T0137632

Forests and Insect Conservation in Australia

Tim R. New

Forests and Insect Conservation in Australia

 Springer

Tim R. New
Department of Ecology, Environment & Evolution
La Trobe University
Melbourne, VIC, Australia

ISBN 978-3-030-06386-3   ISBN 978-3-319-92222-5   (eBook)
https://doi.org/10.1007/978-3-319-92222-5

© Springer International Publishing AG, part of Springer Nature 2018
Softcover re-print of the Hardcover 1st edition 2018
This work is subject to copyright. All rights are reserved by the Publisher, whether the whole or part of the material is concerned, specifically the rights of translation, reprinting, reuse of illustrations, recitation, broadcasting, reproduction on microfilms or in any other physical way, and transmission or information storage and retrieval, electronic adaptation, computer software, or by similar or dissimilar methodology now known or hereafter developed.
The use of general descriptive names, registered names, trademarks, service marks, etc. in this publication does not imply, even in the absence of a specific statement, that such names are exempt from the relevant protective laws and regulations and therefore free for general use.
The publisher, the authors and the editors are safe to assume that the advice and information in this book are believed to be true and accurate at the date of publication. Neither the publisher nor the authors or the editors give a warranty, express or implied, with respect to the material contained herein or for any errors or omissions that may have been made. The publisher remains neutral with regard to jurisdictional claims in published maps and institutional affiliations.

Printed on acid-free paper

This Springer imprint is published by the registered company Springer International Publishing AG part of Springer Nature.
The registered company address is: Gewerbestrasse 11, 6330 Cham, Switzerland

Preface

Forests throughout the world are under threat. For decades, slogans such as 'Save the tropical rainforest' have been clarion calls for conservation, as large-scale clearing and destruction of tree-based vegetation continues to change the appearance and ecological integrity of landscapes in many parts of the world, and is associated with extirpations or extinctions of myriad species that depend on the forest environment. Those changes, however, are not limited to the tropics – forests and woodlands in temperate regions are also suffering in the interests of human need for resources, greed, and continuing population increase. Much of this loss continues unabated and even unheralded. Ciesla (2011) commented that 'The world's forests have been subject to human pressure since the beginning of civilisation'. For much of that time, those pressures were essentially sustainable and most were limited in extent and severity. Recent intensification of forest clearing for exploitation of timber and wood products, and conversion for land uses such as agriculture or urban development, generates a sense of considerable urgency and concern over the future of these, often ancient, environments. As Hanks (2008) put it, 'People have a natural appreciation for trees and forest ecosystems, and degradation of forests by any agent elicits a knee-jerk public reaction'. The continuing, and increasing, high deforestation rates, in particular of tropical forests, affect their characteristically high insect species richness and have led to them being projected as 'extinction hotspots' (May et al. 1995).

The importance of forests for 'biodiversity', including the associated ecological functions and processes so intricately dependent on animals and plants, is acknowledged globally. As Dajoz (2000) put it, 'Forest is – or ought to be – regarded as an ecosystem having multiple roles and which should be conserved or restored'. Primary forests are regarded widely as globally irreplaceable, the status reflecting the high proportions of resident plants and animals recorded only from those habitats. Many of these species have very restricted and specialised ecological needs.

Surveys of forest area and status aim to document primary forest area, protected forest areas, and areas designated for the conservation of biodiversity, through the Global Forest Resources Assessment (Morales-Hidalgo et al. 2015). By 2020, under the Strategic Plan for Biodiversity 2011-2012, and as agreed through the Convention

on Biological Diversity, at least 17 % of the areas of particular importance for biodiversity and ecosystem services are to be conserved by means of 'ecologically
representative and well-connected systems of protected areas and other effective
area-based conservation measures'. Forests are a predominant and vital component
of any such endeavour, with increased documentation of their biodiversity, increased
reservation of key areas, and increasingly sympathetic management regimes necessary to move towards any such laudable target. Such ambitious policy depends on
vastly increased understanding and capability, in part derived from comparisons
across forest treatment regimes and their departures from natural forests. Thus, for
boreal forests in Norway, comparisons of saproxylic fungi and beetles on retention
patches (from single trees to groups), 'woodland key habitats' (habitat patches
exempted from logging because of perceived biodiversity values), and forest
reserves showed that *all* should be part of future forest conservation in filling complementary roles for these organisms (Sverdrup-Thygeson et al. 2014). Assemblages
of aspen-associated generalist beetles (in this study of 153 species of the total 363
saproxylic beetle species obtained) tended to be richest in woodland key habitats
and differed clearly between the three treatments.

More simplistically, forests harbour massive organismal diversity, amongst
which many species are highly localised or ecologically restricted. As late succession climax or near-climax ecosystems, they are botanically diverse and, in addition
to their better-known vertebrate fauna, support enormous numbers of ecologically
specialised invertebrates. Many of these are narrowly endemic to both geographical
region and forest type, poorly documented, and with restricted non-flexible ecological needs flowing from long evolutionary associations with their host plants or
enveloping environments. Insects are predominant amongst these, and changes to
the forests are associated with losses of numerous insect species as their vulnerability increases through erosion of key resources and suitable habitat. Their conservation need, with both ethical and practical dimensions in sustaining biodiversity and
ecological integrity of forest ecosystems, is increasingly acknowledged. However,
human use of forests has generated two major concerns for native biota from the
ensuing changes, as (1) direct loss of forest cover and (2) progressive isolation of
natural forest patches as residual fragments in the increasingly deforested landscape. Allied to these, use of alien species, such as for commercial tree plantations,
can add ecological impacts. Associated management issues – including the need to
control insect pests – also have substantial impacts and implications. Whether of
native or alien tree species, commercially managed plantations are usually monoculture crops that displace and replace the natural high biodiversity of parental
forests.

The vast recent literature on forest ecology and forestry management practices
naturally incorporates much information relevant to conservation of forest inhabitants. However, whilst clarifying many ecological contexts and principles, allusions
to insects are overwhelmingly about the dynamics and management of the relatively
small proportion of pest species rather than wellbeing of the far more diverse taxa
that do not intrude greatly onto 'forester consciousness'. It is sobering to reflect that
strong and persistent campaigns to conserve many global flagship vertebrates (even

primates such as the three species of orang-utans in south-east Asia) are not achieving their aims of protecting forest habitats from loss without considerable opposition and antagonism, or at all. Cases for conserving even the most notable forest insects will appear to many people to be trivial in comparison, but as speciose and functional components of global forest diversity, insects and their relatives are indeed vital considerations in wider conservation endeavour. They add perspective and understanding for the generally smaller scales on which they operate and to the importance of ecological processes in these complex terrestrial environments which are so widely under threat.

In this book, I outline how insects fare in Australia's forests, and the prospects for conserving them constructively in the diverse vegetation associations they collectively inhabit, and as those forests are changed. The history of forest exploitation in Australia is varied and complex, but has included massive losses and changes over only about two centuries of European occupation – so it gives some expectation that the country's 'more natural' forests indeed still harbour substantial endemic biota from the pre-European era. Forest exploitation issues remain amongst the forefront of concerns for the Australian environment and the future of its biota, with realisation that loss of forest and degradation of the forest environments go hand in hand. As elsewhere, measures for conservation of forest-dependent native insects can come into conflict with measures to control pest insects that damage native trees grown for commercial purposes, and the impacts of imported alien species – both in plantation trees and adventive insects – become superimposed on the ecology of native biota alone.

More broadly, the vast majority of literature on the world's forest insects deals with pest management or suppression in production forests – including much elegant biological research to clarify the life histories and dynamics of many notable pest species – or basic faunal documentation. The natural history and ecological variety of insects in forests, together with synopses of their importance in forestry and how to study and control key damaging taxa, is covered in several early entomological texts – that by Barbosa and Wagner (1989) contains much such background that is of current interest in appreciating the development and scope of modern 'forest entomology'. The edited volume by Watt et al. (1997) was innovative in including a titled section on 'Insect Conservation', and a contemporary compilation on forest canopy arthropods (Stork et al. 1997) helped to emphasise the diversity and ecological variety of insects present in that upper, and for long inaccessible and neglected, forest stratum. Both those books were pivotal contributions to realising the need for insect conservation in forests, and documentations of species richness and change continue to aid understanding of ecological impacts and community dynamics from local to landscape and regional levels. A later edited volume (Basset et al. 2003) explored the dynamics of tropical canopy arthropods, illustrating the advances in understanding and documentation made over the previous decade or so, and progress towards functional interpretation of forest insect diversity in many parts of the world. More specifically for Australia, Lunney's (1991) edited book on conservation of forest fauna contains little specific information on invertebrates, and the authors of the only chapter on invertebrates (Postle et al. 1991, on soil and litter fauna in Western Australia) emphasised the importance of species-level interpretations in concluding that 'there is no rational reason for excluding invertebrates from faunal studies which are carried out in forest ecosystems'. However, there are formidable practical difficulties in achieving this.

As for most practical understanding of the ecological bases for insect conservation in Australia, lessons from forest entomology in other parts of the world are amongst the best examples for possible emulation. The long history of attention to conservation of saproxylic insects in production forests of Northern Europe and North America, for example, has elucidated many of the principal needs that might provide similar benefits elsewhere. Australia's varied forests span tropical to cool temperate regimes, across which levels of threats and knowledge of the largely endemic and often localised insect faunas and how they are influenced by change vary greatly, so that parallels elsewhere are relevant. Those parallels draw from a wide range of contexts, from the relatively well-known insect faunas of boreal forests to the daunting 'black holes' amongst tropical forest faunas and attempts to enumerate and understand these. To aid this, I bring together many such studies from amongst the far greater number available (up to late 2017) from different forest types and regions as a basis to compare and consider Australian needs. The publisher's need for each chapter to be read independently has led to some inevitable overlaps in content, as complex themes are treated in different ecological and management contexts; page cross-references are inserted to facilitate more integrated use. The book is intended to provide insights into insect ecology and diversity as background for the many people involved with forests, and concerned about their future, but who are not themselves entomologists. The book thereby encapsulates much recent work on forest insect ecology, and how insect conservation may be rendered more effective in environments ranging from intensive production forests to increasingly isolated natural forest areas in which changes may be far more insidious, and undocumented in any meaningful detail. Without attempts to redress this lack of knowledge, the future for numerous endemic insects in Australian forests is parlous.

Melbourne, Australia   Tim R. New

References

Basset Y, Novotny V, Miller SE, Kitching RL (eds) (2003) Arthropods of tropical forest. Cambridge University Press, Cambridge

Ciesla W (2011) Forest entomology: a global perspective. Wiley-Blackwell, Oxford

Dajoz R (2000) Insects and forests. Intercept, Andover

Hanks LM (2008) Changing relationships among biodiversity, management, and biosecurity in managed and unmanaged forests. In: Paine TD (ed) Invasive forest insects, introduced forest trees, and altered ecosystems. Springer, Dordrecht, pp 153–159

Lunney D (ed) (1991) Conservation of Australia's forest fauna, Ist edn. Royal Zoological Society of New South Wales, Mosman

May RM, Lawton JH, Stork NE (1995) Assessing extinction rates. In: Lawton JH, May RM (eds) Extinction rates. Oxford University Press, Oxford, pp 1–24

Morales-Hidalgo D, Oswalt SN, Somanathan E (2015) Status and trends in global primary forest protected areas, and areas designated for conservation of biodiversity from the Global Forest Resources Assessment 2015. For Ecol Manag 352:68–77

Postle A, Majer J, Bell D (1991) A survey of selected soil and litter invertebrate species from the northern jarrah (*Eucalyptus marginata*) forest of Western Australia, with particular reference to soil-type, stratum, seasonality and the conservation of forest fauna. pp. 193–203 in Lunney D (ed) Conservation of Australia's forest fauna (Ist edn). Royal Zoological Society of New South Wales, Mosman

Stork NE, Adis J, Didham RK (eds) (1997) Canopy arthropods. Chapman and Hall, London

Sverdrup-Thygeson A, Bendiksen E, Birkemoe T, Larsson KH (2014) Do conservation measures in forest work? A comparison of three area-based conservation tools for wood-living species in boreal forests. For Ecol Manag 330:8–16

Watt AD, Stork NE, Hunter MD (eds) (1997) Forests and insects. Chapman and Hall, London

Acknowledgements

The following publishers, organisations, and individuals generously advised on or granted permission for my use of material under their control and are gratefully acknowledged: Acta Forestalia Fennica (Dr. E. Korpilahti); AgriFutures Australia; Cambridge University Press, Cambridge; CSIRO Publishing, Melbourne; Elsevier; *European Journal of Entomology*; Finnish Zoological and Botanical Publishing Board (Dr. K. Raciborski); John Wiley & Sons; IUCN, the World Conservation Union; Marie Selby Botanical Gardens, Sarasota (Dr. B. Holst); National Academy of Sciences, USA; Oxford University Press, Oxford; Taylor & Francis Group, Abingdon; United States Department of Agriculture, Forest Service; Dr. B. Wermelinger. Every effort has been made to obtain permissions for use of previously published material, and the publishers would welcome advice on any inadvertent omissions or corrections that should be included in any future editions or imprints. Most figures have been redrawn to enable standardisation of lettering, and some modified (as noted in individual legends) by omissions of some details or parts.

I have greatly appreciated the enthusiastic reception of my initial proposal for this book by Zuzana Bernhart at Springer. Mariska van der Stigchel's humour, constructive advice, and practical help during its preparation continue to make working with her an enjoyable and rewarding experience; I thank her most sincerely for her patience and support. I very much appreciate the careful preparation of the book by my Project Manager, Mr. Pandurangan Krishna Kumar.

Contents

Contents

Chapter 1
Forests and Their Insect Inhabitants

Keywords Deforestation · Edge effects · Extinctions · Forest fragmentation · Insect functional groups · Insect species richness · Logging

1.1 Introduction: The Ecological Milieu

The world's forests harbour vast numbers of insects and other species not found anywhere else. Defining forests, in view of their great variety of structures and complexity, and describing the major anthropogenic contributions to their loss and change, constitutes a basis for characterising that variety. That information also contributes to assessing the consequences of human impacts on forests for their often highly characteristic insect faunas. This introductory chapter sets out this initial framework, and introduces some key themes to be discussed in more detail later in the book.

By whatever measure is used to define them, forests are a major component of Earth's vegetation. One definition (FAO 2009) declares forests to be 'land spanning more than 0.5 hectares with trees higher than 5 m and a canopy of more than 10%'. That concept thereby incorporates aspects of area and architecture, but is a very broad delimitation that incorporates enormous variety. On that basis, however, forests cover about 3.952 billion hectares, or 30.3% of the world's land surface (Ciesla 2011), with other wooded lands covering a further 1.3 billion hectares. A somewhat different definition for Australia (Barson et al. 2000, RAC 1992) regarded a forest as 'an area, incorporating all living and non-living components, that is dominated by trees having usually a single stem and a mature or potentially mature stand height > 2 m and with existing or potential crown cover of overstorey structure equal to or > 20 %'. Within such structural bounds, vegetation differences impose massive biological diversity based on predominant tree taxa and variety, as the substrates and resources for numerous other taxa.

In Tasmania, alone, Jarman et al. (1991) recognised 46 plant communities, divisible into four main categories of forest types, and parallel subtleties and implications of ecological difference and peculiarity are universal.

© Springer International Publishing AG, part of Springer Nature 2018
T. R. New, *Forests and Insect Conservation in Australia*,
https://doi.org/10.1007/978-3-319-92222-5_1

Despite long recognition of 'forests' in Australia based on observable structural features (as emphasised by Specht 1970), the early definitions of forest varied considerably, and some largely emphasised presence of 'harvestable timber': Hnatiuk et al. (2003) thus noted data from around the middle of the twentieth century claiming that Australia had somewhere between 10 and 15 million hectares of such forest, and traced later developments toward RAC (1992), and generating a more rigorous structural and uniform definition, as above. That definition, used as the basis for later 'State of the Forest Reports' (p. 24), included natural forests and plantations, regardless of species, and was sufficiently broad to encompass areas of trees designated elsewhere as 'woodland'.

Eight of the major biomes recognised by Olson et al. (2001) were regarded as 'forest' by Ciesla (2011), with the primary division of tropical/subtropical (moist broadleaf forests; dry broadleaf forests; coniferous forests; mangroves) and temperate (broadleaf and mixed forests; coniferous forests; Mediterranean forests, woodlands and scrub), and boreal forests/taiga. The last is absent from Australia, and mangroves are not conventionally included in national concepts of the forest estate. Almost all emphasis on entomology devolves on broadleaf forests, and situations that arise from plantations of alien conifers as softwood crops or of native hardwoods, some planted far beyond their natural ranges. Within that more restricted ambit of native broadleaf forest, Bradshaw (2012) categorised eight major dominant forest types, as *Acacia*, *Calitris*, *Casuarina*, tall *Eucalyptus*, low *Eucalyptus*, mallee, other shrublands, and rain forests. Collectively, the temperate region forests are dominated by Myrtaceae, essentially *Eucalyptus* and its close allies, which comprise about 78% of remaining forest vegetation. Tropical rainforest is botanically far more diverse. Each category is geographically circumscribed, and supports different, characteristic insects, many of which are restricted to that particular forest biome and, in many cases, locality – and, so, are vulnerable to change or loss. More recent developments in mapping vegetation at national level contribute continually to the forest data needed for optimal planning and management for conservation and forest use (Thackway et al. 2007).

1.2   Deforestation

Deforestation is recognised widely as the most significant threat to terrestrial biodiversity in Australia (SoE 2011). Whilst the primary concerns over forest losses and biodiversity devolve largely on the extinctions and wider vulnerability of animals and plants restricted to forest ecosystems, the wider importance of forests in providing ecosystem services gives abundant motivation and support for forest conservation in demonstrating widespread benefit to humanity (Brockerhoff et al. 2017). That variety was indicated by the listing included by Brockerhoff et al., which may be summarised as spanning the three major categories of (1) provisionary (nutrition, material [wood biomass], energy [fuel wood]); (2) regulatory (mediation of toxins or nuisances [filtration or sequestration, moderation of noise or visual impacts], mediation of flows [protection from erosion, flood, storms], maintenance of physical, chemical and biological conditions [pollination and seed dispersal, providing habitat, pest and disease control, soil formation, climate regulation]); and (3) cultural (physical and intellectual interactions with nature, spatial and symbolic interactions with nature). 'Biodiversity' is clearly central to many of these, and the main objectives of forest ecosystem management are 'to provide ecosystem services sustainably to society and maintain the biological diversity of forests' (Thom and Seidl 2016). That objective necessitates understanding of the impacts of natural and anthropogenic disturbances of all kinds.

Regulatory controls on permitted levels of deforestation are common, as elsewhere, but many are difficult to enforce and their intent weakened by a variety of exemptions. Thus, in Queensland, exemptions are available for a range of activities – examples noted by Rhodes et al. (2017) include those with impacts less than a given threshold size, essential management activity (such as clearing of firebreaks or power line easements, and fencing), and case-by-case appraisals for larger impacts such as for urban development or mining (Chap. 10). In some other parts of the world, especially in some less-developed countries, weak regulatory policies, needs for trade, and linked industrial logging are key drivers of forest destruction (Laurance 1999). Those impacts can then become far wider than to 'general biodiversity' alone to induce severe dislocations ('transmigrations', or even loss) of indigenous native tribes in Amazonia, Borneo and New Guinea, to whom tropical forest has been their primary dwelling place and resource supply. Laurance (1999) commented that, in the simplest terms 'tropical forests are being cleared, logged, fragmented, and overhunted on scales that lack historical precedent', collectively from a wide range of pressures flowing from increasing numbers and demands of people. Whilst impacts of commercial forestry operations in Australia are often highlighted, deforestation for urban, industrial and agricultural development on private land has had much more widespread impacts.

For any part of the world, published figures for extent and rates of forest loss and degradation, although universally alarming, are not always easy to evaluate – for example, they may be overstated (by environmentalists) or understated (by forestry interests) and so biased to support a sometimes polarised viewpoint or policy. Nevertheless, even very approximate data can be sobering in their implications. For the Amazon region of Brazil, for example, annual clearing of forest reached 29,059 Km^2 in the year 1994–1995, and Rylands and Brandon (2005) quoted cumulative deforestation to 2004 of about 17% of the total area, or 690,000 Km^2. Davis et al. (2001) quoted loss of tropical forest from 1981–1990 to be 154 million hectares, with annual loss of 0.81% of the FAO (1995) estimate of total forest cover in 1980. Implications for biodiversity are inevitably serious. One recent global survey (Keenan et al. 2015) noted that forests cover about 30% of land area, and harbour more than half the known terrestrial animal and plant species.

Although clear-felling of forests remains the most prominent topic in conservation debate, selective logging of remaining forests involves further environmental filtering by removal of valuable timber trees (mainly Dipterocarpaceae in south east Asia, for example). In eastern Malaysia, about 30% of forest in Sabah was lost or degraded over 1974–1985, leading to undisturbed lowland dipterocarp forest becoming very scarce (Davis et al. 2001). In that region, the consequences of reducing those widely exploited dipterocarp forests have received far more attention than other kinds of forest. However, some other forests, such as the peat-swamp forests, are also affected (Houlihan et al. 2013) from selective logging, fires and conversion to plantations. Generalist nymphalid butterflies in Borneo have become predominant in disturbed peat-swamp forests, and richness of forest fruit-baited nymphalids was correlated closely with extent of canopy cover.

Globally, FAO (2006) estimated annual forest area loss of about 130,000 Km^2 over 2000–2005, with about half this being offset in some way through activities such as reforestation or afforestation.

Deforestation has historically been considered as the major threat in forest conservation, with strenuous remedial efforts aiming to reduce levels of loss dominating many conservation agendas. However, simply retaining forest cover – however fundamental and desirable this may be – may not wholly reduce the impacts of wider forest disturbances such as fire and selective logging. For these, numerous published studies document impacts of individual disturbances and on particular sites. As emphasised by Barlow et al. (2016), the combined impacts of various disturbances and forest loss on native biodiversity are vastly more complex to assess and quantify. The impacts of disturbances comprise two main categories, which these authors differentiated as 'landscape disturbance' (deforestation itself, and the effects on condition of the remaining forest) and 'within-forest disturbance' (fire, logging and others that change forest structure and species representations). For two large areas of the Brazil Amazon region, Barlow et al. included forest dung beetles (156 species) in their estimations of the combined effects of anthropogenic disturbances on the conservation values of remnant primary forest, across 175 primary forest patches over 31 catchments, with 30 of those plots having no evidence of internal forest disturbance. Collectively 1575 pitfall traps were deployed to sample the beetles, and the results complemented those for birds and plants. Based on these taxa, catchments with 69–80% of forest cover remaining still lost more 'conservation value' from disturbance than from direct forest loss, emphasising the needs for policies that can counter and prevent disturbance-related losses of biodiversity. Models from that study implied that, even with Brazil's 'Forest Code' (that mandates that 50% of primary forest cover must be retained in Amazonas) successfully meeting that target, those forests may only retain around 46–61% of their original conservation values, with loss of many significant species. Landscape-level measures, with emphasis on protecting large blocks of the remaining forest and promoting connectivity between these, are thus a conservation priority.

About 98% of the world's primary forest occurs in only 25 countries, with about half this being in 'developed countries' (listed as Australia, Canada, Japan, United States, Republic of Korea, Russian Federation) (Morales-Hidalgo et al. 2015). Those countries may thereby have a substantial burden of responsibility in managing their primary forests for conservation and sustainability. The Canadian inclusion, for example, consists largely of boreal forest, which (with that from northern

Table 1.1 The reality of need for tropical forest conservation and the practical problems that underlie this: four factors emphasised by Basset et al. (2004)

1.	The high rate of tropical forest disappearance: Basset et al. cited this exceeding 1–4% annually of current area.
2.	Rate of inventory of biodiversity, especially arthropods, in those forests is not accelerating and many species are likely to go extinct before they are known.
3.	Almost nothing is known of the interactions of most of those species in tropical rainforest environments, so assessing their vulnerability to anthropogenic disturbance and defining measures to slow those extinctions is also unknown.
4.	Efforts to assess and stem losses of biodiversity are manifestly inadequate: Basset et al. noted that more is spent on 'the search for extraterrestrial life' than on countering losses of biodiversity on earth.

Europe) has been claimed to be the 'most extensive terrestrial biome on earth' (Komonen et al. 2000), as covering about 10% of all land area and comprising about 45% of all forest. Both in Fennoscandia and Canada, these forests are extremely rich in species, but used increasingly for timber extraction and production. They contain the most intensively studied forest insect faunas, and the principles harmonising conservation and sustainable foresty for long considered as a partnership have provided a study arena furnishing invaluable examples and lessons that can aid conservation of insects in forests elsewhere. The extensive literature on boreal forest insect needs and conservation has much value to interpreting impacts and changes in Australia.

Destruction of lowland tropical rainforest has been termed 'the greatest threat to the biological diversity of the planet' (Turner and Corlett 1996), with their diversity the most species-rich of all terrestrial ecosystems. However, no complete inventory of insect species for any such area has yet been made. The state of knowledge of impacts of disturbances on tropical forests noted by Bassett et al. (1998, 2004) remains highly relevant more than a decade later: they listed (2004) four factors (Table 1.1) that are individually alarming and collectively formidable to counter and remedy as facilitating practical conservation.

Remaining forest fragments decline in species richness over time (Turner 1996), related to the extent of isolation from the parental contiguous forest, but also retain considerable conservation value in the shorter term, as facilitating persistence of numerous species that could not exist in the deforested landscape. The conservation attention paid to tropical rainforest has partially distracted attention from the needs of the many other tropical forests, such as the seasonally dry Caatinga forests that occur across about 11% of Brazil's land area and have also undergone large-scale destruction (Oliveira et al. 2016). That destruction is, in principle, countered by designation of 'conservation units' through which (in 2012) about 7.5% of the Caatinga area was protected. Comparative surveys of drosophilid flies in and outside conservation units yielded 32 species, with native flies far more abundant in conservation units and alien species more abundant outside. Differences were not always large, but demonstrated the importance of the designated zones for native species.

Impacts of such scale of forest loss as noted above on biodiversity are difficult to define precisely, but are inevitably severe. Most insect groups have simply not been appraised. Strenuous efforts to prevent losses of forest fragments and to protect them from further degradation may be especially important for many invertebrates that 'normally' occupy only relatively small areas, either as local endemics, or existing in small localised populations or metapopulation units. Attempts to quantify impacts have generally drawn on (1) extrapolations from known relationships between species numbers and forest area, and (2) actual or projected deforestation levels (Watt et al. 1997). Forest clearance may be total, partial in order to allow plantations to be established, or include only selective logging based on tree size, condition or species. A frequent managerial outcome is replacement of an array of native species by other vegetation, which may comprise monocultures or near-monoculture stands of native or alien tree species or dramatically different vegetation such as field crops or grassland. Any such change can constitute a 'treatment' for comparison with natural forest in studies that explore the extent and significance of change.

Forests also encompass the dichotomy of (1) natural tree cover and (2) plantations of native or alien tree species and, as Speight (1997) discussed, that cover need not be continuous but may incorporate agroforestry or urban forests allied with human settlements. Speight also noted the managerial dichotomy for tropical forests as (1) forests where the trees are grown for commercial gain or (2) where they form parts of smaller scale systems from which local people benefit, perhaps extending to small scale localised 'village industries' but without designated wider commercial purpose. The difference, bridged by a gradient of intermediate scales of use, is relevant to insects in that commercial forests can demand high intensity of pest management (commonly involving pesticide use) and other efforts to promote uniformity and standardisation. Silvicultural practices may not always be compatible with wider aims to sustain natural variety and varied growth patterns. That variety encompasses the range of tree species, ages, and structural variety included, and the patterns of these across the wider landscape.

In both conservation and forestry interests, three very broad categories of forest states are sometimes recognised, as (1) undisturbed natural forest; (2) disturbed or degraded natural forests, with the severity of changes very variable in intensity and impact; and (3) plantation forests, either as monocultures or with very few species, composed of native or alien species, and the latter posing an ecological scenario not historically encountered by insects native to the plantation region. That many of the alien tree species rapidly acquire pest insects from the local environment, rather than those pests being associated previously with that plant, has several theoretical explanations. Nair (2007) distinguished possible origins from (1) local generalist species that incorporate the 'new' food species into their dietary range, perhaps after accidentally arriving on the plant; (2) newly adapted insects that, over time, change to adopt the new host; and (3) specialised oligophagous insects that are pre-adapted to feeding on closely related tree species, and for which expansion of host range is not difficult. These categories are confounded by a wider variety of alien pests that are introduced unintentionally or arrive by other means, and whose colonisation is facilitated by availability of the plantation species as food, and in an environment that may be largely free of the insect's normal parasitoids and predators.

Writing on the ecological importance of insects in a dry forest in Costa Rica, Janzen (1987) emphasised that 'insects are an essential "glue" and act as building blocks' for much of the habitat. That evocation parallels Wilson's (1987) often-repeated reference to arthropods as 'the little things that run the world', but perhaps nowhere is this significance more overt than amongst the diverse forest insect faunas throughout the world. Many ecosystem processes in forests depend largely on insects for maintenances and effectiveness, and each may thus be influenced by changes to those forests, including loss of area, edge effects, progressive patch isolation, and the shape and connectivity of fragments within the landscape, all in conjunction with the uses of cleared or changed lands within the former forest areas. In exemplifying this enormous variety, Didham et al. (1996) discussed four functional groups of insects – pollinators, seed predators, parasitoids, and decomposers – to illustrate possible functional consequences of imposed changes. Dispersal capability of pollinators, for example, becomes increasingly important as forests become fragmented. Appraisal of the euglossine bees attracted to chemical baits in Brazil indicated that creating patches in contiguous forest alters the dry season populations of most species of this important pollinator group (Powell and Powell 1987). A cleared area of 100 m width was an effective barrier for movement of four species, but bee assemblages in the clearings included species absent from the contiguous forest, or rarely found there. The Powells' survey demonstrated rapid declines in richness and abundance of euglossines from even large forest fragments, implying that large blocks of forest are necessary to support viable populations of some key pollinators and that even a mosaic of fragments might not suffice for this.

Sizes of clear-cut areas may be restricted formally in the interests of conservation. In Poland, for example, the largest permitted clearcut size is five hectares, with the opening width no more than twice the height of the stand (Sklodowski 2017).

Nair (2007) commented that temperate forest insects have received far more attention than tropical forest insects, with few exceptions (such as Speight and Wylie 2001), and that attention has generally emphasised pest management needs. Nair's book discusses many major issues in tropical forest entomology, and underlines the central roles of ecological understanding in effective pest management. Bringing ecological theory to major practical contexts is equally important in insect conservation, in which that same body of information may be directed toward enhancing, rather than suppressing, insect species or guild impacts. However, 'Forest entomology is rich in theory' (Nair 2007), and the scope of the discipline has been transformed as this theory has been translated gradually into practice. Before the middle of the last century, forest entomology focused on three main themes, listed by Liebhold (2012) as (1) identifying the insects causing damage to forest resources; (2) describing their biology and the damage they cause; and (3) developing methods for killing them. Since then the scope of insect pest management has changed dramatically, based in greater understanding and concerns, and changes in social values. In Liebhold's words 'the field of forest pest management has become more complex' as it has progressively incorporated considerations of themes such as proliferation of alien species invasions, climate change and the well-being of non-pest species in the affected forest environments.

The extent of insect species loss from deforestation, although purportedly large from direct loss of habitat and resources, is difficult to quantify, even in those insect groups that have been studied directly (and, in some cases, extrapolated as 'surrogates' to indicate likely wider losses). Examples from tropical forest loss in Cameroon, discussed by Watt et al. (1997), included (1) that complete forest clearance reduced the number of termite species by about half, with soil-feeding taxa the most affected guild; (2) complete forest clearance and plantation establishmant had negative effects on diversity of leaf-litter ants; and (3) butterfly diversity and abundance was negatively affected by forest clearance and conversion to farm fallow, but stayed higher across most afforested plots surveyed. These, and other studies in the Mbalmayo Forest Reserve, were based on samples from relatively small (one hectare) plots that Watt et al. admit as probably too small to uncritically accept the inferences on extent of changes, but generally endorsed that forest clearance has 'a profound negative impact on insect diversity'. Although relatively rarely assessed directly, impacts of logging and similar practices on insect assemblages can be very long-lasting, even when the forests are subsequently allowed to regenerate. Thus, in the Kimbale National Park, Uganda, species richness and larval density of Lepidoptera were lower in logged compartments than in natural forest after 40 years of regeneration (Savilaakso et al. 2009).

Changes in forests and their embedding landscapes affect insects (and other forest inhabitants) in many ways. Habitat and resource extent, structure, accessibility and quality are modified by processes such as deforestation, fragmentation, connectivity and level of spatial and temporal heterogeneity imposed by both natural disturbances and human activity and intervention. The major drivers of change, reviewed by Saura et al. (2014), are deforestation (historically with far more conservation focus than most other factors), forest fires, climate changes, establishment of plantation forests (including those of alien species), wider spread of invasive species, intensification of land uses, abandonment of rural land, and aspects of forest management. Much 'traditional' forestry has strong functional parallels with intensive agriculture, with the major aim being to simplify complex natural communities to form easily managed, uniform and low diversity formats, essentially transforming forest structure and composition from their natural states, with their ecosystem processes and functions correspondingly eroded.

Whilst harvest of timber and wood products is the primary focus of considering impacts of forest loss, harvesting of 'non-timber products' from forests can also have ecological impacts. Excessive collection of seeds and fruits, foliage, inflorescences and entire plants for food, medical purposes or horticultural trading may all lead to changes, many of them poorly documented and inadequately understood. Ticktin (2004) noted that many such materials are currently over-harvested, and that future studies of impacts should consider both population survival and wider community processes, including nutrient recycling. Depletion of critical resources for specialist insect herbivores may also occur unwittingly – declines of some Australian butterflies, for example, have been attributed in part to excessive removal of mistletoes from eucalypts, either as weeds or for commercial sales.

1.3 Fragmentation

Implications and outcomes of forest fragmentation, whereby formerly large continuous forests are reduced to small patches across the landscape, and separated by land otherwise used and comprising areas that are variously unsuitable for forest dependent species, are often severe – reflecting both direct loss of forest and changed conditions within patches. Outcomes may include (1) severe population declines of many specialised resident species; (2) increasing chances of local extinctions; (3) increased more general disturbance, such as by alien species invasions and edge effects; and (4) disruption of key ecological processes, such as pollination and dispersal of plant propagules. One of the general postulated impacts of habitat fragmentation is that the small populations left on patches may, through their isolation and enforced inbreeding, undergo genetic declines and become more vulnerable to stochastic losses. Also generally, genetic diversity equates to evolutionary potential, so that considerations of conserving genetic diversity are a general aim in planning conservation of threatened species. As noted above, four major processes contribute to fragmentation, and are summarised in Fig. 1.1 (Didham 1997). Of these, the first two (effects of fragment area, either in total or as the core area unaffected by edge changes; edge effects) are most easily and commonly appraised, whilst fragment shape (with changing perimeter length: area ratio, and extent of spatial isolation and functional contact with other fragments) directly links with connectivity through species' dispersal propensity. Distinguishing these effects is not always possible, but acknowledgement that they may occur and may aid planning to counter impacts of disturbances is vital.

Rather few insects have been studied specifically in relation to these impacts in forests, but the coprophagous beetle *Canthon* (*Peltocanthon*) *staigi* (Scarabaeidae) in the Atlantic Forest of eastern South America is one (Ferreira-Neto et al. 2017). Although it is sometimes abundant, *C. staigi* is susceptible to habitat changes, and its genetic structure was compared across series of individuals from six populations in eastern Brazil. Three populations showed high levels of heterozygosity, and were recommended as high priority for in situ conservation as sources of genetic variety. Those populations were also 'healthier', and one was located within a designated ecological reserve. The other three populations were far less diverse, and in the future might need enrichment through introduction of individuals from other populations. One of the 'poor' populations was in a very small forest patch under considerable human pressure, and where practical conservation may be difficult to implement. For this beetle, and somewhat to the surprise of Ferreira-Neto et al., fragmentation did not appear to be the primary driver of changing genetic structure – rather, human pressures appeared to be critical in influencing genetic diversity. In practice, the above population groups dictate two separate management strategies useful in pursu-

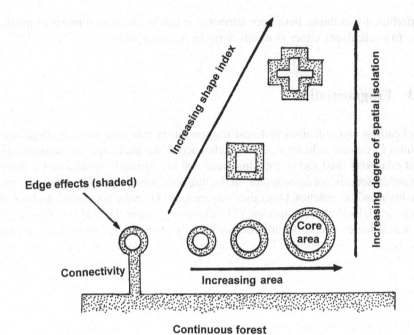

Fig. 1.1 The primary fragmentation processes affecting invertebrates in forests. These are shown as (1) area effects, with either total area or core area free of edge effects; (2) edge effects; (3) shape of the fragment; (4) increasing spatial (and temporal: not shown) isolation; and (5) extent of habitat connectivity in the landscape. (Didham 1997)

ing conservation with the common purpose of maintaining or enhancing genetic diversity through judicious exchanges of individuals across populations.

Forest fragmentation impacts on insects may depend closely on features of the individual remnant patches. Changes amongst ant assemblages in woodland patches in southern Queensland were driven more by such within-patch characteristics than by habitat fragmentation or cover (Debuse et al. 2007), with features such as bare ground, clay content of the soil, and long-term rainfall pattern all influencing competitive ability within the assemblages, and those features explaining more than twice the amount of species variation that was attributed to fragmentation, and four times that explained by habitat loss. Mechanisms underlying apparently straightforward responses to fragmentation may indeed be complex.

Not all forest patches are the outcome of recent or continuing forestry activity, although their historical origins can be unclear. The cool-temperate rainforest patches in south east Tasmania studied by Driscoll et al. (2010), for example, occur naturally surrounded by sedgelands, and patches vary from a few hectares up to several square kilometres in area. Pitfall trap catches of beetles in 31 of those patches comprised 168 morphospecies and confirmed that the forest beetle assemblages were very distinct from those in the enveloping sedgeland (Driscoll 2005).

Fragmentation of any previously continuous habitat area, here forest, partitions that habitat into remnants, or patches. Impacts of fragment isolation are then superimposed on outcomes of direct habitat loss, and further include fragment size and shape, the landscape configuration of the fragments, and the extent of edge effects. Effects on any given insect group vary greatly, with, as Niemela et al. (2007) exemplified from carabid beetles in boreal forests, the responses being affected in many ways that reflect the individual species' environmental and ecological tolerances. In that system, and very broadly, three major responses of Carabidae to forest management were found in strongly managed stands (exemplified by clearcuts), namely (1) open habitat species appear, can become abundant, but disappear as the canopy closes a few decades later; simply because there are usually far more open habitat species colonising these areas than there are forest species lost from them, species richness will often increase; (2) forest generalist species persist throughout the newly founded succession; and (3) species that are true forest specialists requiring closed canopy are affected negatively, and may not recover. This last category, however, included two kinds of recovery process. First, most forest specialists indeed recover following logging, but perhaps not for 30–60 years (Koivula et al. 2002). Sites surveyed 60 years after logging supported carabid assemblages closely similar to those of mature forest, and included almost no open-habitat species. Second, some species do not re-colonise even several decades after logging.

Insect populations in forest fragments may thus decrease simply because of limited area and resources, but can also be due to low rates of immigration which, in some cases, reflect both fragment isolation and the dispersal ability of potential arrivals. Some Australian ground beetles (Carabidae) changed in abundance in small forest patches, in a survey that was one of relatively few monitoring studies that incorporated pre-fragmentation conditions (Davies and Margules 1998). Three of the eight species monitored (from a collective assemblage of 45 species) for six years after fragmentation responded to fragment size, but those responses varied in comparisons across replicated large (3.062 hectare), medium-sized (0.875 ha) and small (0.25 ha) remnants, as part of the 'Wog-Wog habitat fragmentation experiment' near Bombala, New South Wales (p. 196).

Notonomus resplendens was reduced in abundance in all three remnant sizes from its levels in continuous forest, and least abundant in medium-sized remnants. *N. variicollis* was more abundant in small and large remnants than in either medium-sized fragments or continuous forest, and *Eurylychnus blagravei* was most abundant in large remnants, with little difference across the other treatments (Fig. 1.2). Overall richness was not reduced in small fragments but, as in other comparable studies, the beetles showed a wide range of responses to habitat fragmentation.

Much background to the roles and values of tropical forest fragments in conservation has flowed from a major long-term project in the Manaus area of the Brazilian Amazon, on the biodiversity and community structure of fragments over their periods of isolation from contiguous forests, but also in which insect studies have lagged behind those on vertebrates. However, as noted by Turner and Corlett (1996), those fragments differ from parallels studied in several other locations, because they are separated from contiguous forest by only very short distances. The intermediate areas in the Manaus study have also generally developed secondary forest, leading to some level of restoration in functional connectivity and, perhaps, reducing edge

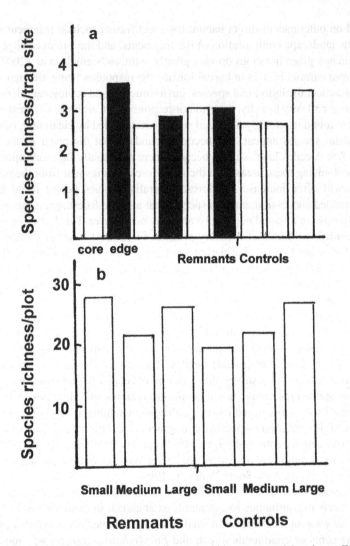

Fig. 1.2 Effects of habitat fragmentation on forest carabids at two spatial scales, at Wog Wog, New South Wales (see p. 196). (**a**) the numbers of species per trap site in core (open bars) and edge sites (black bars) in different sized fragments in remnants and continuous forest controls; (**b**) differences in species richness for small, medium and large remnants and small, medium and large control plots in continuous forest. No significant differences between treatments were found. (Davies and Margules 1998)

effects. 'Edges' of forests or other vegetation types are transitional zones (ecotones, interfaces) between the major habitats they abut. The compositional changes can include features of both neighbouring habitats but also some that are peculiar to the edge itself, providing unique conditions for other species to exploit, as 'edge specialists'. Dajoz (2000) regarded forest edges as having two main structural elements: 'mantle' (shrubs and tree of lesser height than the forest) and 'skirt' (herbaceous, often perennial, understorey plants).

'Edge effects' are sometime a misleading term, as the impacts can penetrate deep into forest fragments, to the extent that smaller fragments may be entirely affected and in essence have their ecological integrity removed (Chap. 6). The processes of enhanced edge effects in isolated forest fragments, discussed by Gascon et al. (2000), include predictable phases such as (1) the initial abrupt forest edge left after isolation producing microclimate changes through wind and light penetration; (2) regenerating vegetation after several years, both at the edge and in the nearby understorey, progressively buffering the forest interior against those initial changes; and (3) in more harsh landscapes, that process breaking down, leading to greater penetration of edge effects, and the original edge being replaced by ruderal scrub vegetation progressively invading the forest fragment. Gascon et al. claimed that edge-affected zones of 1 Km wide are not unusual, so that fragments of up to 1000 hectares may be wholly influenced. Again, the impacts of timber extraction in plantations can extend deeply into the surrounding natural forest (Swart et al. 2017). 'Harshness' of the intervening matrix reflects intensity of use, the presence and abundance of alien species, use of fire, and the vegetation structure, and minimising that harshness may be a crucial need in preserving or restoring forest integrity. Forest fragment edges, especially in harsher landscapes in which they border with non-forest vegetation, are major transition zones or ecotones, across which abrupt changes in microclimate augment the more visually obvious alterations to structure and composition. Microclimates near a forest edge are commonly hotter, drier and brighter than in a forest interior, and tree mortality may also be greater. These changes are reflected in insect assemblages, sampled at intervals across the boundary as spatial 'gradient surveys'.

The factors influencing edge effects are very varied. For Central America, Didham et al. (1998) noted six important and influential environmental variables, as (1) air temperature; (2) canopy height; (3) percent ground cover of twigs; (4) litter biomass; (5) litter moisture content; and (6) an 'air temperature times distance from forest edge' interaction effect reflecting different edge vegetation density and temperature profiles of those edges. Those factors were derived from studies on leaf-litter beetles, based on samples obtained by sieving litter and Winkler bag extraction, in which 15 of the 32 most abundant species (of 993 species obtained) were affected significantly by forest fragmentation. Four categories of species' responses were distinguished amongst those beetles, as (1) edge sensitive, area insensitive; (2) area sensitive, edge insensitive; (3) both edge and area sensitive; and (4) both edge and area insensitive. Didham et al. estimated rates of loss (by absence from samples taken from the fragments) of common beetle species, as follows: 49.8% from 1-hectare fragments, decreasing to 29.8% for 10-hectare fragments and 13.8% from 100-hectre fragments. Intriguingly, common species were more likely to be lost from small fragments than were rarer species. Habitat destruction here seemed to harm poorly dispersing species that were also dominant competitors. More generally, beetle species assemblage composition changed substantially with decreased fragment area and decreasing distance from the forest edge, with assemblage composition more variable among edge sites than among deep-forest sites. Most edge species were rare to reasonably common.

1.4   Selective Logging

In selective logging, individual trees of commercial value are harvested, after which the forest is left to regenerate and the process repeated as the desirable trees again become available. It is thereby a practice that (1) avoids the drastic impacts of less discriminating or broad scale clearing (such as clear-felling), and (2) implies a long-term need for sustainability, but also (3) may increase vulnerability of insects and other specialists that depend on the target tree species. Generally, the process is considered far preferable to clear-cutting (in which all trees are removed), but relatively little information is available on impacts of selective logging on rain forest or other insects. However, many ecologists consider that selective logging, rather than helping forest conservation, may facilitate further destruction, such as by creation of networks of access roads for machinery and log extraction and allowing hunter and slash-and-burn farmers deeper access into forest, and leads to increasing levels of fragmentation.

Comparative surveys imply losses of species: selectively logged forests, surveyed five years later on Buru, Indonesia, supported fewer butterfly species than unlogged forest (Hill et al. 1995, whose surveys did not include the diverse family Lycaenidae). In Sabah, impacts of selective logging disturbance were still clear after 8–9 years, and the detrimental impacts of open areas on the subcanopy butterfly *Ragadia makuta* (Nymphalidae: Satyrinae) exemplified how habitat features contribute to understanding responses to disturbance (Hill 1999).

This 'knowledge gap' is perhaps most severe for the Neotropics, the region subject to the most intensive pressures on forests from this practice (Vasconcelos et al. 2000). Collectively, however, the variety of responses reported endorse the need for taxon-specific appraisals, notwithstanding some tentative generalisations that impacts reflect the extent of structural changes to the forest.

Selective logging is associated with a variety of changes, including decreased canopy cover (with gaps in coverage), increased density of understorey vegetation, and increased litter depth. Several studies on responses of ants have suggested that they are relatively resistant to low levels of logging, but increased intensity of logging can impose more severe effects (Edwards et al. 2012, for Borneo).

Comparison of litter ants across logged and control plots in a central Amazonian forest (near Manaus, Brazil), involved the three treatments of unchanged forest (control) and plots selectively logged 10 and four years previously, each surveyed on replicated four hectare plots. The selective harvesting had removed half the basal area of the selected tree species, for an average removal rate of eight trees/block. Ants were collected by sieving leaf litter and deployment of sardine baits. Selective logging did not affect ant species richness or the general abundance of ants on the forest floor: of the overall 143 species from litter samples, 97 occurred in control plots, 97 in the 10-year logging plots, and 100 in the four-year logging plots. Most of the species in undisturbed forest were able to withstand changes from logging operations. However, some taxa changed in abundance, with that effect persisting at least for the decade after logging – and in accord with studies elsewhere that implied that more than 20 years may be needed for recovery of an insect community. Several ant species rare in undisturbed forest, for example, were common in logged plots. Vasconcelos et al. (2000) also commented on indirect impacts, such as construction of access roads, including compaction from heavy machinery, and noted that in some logged areas up to 20% of ground surface can be affected by logging roads and bulldozer activity, and forest recovery retarded considerably. Minimisation of heavy machinery use could be important in improving conservation of logged forests for ground-dwelling insects, and associated developments of roads and tracks should also be minimised because they can provide entry routes and promote invasions – in this study, of non-forest ant species.

A second study in Brazil, involving pitfall trapping and beating samples of ants in different forest regimes in southern Acre, yielded 263 species – 200 species (46 genera) from control forests and 212 species (41 genera) from managed forests (Miranda et al. 2017) – and only limited effects on functional group composition occurred in the low logging intensity areas. Understorey ant assemblages (the 'beating samples') appeared to be more resilient than the ground-dwelling taxa, the latter differing more between intensity of treatments and with time since logging. Only arboreal predators (mainly *Pseudomyrmex* spp.) were richer in managed forest areas, possibly responding to associated plant abundance including pioneer species favoured by selective logging.

Selective logging can benefit components of the complex saproxylic beetle assemblages (p. 96), even though the overall assemblage composition may not differ between selectively felled, uncut and old-growth stands (Sweden: Hjalten et al. 2017). Assemblages of clear-cut and old-growth stands were similar, with groups of 'cambivores' and 'obligate saproxylics' differing from those of uncut stands, and related to greater abundance of the latter in old growth habitats with large volumes of suitable dead wood. Differences, or possible differences, between saproxylic beetles of even-aged and mixed-age forest stands may provide constructive background for management, but may not be great. Samples comprising 451 species of beetles along a chronosequence spanning 50 years in Sweden, showed that the assemblage had not regained its pre-disturbance complexity, although approaching those conditions (Joelsson et al. 2017). Uneven-aged silviculture may be an important means of sustaining habitat heterogeneity, with selective felling a contributor to this.

1.5   Losses of Insects

In Australia, another quotation from Janzen's (1987) comments on Costa Rican insects is highly pertinent, namely '[Insects] are a diverse, puzzling, complex, inherently attractive and major part of the intellectual display offered by tropical wildlands'. Largely overshadowed in the public and political 'eye' by the remarkably charismatic and unique endemic and emblematic marsupials and birds, Australia's forest insects are no less unusual than the more popular vertebrates, and their conservation needs are complex in both tropical and temperate forest environments.

Numerous studies have demonstrated the susceptibility of insects to loss and degradation of forests, whether tropical or temperate, and from a variety of causes. Probable loss of about 43% of endemic forest dung beetles in Madagascar was linked with elimination of about half the forest (Hanski et al. 2007). The higher level of forest loss in Singapore (95%) was associated with extirpation of phasmids (20% of species) and butterflies (38%) (Brook et al. 2003). Both studies are informative.

In Madagascar, the forest loss (then leaving only about 10% of original forest cover) affected the historical recorded ranges of previously recorded helictopleurine dung beetles (51 species), of which only 29 species were recovered during surveys spanning 2002–2006. Many of the non-recovered species were known from one (n = 9) or two (n = 6) localities, so were probably restricted and/or rare, but Hanski et al.'s (2007) surveys added only four additional species, suggesting that the early inventory was indeed quite comprehensive and reliable. Species in areas that (in 2000) had less than a third of the 1953 forest cover tended to be 'apparently extinct', and many of these had not been collected for 50 years or more. Such 'inferred extinctions' can by far exceed true documented extinctions, with the rarity of numerous insect species inevitably creating doubt over their fate. Relatively reliable information can come only from the most thoroughly documented insect groups.

Butterflies are paramount among these taxa because they are largely diurnal, visible, and well-studied as well as being (despite some ambiguities) easily identifiable and assessable by standard methods. Studies of forest butterflies in Singapore and elsewhere in south east Asia have demonstrated a variety of responses to deforestation and accompanying land use changes (Koh 2007).

Koh (2007) cited several traits that are potentially important in predicting butterfly responses to deforestation, amongst which (1) butterfly species that are restricted to primary or undisturbed forest can have much narrower distribution ranges than species that occur in disturbed areas, and the latter are often much more adaptable and are relative generalists; (2) forest butterflies, as relative specialists that depend on particular forest-limited reources, are thus likely to be affected by forest loss or change far more than are disturbed habitat species; (3) the vertical stratification (p. 82) characteristic of tropical rainforest butterflies may affect their vulnerability, with understorey butterflies undergoing greater losses than either canopy or gap-frequenting species. Additional features, such as flight behaviour, adult feeding preferences, territoriality and thermoregulatory behaviour may also be involved. Adult feeding, for example, can be infuenced by floral composition changes as mature forest is changed to cropland, when larval foodplants and symbionts may also be lost. Again, relative specialists may become especially susceptible.

Loss of forest habitat, largely due to forestry, is the primary threat to many red-listed insect species in northern Europe (Berg et al. 1994), and only very small proportions of production forest areas are effectively protected. For Sweden, Gardenfors (2000) noted that 474 of the approximately 1000 saproxylic beetle species are red-listed – either because of observed declines in abundance or from forest habitat destruction increasing risks to them. Many of those beetles are restricted to particular tree species and decay stages (p. 158) but, in addition to changes in the supply of coarse woody debris, northern forestry has also markedly changed natural disturbance regimes (such as incidence of fire), with impacts of such changes demonstrated repeatedly by surveys of beetles in boreal forests. Amongst others, Gibb et al. (2006a, b) emphasised the need for long-term studies that can assess changes in the beetle assemblages over extended decay periods for their dead wood habitat.

That numerous insects and other species have become extinct as forests are lost or forest land converted for other use cannot be doubted – but the rates of extinction are largely impossible to measure, and difficulties are confounded by the lack of basic inventory studies on most forest insect faunas, and of other hyperdiverse biota. Many species have assuredly become extinct before they have been discovered and recorded – a scenario termed 'Centinelan extinction' after a forested ridge in Ecuador that was stripped of forest before it could be surveyed in any detail despite tantalisng glimpses of its flora, to designate extinctions of species that were unknown before they were lost and, hence, can never be recorded (Wilson 1992). That particular 'haemorrhaging of biological diversity' was fortuitously witnessed, but many apparent parallels have not been seen, and Wilson's (1992) evocative description of those species 'disappearing just beyond the edge of our attention' deservedly brought home to many people that extinctions are indeed far more than are actually documented. Wilson (1992, his p. 244) also wrote 'They enter oblivion like the dead of Gray's *Elegy*, leaving at most a name, a fading echo in a far corner of the world, their genius unused'.

Centinelan extinctions are by no means confined to the species-rich tropical forests. The terrestrial arthropods of old-growth conifer forests of Vancouver Island, Canada, were assessed by analysis of two surrogate groups, staphylinid beetles and oribatid mites, amongst a wide range of taxa collected by a variety of different methods (Winchester and Ring 1996), and both of which contained undescribed species. For both Staphylinidae (Fig. 1.3) and Oribatoidea, new species were found in forest canopy and interior, but not in clearcut areas or the transition zones between these. Winchester and Ring suggested that the faunas of those forests had not been documented sufficiently to overcome likelihood of Centinelan extinctions occurring, and inferences from many other forest insect studies – including in Australia – are similar.

The reality of Centinelan extinctions emphasises the need to understand and document more fully the impacts of forest changes on natural communities. But, as Lamarre et al. (2016) commented 'Entomological surveys in tropical rainforest are notoriously challenging'. In their short survey (two nights of light trapping for larger moths at sites in Brazilian Amazon forest fragments), Lamarre et al. collected numerous new (91) and endemic (90) species amongst 601 recognised barcode identities. The forest fragments were in an area already cleared of about 70% of the natural forest and the authors suggested that their survey implied that many unknown species may already have been lost, and that their small 'snapshot' of local moth diversity and endemism exemplified the lack of basic documentation needed to appraise this.

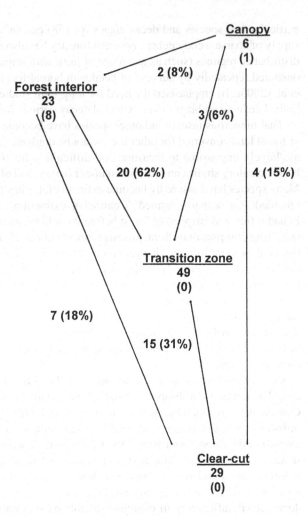

Fig. 1.3 The relationships between staphylinid beetles from four study sites in forests, from pooled samples in the Upper Carmanah Valley (Vancouver, Canada). Numbers along lines are species in common between linked sites; numbers below sites are numbers of species present, and numbers in parentheses mark presence of new species. (Winchester and Ring 1996, with kind permission from the Marie Selby Botanical Gardens)

However, and as for moth surveys in the forests of south-east Asia, the relatively few surveys 'must be regarded very much as pilot observations' (Holloway et al. 1992) and, despite augmentation since then, wider implications of losses remain undetailed within the reality of large scale forest losses that continue largely unabated. Many Australian forest types lack even such 'pilot studies' of their insect fauna, despite tantalising glimpses of the variety from sporadic collection records. Some, however, have been surveyed in far more detail, most studies focusing on particular insect groups rather than the entire communities. Trends revealed in those surveys commonly comparing forest fauna with changed local vegetation types can imply the urgent need for conservation if relatively complete local assemblages are to persist. Those key insect groups, selected for the amount of basic ecological and taxonomic knowledge and logistic grounds such as ease of recognition and sampling, are a major focus in this book, and understanding their ecology in Australian forests can in many cases draw heavily from parallel experiences in other parts of the world. Susceptibility of forest insects to forest management or disturbance is often implied, but equally often is difficult to prove. Loss of forest litter, for example, may affect many insects but details are simply unknown in most cases. Litter loss caused by frequent burning in *Eucalyptus* forests might be responsible for the scarcity of some large forest Swift moths (the spectacular green *Aenetus* spp., Hepialidae), for example (Grehan, in Tobi et al. 1993).

Protecting pristine and near-pristine forests is clearly needed to conserve many otherwise threatened or potentially threatened species. Many conservation biologists, and others, urge increased reservation of forests, as a general laudable principle of 'saving habitat'. A major United Nations biodiversity summit in 2010 set a target of 17% of forested land areas to be reserved (Hanski 2011), a level far higher than current levels for many parts of the world. Even for forest reserves in Europe, Bouget and Parmain (2016), based on their surveys of saproxylic beetles, recommended needs to at least double the current proportion on reserves across European forested landscapes.

References

Barlow J, Lennox GD, Ferreira J, Berenguer E, Lees AC et al (2016) Anthropogenic disturbance in tropical forest can double biodiversity loss from deforestation. Nature 535:144–149

Barson MM, Randall LA, Bordas V (2000) Land cover change in Australia. Bureau of Rural Sciences, Canberra

Basset Y, Novotny V, Miller SE, Weiblens GD, Missa O, Stewart AJA (2004) Conservation and biological monitoring of tropical forests: the role of parataxonomists. J Appl Ecol 41:163–174

Bassett Y, Novotny V, Miller SE, Springate ND (1998) Assessing the impact of forest disturbance on tropical invertebrates: some comments. J Appl Ecol 35:461–466

Berg A, Ehnstrom B, Gustafsson L, Hallingback T, Jonsell M, Weslien J (1994) Threatened plant, animal and fungus species in Swedish forests; distribution and habitat associations. Conserv Biol 8:718–731

Bouget C, Parmain G (2016) Effects of landscape design of forest reserves on saproxylic beetle diversity. Conserv Biol 30:92–102

Bradshaw CJA (2012) Little left to lose: deforestation and forest degradation in Australia since European colonization. J Plant Ecol 5:109–120

Brockerhoff EG, Barbaro L, Castagneyrol B, Forrester DI, Gardiner B et al (2017) Forest biodiversity, ecosystem function and the provision of ecosystem services. Biodiv Conserv 26:3005–3035

Brook BW, Sodhi NS, Ng PKL (2003) Catastrophic extinctions follow deforestation in Singapore. Nature 424:420–423

Ciesla W (2011) Forest entomology: a global perspective. Wiley-Blackwell, Oxford

Dajoz R (2000) Insects and forests. Intercept, Andover

Davies KF, Margules CR (1998) Effects of habitat fragmentation on carabid beetles: experimental evidence. J Anim Ecol 67:460–471

Davis AJ, Holloway JD, Huijbregts H, Krikken J, Kirk-Spriggs AJH, Sutton SL (2001) Dung beetles as indicators of change in the forests of northern Borneo. J Appl Ecol 38:593–616

Debuse VJ, King J, House APN (2007) Effect of fragmentation, habitat loss and within-patch habitat characteristics on ant assemblages in semi-arid woodlands of eastern Australia. Landsc Ecol 22:731–745

Didham RK (1997) An overview of invertebrate responses to forest fragmentation. In: Watt AD, Stork NE, Hunter MD (eds) Forests and insects. Chapman and Hall, London, pp 303–320

Didham RK, Ghazoul J, Stork NE, Davis AJ (1996) Insects in fragmented forests: a functional approach. Trends Ecol Evol 11:255–260

Didham RK, Hammond PM, Lawton JH, Eggleton P, Stork NE (1998) Beetle species responses to tropical forest fragmentation. Ecol Monogr 68:295–323

Driscoll DA (2005) Is the matrix a sea? Habitat specificity in a naturally fragmented landscape. Ecol Entomol 30:8–16

Driscoll DA, Kirkpatrick JB, McQuillan PB, Bonham KJ (2010) Classic metapopulations are rare among common beetle species from naturally fragmented landscape. J Anim Ecol 79:294–303

Edwards DP, Woodcock P, Edwards FA, Larsen TH, Hsu WW, Benedick S, Wilcove DS (2012) Reduced-impact logging and biodiversity conservation: a case study from Borneo. Ecol Appl 22:561–571

FAO (1995) Forest resource assessment 1990, Global synthesis. United Nations Food and Agriculture Organisation, Rome

FAO (2006) Global forest recovery assessment 2005. Progress towards sustainable forest management. FAO forestry paper no. 147. United Nations Food and Agriculture Organisation, Rome

FAO (2009) Global review of forest pests and diseases. FAO forestry paper no. 156. United Nations Food and Agriculture Organisation, Rome

Ferreira-Neto CSA, dos Santos Cruz GA, de Amorin IC, Balbino VQ, Cassia de Moura R (2017) Effects of fragmentation and anthropic pressure on the genetic structure of *Canthon* (*Peltecanthon*) *staigi* (Coleoptera: Scarabaeidae) populations in the Atlantic Forest domain. J Insect Conserv 21:267–276

Gardenfors U (2000) The 2000 red list of Swedish species. ArtDatabanken SLU, Uppsala

Gascon C, Williamson GB, da Fonseca GAB (2000) Receding forest edges and vanishing reserves. Science 288:1356–1358

Gibb H, Hjalten J, Ball JP, Atlegrim O, Pettersson RB et al (2006a) Effects of landscape composition and substrate availability on saproxylic beetles in boreal forests: a study using experimental logs for monitoring assemblages. Ecography 29:191–204

Gibb H, Pettersson RB, Hjalten J, Hilszczanski J, Ball JP et al (2006b) Conservation-oriented forestry and early sucessional saproxylic beetles: responses of functional groups to manipulated dead wood substrates. Biol Conserv 129:437–450

Hanski I (2011) Habitat loss, the dynamics of biodiversity, and a perspective on conservation. Ambio 40:248–255

Hanski I, Koivulehto H, Camerin A, Rahagalala P (2007) Deforestation and apparent extinctions of endemic forest beetles in Madagascar. Biol Lett 3:344–347

Hill JK (1999) Butterfly spatial distribution and habitat requirements in a tropical forest: impacts of selective logging. J Appl Ecol 36:564–572

Hill JK, Hamer KC, Lace LA, Banham WMT (1995) Effects of selective logging on tropical forest butterflies on Buru, Indonesia. J Appl Ecol 32:754–760

Hjalten J, Joelsson K, Gibb H, Work T, Lofroth T, Roberge J-M (2017) Biodiversity benefits for saproxylic beetles with uneven-aged silviculture. For Ecol Manag 402:37–50

Hnatiuk R, Tickle P, Wood MS, Howell C (2003) Defining Australian forests. Aust For 66:176–183

Holloway JD, Kirk-Spriggs AH, Khen CV (1992) The response of some rain forest insect groups to logging and conversion to plantation. Phil Trans R Soc Lond B 335:425–436

Houlihan PR, Harrison ME, Cheyne SM (2013) Impacts of forest gaps on butterfly diversity in a Bornean peat-swamp forest. J Asia-Pacific Entomol 16:67–73

Janzen DH (1987) Insect diversity of a Costa Rican dry forest: why keep it, and how? Biol J Linn Soc 30:343–356

Jarman SJ, Kantvilas G, Brown MJ (1991) Floristic and ecological studies in Tasmanian rainforest. Tasmanian National Rainforest Conservation Program, Report no 3. Forestry Commission/ Australian Government Printing Service, Hobart/Canberra

Joelsson K, Hjalten J, Work T, Gibb H, Roberge J-M, Lofroth T (2017) Uneven-aged silviculture can reduce negative effects of forest management on beetles. For Ecol Manag 391:436–445

Keenan RJ, Reams GA, Achard F, de Freitas JV, Grainger A, Lindquist E (2015) Dynamics of global forest area: results from the FAO Global Forest Resources Assessment 2015. For Ecol Manag 352:9–20

Koh LP (2007) Impacts of land use change on South-east Asian forest butterflies: a review. J Appl Ecol 44:703–713

Koivula M, Kukkonen J, Niemela J (2002) Boreal carabid-beetle (Coleoptera, Carabidae) assemblages along the clear-cut originated succession gradient. Biodivers Conserv 11:1269–1288

Komonen A, Pentilla R, Lindgren M, Hanski I (2000) Forest fragmentation truncates a food chain based on an old-growth forest bracket fungus. Oikos 90:119–126

Lamarre GPA, Decaens T, Rougerie R, Barbut J, Dewaard JR et al (2016) An integrative taxonomy approach unveils unknown and threatened moth species in Amazonian rainforest fragments. Insect Conserv Div 9:475–479

Laurance WF (1999) Reflections on the tropical deforestation crisis. Biol Conserv 91:109–117

Liebhold AM (2012) Forest pest management in a changing world. Int J Pest Manage 58:289–295

Miranda PN, Baccaro FB, Morato EF, Oliveira MA, Delabie JHC (2017) Limited effects of low-intensity forest management on ant assemblages in southwestern Amazonian forests. Biodivers Conserv 26:2435–2451

Morales-Hidalgo D, Oswalt SN, Somanathan E (2015) Status and trends in global primary forest protected areas, and areas designated for conservation of biodiversity from the Global Forest Resources Assessment 2015. For Ecol Manag 352:68–77

Nair KSS (2007) Tropical forest insect pests: ecology, impact and management. Cambridge University Press, Cambridge

Niemela J, Koivula M, Kotze DJ (2007) The effects of forestry on carabid beetles (Coleoptera: Carabidae) in boreal forests. J Insect Conserv 11:5–18

Oliveira GF, Garcia ACL, Montes MA, De Araujo Luca JCL, Da Silva Valente VL, Rohde C (2016) Are conservation units in the Caatinga biome, Brazil, efficient in the protection of biodiversity? An analysis based on the drosophilid fauna. J Nat Conserv 34:145–150

Olson DM, Dinerstein E. Wikramanayake ED, Burgess ND, Powell GVN (and 13 other authors) (2001) Terrestrial Ecosystems of the world: a new map of life on earth: a new global map of terrestrial ecosystems provides an innovative tool for conserving biodiversity. BioScience 51:933–938

Powell AH, Powell GVN (1987) Population dynamics of male euglossine bees in Amazonian forest fragments. Biotropica 19:176–179

RAC (Resource Assessment Commission) (1992) A survey of Australia's forest resource. Australian Government Printing Service, Canberra

Rhodes JR, Cattarino L, Seabrook L, Maron M (2017) Assessing the effectiveness of regulation to protect threatened forests. Biol Conserv 216:33–42

Rylands AB, Brandon K (2005) Brazilian protected areas. Conserv Biol 19:612–618

Saura S, Martin-Quellen E, Hunter ML (2014) Forest landscape change and biodiversity conservation. In: Azevedo JC, Perera AH, Pinto MA (eds) Forest landscapes and global change: challenges for research and management. Springer, New York, pp 167–198

Savilaakso S, Koivisto J, TO V, Pusenius J, Roininen H (2009) Long lasting impact of forest harvesting on the diversity of herbivorous insects. Biodivers Conserv 18:3931–3948

Sklodowski J (2017) Manual soil preparation and piles of branches can support ground beetles (Coleoptera, Carabidae) better than four different mechanical soil treatments in a clear-cut area of a closed-canopy pine forest in northern Poland. Scand J For Res 32:123–133

Specht RL (1970) Vegetation. In: Leeper GW (ed) The Australian environment, 4th edn. Melbourne University Press, Melbourne, pp 44–67

Speight MR (1997) Forest pests in the tropics: current status and future threats. In: Watt AD, Stork NE, Hunter MD (eds) Forests and insects. Chapman and Hall, London, pp 207–228

State of the Environment (SoE) (2011) Australia, state of the environment 2011. Department of Sustainability, Environment, Water, Population and Communities, Canberra

Swart RC, Prycke JS, Roets F (2017) Arthropod assemblages deep in natural forests show different responses to surrounding land use. Biodivers Conserv 27:583. https://doi.org/10.1007/s10531-017-1451-4

Thackway R, Lee A, Donohue R, Keenen RJ, Wood M (2007) Vegetation information for improved natural resource management in Australia. Landsc Urb Plan 79:127–136

Thom D, Seidl R (2016) Natural disturbance impacts on ecosystem services and biodiversity in temperate and boreal forests. Biol Rev 91:760–781

Ticktin T (2004) The ecological impacts of harvesting non-timber forest products. J Appl Ecol 41:11–21

Tobi DR, Grehan JR, Parker BL (1993) Review of the ecologcal and economic significance of forest Hepialidae (Insecta: Lepidoptera). For Ecol Manag 56:1–12

Turner IM (1996) Species loss in fragments of tropical rain forest: a review of the evidence. J Appl Ecol 33:200–209

Turner IM, Corlett RT (1996) The conservation value of small, isolated fragments of lowland tropical rain forest. Trends Ecol Evol 11:330–333

Vasconcelos HL, Vilhena JMS, Caliri GJA (2000) Responses of ants to selective logging of a central Amazonian forest. J Appl Ecol 37:508–514

Watt AD, Stork NE, Eggleton P, Srivastava D, Bolton B et al (1997) Impact of forest loss and regeneration on insect abundance and diversity. In: Watt AD, Stork NE, Hunter MD (eds) Forests and insects. Chapman and Hall, London, pp 273–286

Wilson EO (1987) The little things that run the world: the importance and conservation of invertebrates. Conserv Biol 1:344–346

Wilson EO (1992) The diversity of life. Harvard University Press, Cambridge

Winchester NN, Ring RA (1996) Centinelan extinctions: extirpation of northern temperate old-growth rainforest arthropod communities. Selbyana 17:50–57

Chapter 2
Australia's Forest Ecosystems: Conservation Perspective for Invertebrates

Keywords Endemism · *Eucalyptus* · Forest estate · Forest loss · Forestry industry·
Myrtaceae · *Pinus radiata*

2.1 Introduction: Extent and Variety of Australia's Forests

The isolated land mass of Australia spans climates from tropical to cool temperate, with an accompanying variety of vegetation types, including forests. Those forests have a long evolutionary history, and the present fire- and drought-adapted sclerophyll forests, replacing rainforests over much of the continent from around 15 million years ago, now comprise the majority of forest vegetation communities.

The importance of the 'forest estate', both natural and silvicultural, to native insects has received relatively little notice – especially (and perhaps understandably!) in relation to concerns for the fate of iconic mammals and birds as forests are lost or exploited. There is little doubt that notable marsupial species such as the endangered Leadbeater's possum (*Gymnobelideus leadbeateri*), Victoria's faunal emblem and restricted to old growth forests in the central part of the state, and its parallels elsewhere across many tropical and temperate forest areas are important 'umbrella species' for the multitude of co-occurring invertebrates. However, even such notable and popularly-acknowledged taxa remain at the centre of continuing heated controversy between the 'conservation lobby' and forest industry bodies over their conservation status and worth in relation to forestry activities (with the human livelihoods and industry they support), whereby biologically significant forest areas may either be exploited or preserved in some way. Proposals for large–scale exploitations of eucalypt forest not within the reserves system (p. 229) include industries such as manufacture of charcoal, and biomass burning for power generation (Lindenmayer et al. 2002), activities that have potential for substantial risks to much native biodiversity. Changes to Australian forests continue to diversify in extent and variety, and the importance of forests to the country's environments and economy are appreciated widely. They are also embedded in formal policy. The eleven goals of Australia's National Forest Policy Statement (1992) embody many principles that have persisted in later documents. They include '1. Conservation: to

© Springer International Publishing AG, part of Springer Nature 2018   23
T. R. New, *Forests and Insect Conservation in Australia*,
https://doi.org/10.1007/978-3-319-92222-5_2

Table 2.1 Australia's State of the Forests Report 2013: the criteria listed in assessing forest condition (SoFR 2013)

1.	Conservation of biological diversity.
2.	Maintenance of production capacity of forest ecosystems.
3.	Maintenance of ecosystem health and vitality.
4.	Conservation and maintenance of soil and water resources.
5.	Maintenance of forest contribution to global carbon cycles.
6.	Maintenance and enhancement of long term multiple socio-economic benefits to meet the needs of societies.
7.	Legal, institutional and economic framework for forest conservation and sustainable management.

maintain an extensive and permanent native forest estate and to manage it in an ecologically sustainable manner for the full range of forest values'.

Australia's most recent 'State of the Forests Report' (SOFR 2013), a document produced at five-year intervals, summarises changes over the period 2007–2011 according to a number of different Criteria and as the most authoritative consensus over forest status and concerns for the future. That fourth report in the series has the major purpose 'to keep the public informed about Australia's forests, their management, use and conservation, and to provide information on how they are changing'. Of the Criteria used (Table 2.1), two (Criteria 1 and 3) are of particular relevance here as indicating (1) the substantial ecological background available on forest diversity and extents, with comment that fragmentation for land-use changes has led to clearing or substantial modification of about a third of Australia's native vegetation in forested areas, so that many ecological communities have become highly fragmented or reduced to a tiny fraction ('less than 1%') of their original extent; and (2) the lack of detailed knowledge of much characteristic forest-dependent or forest–dwelling biota, and its incorporation into meaningful conservation measures.

Data on forest fauna also reveals major biases, implying that it is currently impossible to incorporate invertebrates meaningfully in any general approaches to assessing conservation management needs and priorities. As Wilson (1991) remarked, referring to Victoria but with much wider relevance 'Our knowledge of invertebrate ecology in forests at any level is poor'. Thus, 1101 forest–dependent vertebrates but only 32 invertebrates are included on a national list of threatened taxa under the Environment Protection and Biodiversity Conservation Act 1999 (EPBC Act). These were selected from the total EPBC Act listings and comprise (at July 2017) Critically Endangered (10 species of total 27, of which seven are insects), Endangered (seven of 22, nine insects) and Vulnerable (six of 11, four insects) and, other than for several stag beetles (from Tasmania, p. 131), the level of forest dependence by the very few insects listed is unclear. However, despite increase in numbers since the previous SOFR, of 14 additions (Critically Endangered), three (Endangered) and two (Vulnerable) species toward the total of 32, the lack of inclusions remains disappointing. Little specific comment on any invertebrate is made in SOFR 2013, and the key needs for forest management and conservation have flowed largely from information on vertebrates (predominantly mammals and birds, with less information from other groups) and vascular plants. 'Information remains very limited on forest-dwelling invertebrates' (SOFR 2013, p. 79), with 85% of forest-dwelling insect species regarded as having minimal or inadequate information that could inform management, most of them known by little more than a name or even being undescribed.

The second most relevant criterion (Criterion 3) directly addresses forest management in relation to two universal and continuing themes – the agents and processes affecting forest health, and the impacts of fire.

Australia's native forests, whether tropical or cool temperate, are largely of endemic trees and understorey species, and support large numbers of endemic insect species, many of them apparently scarce and localised. Collectively, the forest estate comprises numerous forest types, with many of the predominant trees themselves having very limited distributions, but also includes increasing areas of plantation forests of either native or alien tree species – and which pose an additional suite of ecological scenarios and conservation considerations. As well as true forest, much savanna country is essentially open woodland in which many of the true forest insects have close relatives living on closely related plant species – and which are also under threat from destruction and wider anthropogenic changes. That open woodland shows many parallels to forest systems, and adds insights into insect diversity and vulnerability in Australia.

The major categories of concern here are rainforest (Fig. 2.1) and the southern forests other than rainforests, each with subdivisions based on predominant floristic components, architecture and climate. The brief synopsis below introduces this massive biological variety and the difficulties of categorising the various components. Within the southern temperate forests, the 'Mediterranean forest' category reflect the characteristic structure of forest related to the well-defined Mediterranean climate of parts of south west Western Australia and the 'green triangle' of far south east South Australia and extending marginally into western Victoria (Fig. 2.2).

Much conservation attention has focused on so-called 'old-growth' forests in Australia, as a term implying lack of anthropogenic disturbance and a relatively natural or pristine condition. Those forests in south east Australia have several key ecological characteristics, noted by Scotts (1991) and applying to all forest types in the region. Their key features were listed as (1) a deep multi-layered canopy resulting from presence of more than one tree age class and/or dominant and subdominant members of one age class; (2) individual live large or old trees; and (3) significant numbers of stags (standing dead trees) and large logs. Elsewhere, the key structural features of old-growth forests in Victoria have been listed as large trees, hollows, dead branches and large woody debris on the ground – with large fallen logs a significant feature of south-east Australian temperate forest. These structural features are all conspicuous indicators for forest condition as 'old-growth' but, as Burgman (1996) noted, each individual forest type will possess a suite of unique characters that define its status and render it distinctive. Burgman emphasised, also, that the wider 'old-growth estate' must incorporate both current old-growth stands and the young forest stands that must be conserved largely free of disturbance in order for them to become 'old', perhaps over several centuries of life.

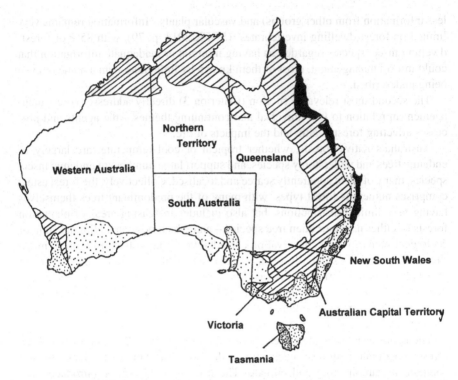

Fig. 2.1 Indicative distribution of major forest and woodland formations in Australia: tropical rainforest (black), temperate rainforest (dense dots), sclerophyll forest (intermediate dots), mallee (sparse dots), open woodland, savanna (diagonal hatch), arid zone (open). From various sources, after AntWiki (https://www.antwiki.org/wiki/Australian_ants)

Treed landscapes in Australia vary greatly, and the concepts of 'woodlands' and 'forests' are often not differentiated clearly. In general, 'woodlands' imply ecosystems in which the trees do not form any continuous canopy cover, and with foliage cover up to only about 30%. Specht (1970) distinguished a range from 'low woodland' (trees <10 m high and with up to only about 10% foliage cover) to 'tall woodland' (trees up to >30 m high, with 10–30% cover), each with understorey ranging from 'grassy' to 'shrubby'. Collectively, woodlands are floristically very varied, and have been influenced by losses and fragmentation processes equivalent to those for forests, and with analogous outcomes and concerns. Losses of the predominant woodland categories in Australia have been high (AUSLIG 1990), and few of the numerous recognised forest and woodland systems escaped substantial levels of land clearing or other exploitation from the nineteenth century on (Australian Government 2007). Conservation concerns for insects are widespread and strongly coincide with those for forests in emphasising needs for sustaining biological and structural variety, by management measures such as coppicing, pollarding, supplementary planting, and preservation of older growth.

Fig. 2.2 The distribution of Mediterranean forest climate areas (black) in southern Australia. (After Nahrung et al. 2016)

In Europe, 'open woodlands' are recognised as critical habitats for conserving biodiversity, with three major faunal elements, each paralleled in Australia. Those elements are (1) their own characteristic elements, many of them endemic, (2) species associated with forests, and (3) species associated with open habitats – and key structural features such as old trees upon which many insects and others may depend. However, woodlands have largely been eroded and diminished in much of Europe (Miklin and Cizek 2014), where rates of loss may even exceed those of tropical forests, as claimed by DeFries et al. (2005). Strong evidence for losses of the conservation capacity of European woodlands in protected areas also emphasises the critical values of remaining tracts and needs to actively conserve these.

'Forest' ecosystems generally imply predominance of taller trees and higher foliage cover. Again from Specht (1970), whose scheme was adopted widely over the next decades and has been influential in ecological interpretation, the most widely adopted categories reflect a trio of 'open forests' (trees 10–30 m high, cover of 30–70%), 'tall open forests' (trees >30 m high, also with cover of 30–70%), and 'closed forests' (trees >10 m high, but cover >70%). Forests also divide into the two geographical categories of northern and eastern rainforests and other southern forests. The former, especially, are botanically diverse and complex, with categories

Table 2.2 The Australian rainforest ecofloristic provinces and the formation group associated with each (as used by Webb et al. 1984)

Province	Formation group
A1	Warm subtropical ombrophilous lowland; cool subtropical ombrophilous lowland and upland
A2	Subtropical lower montane ombrophilous; warm temperate evergreen ombrophilous; warm temperate evergreen seasonal broad-leaved
A3	Cool and warm temperate evergreen ombrophilous lowland and upland, with evergreen conifers in places
B1	Tropical semi-deciduous lowland
B2	Tropical ombrophilous lowland; tropical evergreen seasonal lowland; tropical ombrophilous alluvial; tropical ombrophilous swamp; tropical ombrophilous upper lowland or upland; tropical ombrophilous lower montane with fragments of tropical ombrophilous cloud forest on highest summits
B3	Tropical drought-deciduous lowland
C1	Subtropical evergreen seasonal lowland, characteristically with evergreen conifers
C2	Subtropical semi-deciduous lowland

delimited by floristic surveys (Webb et al. 1984), and temperate region forests are also very diverse.

Webb et al.'s rainforest divisions were based on the presence/absence of 1316 key tree species, leading to a broad classification of rainforest into three 'ecofloristic regions', each separated into 'ecofloristic provinces' (Table 2.2). Many categories of rainforest comprise core areas, but may otherwise occur as outliers well separated from those cores and so constituting distinctive communities. That survey also endorsed that each of the three major vegetation types – tropical, subtropical, temperate – is ancient. In urging the need to conserve remaining rainforest ecosystems, Webb et al. noted 'The rapid disappearance and conversion of Australia's rainforests and their transitional types provide a chastening example of how several technological generations can obliterate complex and irreplaceable natural ecosystems …'.

Temperate region forests are also very varied and, although modifications have been suggested to the Webb et al. (1984) classification scheme, it has largely withstood changes over the period in which parallel attention to insect conservation in

those forests has developed. The southern forests occur as three geographically discrete entities – the south eastern mainland forests, the south western mainland forests, and Tasmania. In addition to closed rainforests in the east (part of Webb et al.'s
'section A'), the open forests and tall open forests contrast with closed forest by
being dominated by *Eucalyptus* s.l. species (Myrtaceae). These include iconic individual 'tall trees' of Mountain ash (*Eucalyptus regnans*). Southern rainforests are
markedly less diverse than those further north, but a number of their resident plants
evolved as understorey in tall open forests.

By 2012, Australia had lost nearly 40% of its forests (Bradshaw 2012), based on
an estimate that about 30% of land area was covered by forest at the time of European
settlement. Indeed, by 1995, Australia already had the lowest total area of remaining
closed forests (4.6% of land mass) relative to the 15 countries compared by Singh
et al. (2001). About 4% of the world's forests then occurred on about 5% of the
world's total land area.

Bradshaw's (2012) figures that Australia's native forest covers 1.474 million
Km^2, about 19% of land area, were used in his estimate of loss, of about 38% of
area. Greatest losses amongst the considerable variety of forest categories have been
in eucalypt forests, for which more than 80% of forests have been lost or degraded
severely by human interventions. Whilst precise figures for the extents of forest loss
will continue to be debated, Lunney's (1991) contention that essentially half of
Australia's forests have been lost and much more is under threat over the next
decade, with the mosaic complexity of these systems then only starting to become
appreciated, is a sobering summary. The most complete early data on losses are for
the southern temperate regions, where 'old growth forests' are critical habitat for
much fauna. Thus, in New South Wales about 35.3 million hectares had been partially cleared by 1921 (Reed 1991), and in Victoria about 56% of original forest
cover has been cleared and 70% of the remainder modified severely (Wilson 1991).
Clear-felling has been a predominant method in forestry in the south east, as in
Western Australia (Wardell-Johnson and Nichols 1991), so that impacts on natural
habitats tend to be far more severe and widespread than processes such as selective
logging (p. 200).

The authoritative Australian Forestry Standard (2007)(AFS) also gives information on forest losses and constitution, quoting that (for 2003) the total forest area
exceeded 164 million hectares, of which more than 162 million hectares was native
forests and the remainder being plantation forests. That varied forest estate included
457 of the 910 Australian plant communities then recognised, and the two main
plantation categories were distinguished as hardwood (mainly native *Eucalyptus*
and its allies) and softwood (largely introduced pines, dominated by *Pinus radiata*).
The report also cited the above figure, that about 40% of total forest area had been
lost, mainly to clearing for agriculture, since European settlement, and noted that
the major growth in plantation area has occurred only from the 1960s–1970s on.
Much of the plantation estate is under private ownership, and its expansion funded
through private investment. AFS noted 58% privately owned, 36% publicly owned,
and the remainder of the total plantation estate in joint ownership. Much of the
plantation expansion also took place on previously cleared agricultural land rather

than from further loss of native forest, as a trend increasing from the 1990s, from which time policies have progressively prohibited conversion of native forest to new plantations on all mainland States and Territories. Plantation increase has been a driver for some regional economies (Gerrand et al. 2003).

In contrast to the modern trend of broad-scale clearing, with its 'significant and irreversible effects on forest and woodland systems' (AFS), the early direct impacts for timber exploitation were less, and often transient. Selected use of preferred and commercially desirable timber tree species, however, has led to their declines, together with their associated insects. Many of these remain difficult to interpret from historical records, but one possible case is the extraction of Kauri (*Agathis robusta*, Araucariaceae), as perhaps the most attractive timber for crafting fine furniture in Australia and which was logged heavily throughout its southern Queensland rainforest range, from which it has been largely removed. Kauri cones are the putative habitat of the primitive moth *Agathiphaga queenslandensis*, with larvae thought to feed on the seeds within (Common 1993). Absence of recent records of the moth implies that it might now have become extinct as its host tree declined.

The wider and less discriminatory forest clearing is almost certain to have led to declines of numerous specialised and locally endemic forest insects, but few of these can now be interpreted reliably, again because of lack of historical records. One example is of several species of large groundbeetles (Carabidae), *Nurus* spp., formerly abundant and widespread in heavily wooded areas of northern New South Wales which have now been cleared extensively. The beetles have become extremely scarce, and are rarely encountered even by experienced collectors - but collection records indicate that they were far more common a century or so ago. *Nurus atlas* and *N. brevis* are both listed formally as 'Endangered' in New South Wales.

2.2   Impetus for Management

Informed management of such forests in the temperate regions of Australia is thus one of the nation's foremost conservation concerns. The major issues have been plagued by controversy and emotion for decades: historical essays contain much information on the development of forestry and the changing priorities, attitudes and impacts of the practices involved (Dargavel and Feary 1993). The Resource Assessment Commission's 'Forest and Timber Inquiry' (1991) had gone far toward quantifying forest losses up to that time, but has not allayed persisting demands for logging much of the remaining accessible forest area either for timber or for woodchips, the latter largely for export at a small fraction of the true commercial worth of prime hardwood timber. Only an 'uneasy truce', at most, is evident between many conservation interests and sectors of the timber industry, with numerous livelihoods (both of individuals and local communities) depending on continued harvesting and processing of forest products, and also with government agencies. The two-tiered agency influences and policies, at Commonwealth and State/Territory government levels, can add both complexity and confusion, sometimes allied to inconsistency. In essence, an administrative mosaic of management obligations and methods is superimposed on a daunting mosaic and variety of biological scenarios that also transcend administrative boundaries and biological gradients. Nevertheless, from the Australian Forestry Standard 2007 'With appropriate harvesting, silvicultural management and conservation strategies, the composition of forest has the ability to be maintained'.

The extent of the management needed is illustrated by the economic values and benefits of forestry for human society, collectively indicating the potential changes at stake. AFS (2007) noted that the value of wood product exports from Australia had more than doubled over the decade to 2003–2004, with softwood production showing a parallel doubling in production, from 1.34 million cubic metres (1990–1991) to 2.98 million cubic metres (2002–2003). Societal implications include citation that that >170,000 people were employed directly in the forestry sector in 2002–2003: a more recent figure of ca 64,300 people in forestry and forest products industry (ABARES 2017) still reflects enormous importance. Ecologically sustainable management is essential to ensure that such benefits continue, and the broad national aim of successful forest conservation (including safeguarding biodiversity) in harmony with integrated production and industry development has been urged repeatedly.

The broad patterns of forest exploitation have led to two major arenas dominating conservation concerns. These may be distinguished simplistically as the northeastern tropical rainforest, largely within a designated World Heritage Area, and the southern temperate forests with particular concerns for the roles and fate of old growth forest on the south-eastern mainland, in Tasmania, and in the far south west of the continent. However, notwithstanding attention to those foci, conservation needs extend to all forest and woodland types in Australia. Extensive local endemism amongst insects, and the ecological specialisations of many taxa ensure that enormous numbers of species may be rendered vulnerable by uncritical changes to their habitats. The figures of 39 million hectares of native forest reported as protected for biodiversity conservation, with this the primary interest for 26.4 million hectares, and 21.5 million hectares being on nature conservation reserve tenure (ABARES 2017) suggest that there is still ample opportunity for effective conservation of forest biota to occur.

References

ABARES (Australian Bureau of Agricultural and Resource Economics and Sciences) (2017) Australia's forests at a glance, with data to 2015–2016. Department of Agriculture and Water Resources, Canberra

AUSLIG (Australian Surveying and Land Information Group) (1990) Atlas of Australian resources. Vegetation Department of Administrative Services 6 Canberra

Australian Forestry Standard (2007) Standards Australia, Sydney

Australian Government (2007) Australia's native vegetation; a summary of Australia's major vegetation groups, 2007. Department of the Environment and Water Resources, Canberra

Bradshaw CJA (2012) Little left to lose: deforestation and forest degradation in Australia since European colonization. J Plant Ecol 5:109–120

Burgman MA (1996) Characterisation and delineation of the eucalypt old-growth forest estate in Australia: a review. For Ecol Manag 83:149–161

Common IFB (1993) Moths of Australia. Melbourne University Press, Carlton

Dargavel J, Feary S (eds) (1993) Australia's ever-changing forests II. In: Proceedings of the 2nd national conference on Australian Forest History. Centre for Resource and Environmental Studies, Australian National University, Canberra

De Fries R, Hansen A, Newton AC, Hansen ML (2005) Increasing isolation of protected areas in tropical forests over the past twenty years. Ecol Appl 15:19–26

Gerrand A, Keenan RJ, Kanowski P, Stanton R (2003) Australian forest plantations: an overview of industry, environmental and community issues and benefits. Aust For 66:1–8

Lindenmayer DB, Claridge AW, Gilmore AM, Michael D, Lindenmayer BR (2002) The ecological roles of logs in Australian forests and the potential impacts of harvesting intensification on log-using biota. Pac Conserv Biol 8:121–140

Lunney D (1991) The future of Australia's forest fauna. In: Lunney D (ed) Conservation of Australia's forest fauna, Ist edn. Royal Zoological Society of New South Wales, Mosman, pp 1–24

Miklin J, Cizek L (2014) Erasing a European biodiversity hot-spot: open woodlands, veteran trees and mature forests succumb to forestry intensification, succession, and logging in a UNESCO Biosphere Reserve. J Nat Conserv 22:35–41

Nahrung HE, Loch AD, Matsuk M (2016) Invasive insects in Mediterranean forest systems: Australia. In Paine TD, Lieutier F (eds) Insects and diseases of Mediterranean forest ecosystems. Springer, Cham, pp 475–498

RAC (Resource Assessment Commission) (1991) Forest and timber inquiry, Final report, overview. Australian Government Printing Service, Canberra

Reed P (1991) An historical analysis of the changes to the forests and woodlands of New South Wales. In: Lunney D (ed) Conservation of Australia's forest fauna, Ist edn. Royal Zoological Society of New South Wales, Mosman, pp 393–406

Scotts DJ (1991) Old growth forests; their ecological characteristics and value to forest-dependent fauna of south-east Australia. In: Lunney D (ed) Conservation of Australia's forest fauna, Ist edn. Royal Zoological Society of New South Wales, Mosman, pp 147–159

Singh A, Shi H, Foresman T, Fosnight EA (2001) Status of the world's remaining closed forests; an assessment using satellite data and policy options. AMBIO 30:67–69

SOFR (Australia's State of the Forests Report) (2013) Department of Agriculture, Australian Bureau of Agricultural and Resource Economics and Sciences. In: Canberra

Specht RL (1970) Vegetation. In: Leeper GW (ed) The Australian environment, 4th edn. Melbourne University Press, Melbourne, pp 44–67

Wardell-Johnson G, Nichols O (1991) Forest wildlife and habitat management in south-western Australia: knowledge, research and direction. pp. 161-192 in Lunney D (ed) Conservation of Australia's forest fauna (Ist edn). Royal Zoological Society of New South Wales, Mosman

Webb LJ, Tracey JG, Williams WT (1984) A floristic framework of Australian rainforests. Aust J Ecol 9:169–198

Wilson BA (1991) Conservation of forest fauna in Victoria. In: Lunney D (ed) Conservation of Australia's forest fauna, Ist edn. Royal Zoological Society of New South Wales, Mosman, pp 281–299

Chapter 3
Changes and Threats to Australia's Forests

Keywords Agroforestry · Alien trees · *Eucalyptus* · Forest insect pests · Forest management principles · Oil palm · *Pinus radiata* · Plantations · Saproxylic insects

3.1 Introduction: Needs for Management

The ways in which forests are managed for production and sustainability have many influences on the resident biota, and some of the effects may not be obvious or predictable before a disturbance is imposed and changes noticed. Thus, many saproxylic beetles are found only in the old-growth areas of boreal forests of Canada. The short rotation times urged by managers for maximum yields may not favour the beetle fauna, and lead to disruptions of their roles in breakdown of coarse woody material (Spence et al. 1997) and, despite incomplete knowledge of many individual species' responses, may prove to be a threat. However, understanding how forest management practices can be honed to minimise loss of insect richness and abundance without unduly compromising production forestry targets is still elusive for many forest types.

The history of forestry in Australia, although relatively short, is complex. It is not detailed here, other than to note some trends relevant to insect wellbeing: valuable sources of early information on that history include books by Carron (1985) and AAS (1981), both produced at a time when considerations of forest biodiversity conservation were very secondary to production issues. Nevertheless, broader recognition that 'Forests generate environmental issues' (Gentle 1981) was well established and, even though those issues traditionally have not involved insects other than in pest management contexts, a growing recognition of insect diversity and conservation needs in forests, and willingness to respond to those needs represents a welcome maturing of perspective.

© Springer International Publishing AG, part of Springer Nature 2018
T. R. New, *Forests and Insect Conservation in Australia*,
https://doi.org/10.1007/978-3-319-92222-5_3

3.2   Management Priorities

Moderation of intensity of interventions – such as replacing clear-felling by more restricted mosaic clearance – is a clear general benefit (Watt et al. 1997). Those insects feeding on or otherwise depending on dead wood have been a major source of information aiding forest management for conservation. Thus, in addition to the features of living trees, recognition that forest management should seek to create and retain a high diversity of dead wood for saproxylic invertebrates is widespread, with Ranius et al. (2015) emphasising that features such as tree species and whether standing or downed wood is involved were also both important, as reflecting substrate specificity or 'preference' for particular taxa. However, simply because it is far easier to determine favourable characteristics for saproxylic insects at a 'dead wood item' scale than at a stand or landscape scale, the full implications of such differences for management are unclear. Thus, for some rare and specialised species, habitat connectivity (ability to reach additional dead wood items) may be more important than the precise substrate available. At any scale, heterogeneity of resources and microhabitats – in both natural and managed forests, including plantations, is likely to favour richness of insects associated with living trees and saproxylic insects, and providing that heterogeneity amongst dead wood includes species, position, decay stage, and extent of shading. The main focus of management is suggested to be combining adequate supply of the basic resource of dead wood with environmental variety. The principles of retaining heterogeneity in plantation management are accepted widely. Hartley (2002) endorsed these in a broader review of how to conserve biodiversity in plantation forests, based on acceptance that these can indeed have substantial conservation values at all stages from establishment to harvesting, and within a variety of landscape contexts. Preservation or retention of some natural vegetation there is a key need (as biological legacy), for example, and tree harvesting later should leave some mature trees or native remnants (p. 200). Hartley gathered plantation management recommendations under four groups (Table 3.1), and many of those measures can have conservation benefits with little or no decline in plantation productivity.

Lindenmayer et al. (2006) set out five 'guiding principles' for biodiversity conservation in forested areas, each associated with a 'check list' of measures that may collectively help evaluation of those principles. The principles, listed here to indicate the broad scope of management needed, are (1) maintenance of connectivity; (2) maintenance of heterogeneity within the landscape; (3) maintenance of structural complexity within stands; (4) maintenance of aquatic ecosystem integrity; and (5) the guidance of human disturbance regimes by natural disturbance regimes. Each applies intuitively to many kinds of ecosystem as a collective and adaptable framework for conservation management, but the last has become a widespread model for silvicultural management.

Forest clearing, the removal of entire stands with associated changes to ground and understorey communities through compaction and removal, is obviously one of the most abrupt and severe environmental changes still taking place in terrestrial

Table 3.1 Some management recommendations for aiding biodiversity conservation in plantation forests. (After Hartley 2002)

Harvest considerations
Leave as many snags and cavity tree as possible during plantation establishment; plan to leave native vegetation throughout second rotation by one or more of (1) dispersed individual retention trees, (2) aggregated clumps, (3) linear strips, (4) buffer strips along aquatic habitats; delimb trees near the stump and leave tops on site; instead of clear-felling, manage others though irregular shelterwood and selective felling.
Species composition
Where possible, use native species rather than exotic species; if exotic species used extensively, increase emphasis on retaining areas of native vegetation; juxtapose exotic and native stands in the landscape; retain or underplant especially important plant species; maintain genetic diversity by using a variety of planting stock; use multispecies stands with appropriate integrative management with native taxa.
Site preparation
Avoid intensive preparation that involves disturbing soil nutrients, promotes soil erosion and reduces/destroys woody debris; leave some snags and logs (coarse woody debris) after site preparation; consider control burning to promote native understorey plants; emulate natural disturbance regimes in assessing fire frequency.
Tending
Thin some plantations earlier and more intensively to stimulate diverse understorey; leave others unthinned to create mosaic conditions; maintain plant variety, and use herbicides sparingly.

ecosystems. The various components of moderation needed to progressively harmonise wider interests broadly incorporate changes to the extent of forest despoliation, and the spatial and temporal patterns over which those changes occur. Forest preservation and restoration contribute to countering habitat loss and fragmentation, and to assuring continuity of key resources and natural variety of native resident biota. The high levels of species extinctions attributed to large-scale forest losses are a widespread concern, even if not documented in detail for many cases. The more imponderable impacts of processes such as microclimate change superimposed on direct disturbances may affect how the various issues are considered, but the complexity of forest environments dictates that priority areas for conservation management may be designated, and also that an 'ecosystem management approach' may be most effective. Nevertheless, and paralleling much insect conservation in other environments, studies on individual focal threatened species in forests contribute to both progress and wider understanding.

Most forestry regulation and management that incorporates conservation in Australia has been developed primarily for vertebrates and plants, with the few demonstrably successful cases helping to draw attention to the values of such 'flagship species', with implications of less publicised 'umbrella roles' in assuring habitat continuity for invertebrates and other less-attended taxa. Many insects are the welcome, if unheralded, beneficiaries of campaigns for the conservation of better-known and more popular forest-dependent biota.

3.3   Plantation Forestry

The history of forest plantation development was initially driven largely by increasing demands for timber and wood products, with realisation that concentrating on selected tree species for these resources and using them to replace mixed forest with low density of the most suitable timber trees (so, much commercial redundancy) would be commercially beneficial. In the controlled conditions of commercial plantation forestry, selection for features such as increased growth rate and practical modifications of spacing, fertiliser applications and other measures to promote productivity and uniformity – including mechanical thinning and harvesting, and suppression of competing plant species – can occur. These needs have led to focus on extensive plantations of various conifers (as softwoods) and hardwoods such as *Eucalyptus*, in many parts of the world. Local demand, often accompanied by export needs, for construction timber, pulpwood or other purposes, dictate the taxa used. Many different trees can become involved – Nair (2007) noted 'about 170 species' in plantations in India, and 80 in Malaysia, for example, and also that increasing numbers of species are progressively being incorporated. More than 400 tree species were then being exploited in Cameroon alone. However, most forestry plantations have focussed largely on pines, eucalypts and acacias, with the greatest needs for pulpwood production based on short rotation periods. Each plantation of an alien tree species represents both (1) an intrusion into the region's natural environment, with possible displacement of native species, and (2) a possible resource for native insect species to exploit, many of these then viewed as pests and targets for suppression. Each alien tree species may also invade native vegetation, with more widespread impacts on the insects present.

In addition, replacement of forest by crops, ranging from local subsistence needs to massive commercial expansions of commodity crops such as oil-palm (p. 47) poses another suite of concerns for conservation. Each plantation, or crop, represents major change from the parental forest, with consequences including incidence of new insect species (many of them alien, and some likely to become pests), and applications of fertilisers and pesticides in the broad interests of crop protection and enhancement. Many of the impacts are on native biota, and range from direct destruction to more subtle alien species' influences, novel interactions, and changing food webs and resource availability and synchrony.

The roles of forestry plantations in conservation continue to be debated. As Carnus et al. (2006) noted 'there is no single, or simple, answer to the question of whether planted forests are good or bad for biodiversity', with features such as area, species and ecological context interacting also with local social and economic influences on land use. Large-scale monoculture tree plantations have drawn much comment on their impacts on the biodiversity present earlier on those sites, but in wider context tree plantations can also provide many of the ecosystem services and conservation roles attributed to more natural forests (Paquette and Messier 2010).

The term 'plantation' thus covers a very varied array of ecological regimes, and the formerly widespread implications that they are 'biological deserts' is by no means unchallenged. Characteristics of some plantations may accord them unexpectedly high biodiversity, and the parameters of comparing plantations with parental or 'natural' forest also vary greatly across studies. Stephens and Wagner (2007) noted comparisons with natural forest, semi-natural forest, natural non-forested ecosystems (such as grassland), logged forest, agricultural lands, and other plantations – with most studies they reviewed purportedly comparing plantations with natural forest, notwithstanding that many of the plantations had been established on non-forested lands. However, the above comparison categories each included studies in which invertebrate diversity was equal or greater in the plantations than the environments with which they were compared. Stephens and Wagner regarded many of the studies as making inappropriate comparisons, such as comparing alien monoculture plantations with natural forest conditions, or ensuring that comparisons are with the land use they replace.

Forest plantations in Australia fall into two broad categories – based repectively on use of native species, and use of alien tree taxa. Both are primarily to supply timber (or fibre derivatives such as woodchips) in crop form. The first focusses on hardwood eucalypts, notably *Eucalyptus nitens* (Shining gum) and Blue gum (*E. globulus*), but debate continues over the proper epithet for such native species planted well beyond their natural range but in the same country – the widespread use of Tasmanian blue gum (*Eucalyptus g. globulus*) on the Australian mainland, for example, is at one level the introduction of a previously absent alien species to that region, with opportunity for novel ecological interactions to be generated, or arise. By 2013, eucalyptus plantations comprised about 49% of total Australian timber plantation area (Frew et al. 2013). Most eucalypt plantations in Australia have been founded on *E. globulus*, *E.grandis* or *E. nitens* and, from one perspective, can be viewed as complementing the resources potentially available to native insects using other eucalypts in the plantation areas. However, plantations and other production forests are anticipated widely to support less diverse insect communities than do natural forests, reflecting reduced plant richness and lack of structural or age-related variety, such as understorey and woody debris. An important 'test case' for this premise has been the extensive plantations of *E. globulus* on cleared agricultural land in southern Australia. Congeneric trees are distributed widely and naturally across the region, and support complex insect communities as remnant sources for colonisation of plantations.

In Western Australia, the insects of *E. globulus* and remnant native *E. marginata* were compared, together with insects of open parks near each of the eucalypt treatments (Cunningham et al. 2005). Although plantations contained many of the natural forest species of Coleoptera, Hymenoptera and Lepidoptera, they had lower richness and a few more abundant species – some of these known forestry pests. Much of the plantation insect fauna was essentially a subset of the remnant forest fauna, but a relatively few species were not found in forests. Rather than these being claimed as plantation specialists they could alternatively simply be rare in forests and so unrepresented in the samples examined. Nevertheless, the acceleration of eucalypt plantation establishment over recent decades, with concomitant shift away from expansion of *Pinus radiata*, has enabled far greater interchange of native biota between plantations and native forests. Of the 85 or so significant insect pests of *Eucalyptus* plantations in Australia, the great majority normally use the same or related host trees in native forests (Strauss 2001). Strauss noted also that 'exchanges' occur in both directions, from forest to plantation and the converse, so that *E. globulus* plantations could act as a 'conduit' for pest insects to reach native forests in which suitable hosts may ocur. The latter may reflect the large insect populations that can build up under ecologically favourable conditions in plantations and their spillover into nearby native forest remnants – perhaps then disadvantaging less abundant native resident species there.

A few native herbivorous insects have become pests in Australian plantations: Steinbauer et al. (2001) noted, for example, that the geometrid moth *Mnesampela privata* is often quite rare in native eucalypt forest but larvae can cause immense damage to plantations of young *Eucalyptus globulus*, in part because of the large amounts of young foliage available. One widely claimed benefit for plantations is the reservoir of natural enemies of native pests that remain in native forests and that can help to control their hosts/prey species should they invade plantations. Outbreaks of *M. privata* occur only in plantations, and this is perhaps aided by the ease with which larvae, having defoliated their natal tree (where large, multiparental egg masses can lead to high local damage) can disperse and colonise other trees nearby. In native forest, a variety of host and non-host eucalypt species might reduce this capability to build up populations, in contrast to monoculture crops in which populations of natural enemies may also be lower.

Attractions of eucalypts for plantations include their rapid growth and production of high quality wood useful for both timber and paper manufacture, as well as their wider uses for shade, windbreaks and ornamentals in many parts of the world. At least 90 species of eucalypts have been planted in California alone as landscape and windbreak trees since around 1850 (Doughty 2000), for example. An additional attraction in the early years of such use was the low levels of insect pests and diseases, but this pattern has changed radically since than – Paine et al. (2010) noted that 16 Australian insect pests established in California from 1983–2008, adding to the two species present before then. In California, no native insect species have shifted on to *Eucalyptus* and, as the stocks were initiated from seeds, no introductions of Australian insects occurred with the plants when they were established. No native Myrtaceae occur in California so that, as Paine (2016) remarked, there is no local pool of host-related insects that could transfer readily to eucalypts. The history of later colonisation by Australian insects has represented mostly recent arrivals (Paine 2016). Many of those adventive species occur in several regions in which *Eucalyptus* plantations have been established: for example, some of the Coleoptera now thrive in North America, Europe and the Mediterranean area, South America and southern Africa. Paine et al. also suggested the possibilty that some of the introductions to California may have been intentional, with opponents of eucalypts perhaps advocating use of alien insect pests as biological control agents to suppress the trees. *Eucalyptus* plantations in Chile are likely to increase in extent, with the small array of introduced Australian insect pests discussed by Lanfranco and Dungey (2001) also likely to increase, perhaps especially in climatically marginal areas where monitoring and management may need to be planned carefully. At that time, none of the alien insect species posed any economic threat in Chile, but Lanfranco and Dungey raised the practical matter that many plantations were owned by small or medium-scale proprieters, who might not be able to afford to control serious pest outbreaks.

Historical recognition of the plantation advantages and potential of eucalypts flows largely from early importations to arboreta and botanical gardens in Europe (Turnbull 1999), and global plantation areas continue to increase from the 'at least 12 million hectares' quoted by Turnbull, with increase in large scale industrial plantations, and many more local use plantations across rural landscapes in many places.

One common feature of native tree plantations is thus that rotation times are relatively short, in the interest of harvestable productivity, so that mature trees do not develop. Longer rotation times are a commonly recommended conservation measure (Harris 1984) in order to help diversity conservation, but such older plantings may still lack many of the attributes of true old-growth forests, most notably dead wood. Nevertheless, many declining saproxylic beetles not found in mature forests can occur in 'over mature' plantation stands (Martikainen et al. 2000).

More usually, alien trees are represented by introduced softwoods to provide a supply of timber otherwise not naturally available within Australia. Extensive conifer plantations, predominantly of Monterey pine (*Pinus radiata*) in the south east dominate that industry, many of them on ground occupied previously by natural eucalypt forests. In addition, and posing rather different conservation scenarios, long-cleared agricultural land may be converted to pine or eucalypt plantations. This practice, which is more widespread elsewhere, led to establishment of pine plantations on former heathland in the northern hemisphere, largely in response to needs for increased wood production since the nineteenth century, and represents replacement of a complex natural non-forested biotope by the plantation (Vangansbeke et al. 2017).

Eucalypt plantations in Australia continue to increase in importance, but their extent is still smaller than that of exotic conifer plantations (dominated by *P.radiata*) – but they are seen as more environmentally acceptable. Pine plantings commenced around 1876, and their later establishment on sites converted from native forests led to considerable conservationist opposition from the 1970s on, augmented by claims that lasting harm resulted to the ground involved. Reviewed by Turner et al. (1999), soil changes and increased management intensity of conifer plantations led to impacts on the sites, as well as potential for increased soil compaction and erosion, together with increased water use. Expansion of eucalypt plantations seems assured – one earlier 'vision' for Australia's plantation future is to have three million hectares of *Eucalyptus globulus* and other eucalypts by 2020 (Anonymous 1997, Australian Government 2007), spanning both the species' natural and 'extended Australian' ranges, so that the collective range of sites encountered will differ enormously in features and suitability, as well as in their local insect faunas.

The Australian Forestry Standard (2003, 2007) defined plantations as 'Stands of trees of native or exotic species, created by the regular placement of cuttings, seed-

Table 3.2 Principles for enhancing wildlife conservation in plantations (Lindenmayer and Hobbs 2007)

General principles
1. Habitat suitability is critical
2. Some fauna may not persist in plantations, so replacement of native vegetation by plantations is a negative outcome

Landscape level principles
1. Landscape heterogeneity/diversity can promote conservation values; different age structure and composition are important components of this
2. Remnant native vegetation: Larger remnants within planted areas more likely to persist and be viable
3. Remnant native vegetation: Patch shape can influence some edge effects
4. Riparian areas can have considerable conservation value as habitats and may function as dispersal routes
5. Connectivity: Physical links between patches (both remnant vegetation and planted areas) may assist dispersal
6. Adjacency and landscape context: Conservation value of plantations likely to be greater if near native vegetation. Nature of surrounding landscape important

Stand level principles
1. Tree species diversity: Greater diversity provides greater range of resources and can support more species
2. Stand structure: Range of features needed, including logs, large trees, hollows and others
3. Stand age: Older trees likely to develop features needed by some species

lings or seed selected for their wood-producing properties and managed intensively for the purposes of future timber harvesting'. Writing generally on the importance of the Australian plantation estate for wildlife conservation, Lindenmayer and Hobbs (2007) listed five key reasons for considering this theme. They noted (1) that the plantation estate, whether based on native or alien tree species, is likely to triple within a few decades: DPIE (1997), for example, projected that plantations could reach three million hectares by 2020, so that siting and management of new plantations may have severe impacts on biota currently thriving in those pre-plantation landscapes: (2) key ecosystem processes, including pest control, may be improved and sustained by the natural biota; (3) that some species cannot be conserved in plantation-dominated landscapes, but many others may be amenable to conservation through modifying forest management – in some cases to only minor extents; (4) maintenance (or loss) of native biota in plantations is increasingly relevant in setting ecological standards for plantations; and (5) simple plantation forestry may itself not achieve societal demands for plantation outputs or be compatible with ecological sustainability.

Key studies on the biota of Australian forestry plantations, with far more appraisal still needed, led Lindenmayer and Hobbs to list important findings then to hand (Table 3.2). The first of these counters the often-claimed theme of pine forests being 'biological deserts', whilst the second topic emphasises that the assemblages of native species present differ in detail from those of parental forests, or of landscapes

Table 3.3 Main methods to improve levels of biodiversity in plantation forests (Spellerberg and Sawyer 1996)

1 Diversify age structure of the plantation
2 Provide open space habitats
3 Provide dying and dead wood
4 Increase tree species richness
5 Maintain stands of mature tree beyond normal economic maturity
6 Sensitive management of riparian zones, woodland set-asides and remnants of native vegetation; incorporate buffer zones
7 Increase resources for wildlife - for example, nest boxes
8 Landscape design
9 Sensitive management of understorey vegetation

dominated by cleared land. Dominant vegetation cover within and near plantations can enhance native species richness, and the sizes and shapes of remnant patches affect the array of native species that can thrive. Much of that account deals primarily with vertebrates, and Lindenmayer and Hobbs (2007) noted the preponderance of studies on birds in published literature. The principles for conservation are equally relevant for invertebrates.

Those principles are also central to the series of management recommendations for enhancing biodiversity in plantations listed by Spellerberg and Sawyer (1996, 1997) (Table 3.3). The three tables above overlap in many details and collectively display the major enduring needs that can safeguard and improve the values of plantations for insect life.

Concerns over clearing native forests for *Pinus radiata* plantations in Australia accelerated from the 1960s on, with initial concerns focusing on the fate of arboreal marsupials. A large number of surveys over the next decades confirmed that vertebrate assemblages were much less diverse in pine plantations than in native forests. Comparisons of invertebrate assemblages in pine plantations and native eucalypt forests implied similar reduction of richness across many groups. Studies in Victoria (Ahern and Yen 1977, Neumann 1979) were amongst the first to proclaim such differences. Comparisons have also been made between pine plantations and eucalypt plantations (Bonham et al. 2002), in which several invertebrate groups (Onychophora, Gastropoda, selected Diplopoda, carabid beetles) were surveyed by hand-collecting across 46 localities in northern Tasmania. At each site a plantation block (one of *Pinus radiata* [26 sites], *Pseudotsuga menziesii* [3 sites], together 'pines', or *Eucalyptus nitens* [17 sites]) was paired with a nearby native forest patch. The 264 individual carabids represented 16 species, all of which were found in plantations. Two species were not found in native forests. Three of the flightless species (*Lestignathus cursor, Percasoma sulcipenne, Theprisa* sp.) were all more common in native plantations. In contrast, the flightless *Prosopogmus chalybeipennis* was more common in pine plantations.

The diverse responses of ground-dwelling insects to pine plantation environments have been assessed mainly by comparative pitfall trap surveys across different

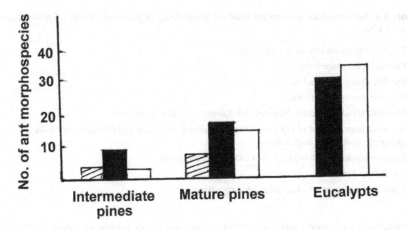

Fig. 3.1 The numbers of ant morphospecies in intermediate-aged and mature pine plantations and eucalypt forests in three sites (hatched, black, open) in Victoria, Australia (Sinclair and New 2004) (note that different sampling effort invalidates direct comparison across sites, but relativity across treatments at each site is more acceptable)

forest regimes. Study of different taxa has sometimes inferred very different impacts. The diverse epigaeic beetles of eucalypt forest and three age stages of *P. radiata* plantations established on land formerly occupied by that forest in Victoria showed rich assemblages in all four treatments (Gunther and New 2003), evidencing that the plantations can support considerable diversity. This was also the case found earlier by Neumann (1979), leading him to assert that 'Australian pine environments can provide habitats for a wide range of insects', among which beetles can be abundant. Neumann's surveys were based on Malaise trap catches, so probably included a far wider ecological spectrum of beetles than pitfall traps provide. However, and acknowledging different sampling effort through bulking the pine treatments, both studies suggested high richness in both vegetation systems - eucalypt: pine morphospecies richness of 199:241 (Neumann), and 114:176 (Gunther and New). Both studies also showed considerably heterogeneity among sites across and within treatments. In contrast to these beetles, epigaeic ant assemblages manifested considerable differences in diversity and composition between eucalypt forests and *P. radiata* plantations in the three sites in Victoria surveyed by Sinclair and New (2004), with the pine ants being an impoverished subset of the local eucalypt forest assemblages. In older, thinned, pines in which understorey vegetation was well-developed, numbers of ants were substantially greater than in unthinned pines, but were still far lower than in eucalypts (Fig. 3.1). Conservation of any localised or specialised ant species in the region may depend on presence of native forest patches as refuges within otherwise altered landscapes. Considerable differences in functional group distribution also ocurred, with opportunists (such as *Rhytidoponera* spp.) predominant in pines but comprising only about 6% of eucalypt ant individuals; dominant Dolichoderinae (notably *Anonychomyrma convexa*) were predominant in eucalypts and absent from most pine sites.

Surveys in other parts of the world endorse that insect assemblages may be transformed by replacing one tree category with another under plantation conditions, whether the plantation is of native or alien plant species. Beetle assemblages in natural oak (*Quercus liaotungensis*) forest in northern China, for example, differed from those in mature conifer plantations replacing that oak parental forest (Yu et al. 2010). Many oak forest species were absent from, or substantially less abundant in, pine plantations, and preservation of oak forest was identified as a key need to conserve the local forest beetle fauna. The principle of conserving native trees as needs and calls for alien tree species increase is perhaps a universal need.

Plantation forest can provide valuable habitats for native biota, including some threatened insect species for which other, more 'natural', habitat has disappeared. The locally endemic New Zealand carabid beetle *Holcaspis brevicula*, for example, now depends entirely on pine plantation forest (Brockerhoff et al. 2005). All 10 known specimens were collected in the same plantation forest (Eyrewell Forest), and extensive trapping in nearby native forest areas on the Canterbury Plains over five years yielded 47 other species of Carabidae. Eyrewell Forest was planted only from 1926, and the original habitat of *H. brevicula* is unknown. However, all pine stands in the plantation have been clearfelled at least once or twice, and this supports that the beetle can withstand that disturbance – whether through refuges, colonisation from some yet unknown source, or both of these. Conservation measures are thereby difficult to suggest, but if future surveys reveal any local 'hotspots' in or near the plantation, those areas could be protected and managed as sympathetically as possible.

Mixed species plantations can lead to increased biodiversity simply through providing wider resources – not only for specialist insects but with wider attributes such as encouraging a broader spectrum of potential natural enemies of value in pest management.

Pitfall trap comparisons of spiders, Carabidae and Staphylinidae in single species stands of native Ash (*Fraxinus excelsior*) and introduced Norway spruce (*Picea abies*) and mixed plantations of these species in Ireland (Oxbrough et al. 2016) revealed that only carabid richness was affected by presence of spruce. The importance of including a native species in mixed plantations in supporting diversity led to recommendation of a 50:50 ratio of native to non-native trees in mixed plantations – in this study, native ash was clearly important for ground beetle conservation.

In a related context, also increasing heterogeneity and resource variety, presence of understorey in pine plantations facilitates colonisation and resilience by displaced local species. The Chilean endemic carabid *Ceroglossus chilensis* was favoured by understorey in *P. radiata* plantations (Russek et al. 2017), so that allowing development of understorey in those plantations established after clear-felling of native forest helped to offset the impacts incurred. Earlier, Cerda et al. (2015) recorded the presence of *C. chilensis* in mature pine plantations, especially in stands with well-developed understorey. The flightless beetles 'preferred' sites with dense understorey, allowing plantations to function as surrogate natural habitats, much as for *H. brevicula* in New Zealand (above). Recognising the values of understorey for insects and fostering its presence in plantations may be an important general contribution to biodiversity conservation, but involves some modification to the more traditional approach of plantation management, of removing all understorey growth. In New Zealand *P. radiata* plantations, the pines formed a 'monoculture canopy', whilst the more diverse understorey shared many species with the native Kunzea (*Kunzea ericoides*) forests they replaced. As Berndt et al. (2008) noted, in such cases the plantation may provide habitat for some species of carabids that can not survive in agricultural or other more open areas. Comparison of a wider range of beetles in New Zealand *P. radiata* plantations and adjacent habitats showed that mature plantations and clear-felled areas had the higher species richness, and native forest and pasture had fewer species (Pawson et al. 2008). This inferred that mature pine stands were the most suitable non-native habitat to augment the beetle fauna of native forest, notwithstanding the higher relative abundance of some native species that characterise the latter fauna. The importance of understorey in forest plantations was emphasised also by Brockerhoff et al. (2013), both for increasing structural complexity of the plantations as a general contribution to enhancing heterogeneity and for providing habitat for species not associated with the canopy tree species.

Parental systems other than native forest can also be placed at risk from plantation establishmant. Conversion of heathlands to pine plantations in parts of Europe, for example, has led to heathland becoming a rare and threatened ecosystem signalled for protection under the European Habitats Directive (Vangansbeke et al. 2017). Plantations of timber trees on former pasture or other agricultural land help to reduce logging pressures on remnant native forests, but (as in forest-based plantings) their establishment may also necessitate measures to counter seedling loss from insects or other herbivores. Local context and effects may be important, and insecticide applications may be necessary to protect young trees (Panama: Plath et al. 2011). In former pasture areas, tree plantations become the prevalent landscape form, with other elements (native forest remnants, cleared pasture patches, scattered trees) embedded within this.

Resumption of agricultural land in Australia is exemplified by a study based near Jugiong, New South Wales, in which pine plantations were established in 1998, with a series of small remnant patches of eucalypts selected for exemption at that time, as the foundation of a planned long-term study of the trajectories of biodiversity changes (Lindenmayer et al. 2001). Most of the 52 remnants were less then five hectares in area. That longitudinal study was designed primarily to examine the responses of vertebrates, notably arboreal marsupials, but with terrestrial mammals and reptiles also appraised across a larger array of 126 sites, over at least 10 years – a period and scope unusual amongst such surveys and which has potential to provide far greater insights on outcomes from successional and fragmentation changes than short-term or 'spot' surveys. Ground-dwelling beetles were sampled by pitfall trapping across 30 of the study sites, six each of the five categories (1) woodland patches near both pine plantations and farmland; (2) farmland; (3) woodland patches close to farmland; (4) woodland patches close to dense pine plantations; and (5) pine plantations. Only a single trapping period was undertaken, of three weeks in late summer 2013, but this yielded 562 beetles (130 morphospecies, 28 families) (Sweaney et al. 2015). These were categorised by trophic role, flight capability and body size. Significantly fewer flightless species, and lower richness and abundance of large-bodied beetles occurred in pines and patches embedded in pine plantations than in other site categories. Those species were considered to have low dispersal capability, but the findings suggested that pine plantation establishment had led to their losses there and in the embedded patches, so contributing to biotic homogenisation. In this context, the pines may also restrict dispersal of grassland and pasture beetles, for which movements are hampered by taller vegetation. The cleared farmland supported more beetles than any other treatment, with many species found only in farmland – perhaps because the species previously sensitive to this change from native vegetation had already become extinct. Sweaney et al. also suggested that plantation expansion in areas where the pines contrast strongly with native vegetation might increase risks of species declines in such fragmented landscapes. The level of contrast between pines and nearby natural eucalypt-associated environmental variables may be linked with homogenisation. Native defoliating insects are a major concern in eucalyptus plantations, and some can become sufficiently abundant to kill large numbers of trees, and contribute to 'dieback' (p. 146). Impacts of plantation management (perhaps most notably influences of irrigation and fertiliser applications) on ecology and population dynamics of those insects can thus contribute to sporadic or more regular 'pest status', and some taxa have been studied in considerable detail. Defoliating scarab beetles pass their larval life feeding underground in the grassy understorey of plantations, where fertiliser applications can induce increased and nutritionally enriched grass food. In plantations of *Eucalyptus saligna* in New South Wales (Frew et al. 2013), fertilisation was associated with increased scarab (notably species of *Sericesthis*) larval populations and might induce 'reservoirs' of scarabs that lead to later increased tree defoliation by the emerging adults.

Numerous parallel exercises elsewhere contribute to ecological understanding of the processes and consequences of such transformation. In Japan, monoculture plantations (mostly of conifers, and notably Japanese cedar [*Cryptomeria japonica*] and Hinoki cypress [*Chamaecyparis obtusa*]) increased markedly after World War II, and their replacement of natural or secondary broad-leaved forests represents one of the largest regional transformations of natural habitats (Makino et al. 2007). The anticipated decline in insect richness was explored by comparisons of timber beetles (Cerambycidae, Disteniidae) as possible surrogates (p. 113), with an overall richness of 99 species in broad-leaved forest (10 plots) and 66 species in eight conifer plots. The survey was essentially a chronosequence in which beetles were compared, by Malaise trap samples, on different age classes of the two forests. Average richness was higher in broad-leaved plots, and in younger trees of both categories (Fig. 3.2). Earlier study of a similar suite of forests in Japan, using baited 'collection traps', had demonstrated that the timber beetle assemblages of old-growth forests (with no record of clearing for at least 120 years) were distinct from those of conifer plantations (30–40 years old), with those of second growth forest (30–70 years old) having intermediate assemblages (Maeto et al. 2002). A number of the more

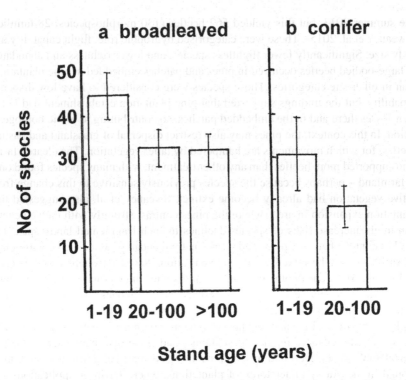

Fig. 3.2 Species richness of longicorn beetles in relation to stand age in (**a**) broadleaved and (**b**) conifer plantations in Japan. (Makino et al. 2007)

common longicorn species (from the collective samples comprising 25,115 individuals and 82 species) were either specific to old-growth forests or closely associated with them. Some had high indicator values, whilst other species had such values for conifer plantations. Six of the 12 species entirely or closely associated with old growth were species of the widely distributed Holarctic genus *Pidonia*, which Maeto et al. (2002) suggested might be valuable indicators for monitoring the processes of forest retention. Nevertheless, species richness was diminished in both conifer and second growth forests, the latter possibly reflecting paucity of larger-diameter trees. Conifer plantations in Japan apparently cannot fully replace or cater for the longicorn beetle diversity of natural forests. In Makino et al.'s (2007) survey, only two of the 468 species for which host tree records were available were true specialists on Japanese cedar or other conifers. Replacement of broadleaved trees by conifers was thus postulated to diminish habitat for numerous native beetles, and lead to impoverished local assemblages.

The roles of alien tree plantations in Ireland are likely to be paralleled elsewhere: most forest plantations in Ireland are of non-native conifers, predominantly Sitka spruce (*Picea sitchensis*) or Norway spruce (*P. abies*) (O'Callaghan et al. 2016) that have values as surrogates of the natural fragmented forest for many generalist insect species and some taxa of conservation concern, but are very poor habitats for native forest specialists. Not surprisingly, the richness of canopy beetles was significantly greater in oak/ash woodlands than in plantation conifer forests (Pedley et al. 2014), whilst the similarity of carabid richness across these forest types was attributed to the general paucity of forest specialist taxa (Irwin et al. 2014). Canopy fogging (p. 81) thus clearly demonstrated differences between native and conifer plantation forests in Ireland.

In general, plantations can support native biodiversity through three major mechanisms: (1) as supplementary or complementary habitat for forest species; (2) promoting habitat connectivity across landscapes; and (3) providing buffering effects along edges. The tree species and relationships in any particular plantation influence those roles, with alien trees often deemed less hospitable than native species. Whatever tree species are involved, the associated insect assemblages will change in character over the plantation's life. Epigaeic arthropod assemblages in plantations of Norway spruce (*Picea abies*) in the Czech Republic were studied by pitfall traps across 15 differently-aged stands, where Purchart et al. (2013) showed that litter depth, cover of ground herbs, the amount of coarse woody debris (p. 156), and extent of canopy closure affected composition of all focal groups studied. The only insect group included in that study, Carabidae, showed the trend typical in other contexts, of increased dominance of forest species in older stands, together with decreased representation of generalist and open-habitat taxa, but with their changes in abundance and total richness not significant among the 29 species represented.

3.4 Agricultural Conversion

In contrast to the replacement of agricultural land by forestry plantations, much agricultural development involves replacement of forests by planted crops, other than forestry trees, or by grazing land. These processes may involve severe and abrupt changes in structure, with removal of most or all preexisting vegetation for establishments of monocultures of alien plant species for which continuing and intensive protective management is needed.

Those monocultures may involve alien grasses favoured for pasture, or more complex cropping systems involving annual and perennial species. Oil-palm plantations are simply one example, attracting particular attention from conservationists because of the preparatory clearing of large areas of tropical forests. Although not currently a major focus in Australia, the expansion of oil palm plantations in nearby regions and elsewhere in the world's tropics and subtropics illustrates many of the more general conservation problems and concerns that can arise. Indeed, oil palm plantation expansion is one of the most extreme manifestations of so-called 'international commodity agriculture' across the wet tropics but, as Livingstone et al. (2013) commented, very few studies have directly examined impacts on biodiversity, especially in relation to ecological functions. Across 13 comparisons of animal taxa in forest and oil palm plantations, species richness in the latter was only 15% of that in forests (Fitzherbert et al. 2008). Studies on impacts of oil palm plantations on insects are predominantly on terrestrial taxa, but a study in Brazilian Amazonas demonstrated the significant changes in aquatic insect assemblages in plantation areas (Luiza-Andrade et al. 2017). Larvae of Ephemeroptera, Plecoptera and Trichoptera were identified to generic level, and all three orders were affected by the plantations. Retention of effective buffers, such as riparian vegetation strips, to counter impacts on stream environments may be a key conservation need.

The needs to minimise additional impacts on biodiversity in establishing further oil palm plantations are recognised widely. In general, expansion of oil palm planta- tions represents a major threat to global forest biodiversity, with substantial losses of old-growth forest taxa, and it has been suggested that impacts within the Neotropics (even though they have not been reported in detail for many groups) may be even more severe than those reported for south-east Asia (Alonso-Rodriguez et al. 2017). Richness and diversity of two important moth groups, Geometridae and Erebidae: Arctiinae, were both reduced in Costa Rican oil palm plantations when compared with old-growth and secondary forest. Light trap surveys in the understo- rey near the Piedras Blancas National Park, where oil palm plantations have been extended considerably in recent years, yielded 170 geometrid species and 139 arc- tiines, with relativity (old-growth forest interior: oil palm plantations) being 113:31 and 81:35 species respectively. These two environments thus supported very differ- ent moth assemblages (Alonso-Rodriguez et al. 2017), but forest margins and young secondary forest did not differ substantially from each other. The assemblages of Geometridae were more distinct than those of Arctiinae, suggesting that they might be a useful indicator group (p. 113) for forest condition.

In addition to direct loss of native forests to oil palm, the permeability of the plantations may affect species' access. A survey of Costa Rican orchid bees showed that their richness, abundance and community similarity in oil palm declined as distance from the forest increased (Foster et al. 2011). Some of the 26 species could form populations in both systems, but only two bee species occurred only with oil palm, in contrast to seven found only in forest. A much lower abundance of orchid bees in oil palm implied that the plantations were not a high quality matrix for their movements, in turn leading to implication that landscape-level management of oil palm is critical for conservation.

The alternative trajectory for oil palm plantation establishment, from pastures rather than from forests, clearly lessens further forest destruction and can incur fewer future impacts on biodiversity – whilst recognising that any earlier forest clearing to prepare the agricultural land now used may have had serious impacts at some earlier stage. In Colombia, the plantations established on pasture land have higher proportions of species of ants and dung beetles characteristic of forest than of cattle pastures (Gilroy et al. 2015), so that agricultural resumption might indeed be preferable. The plantations supported communities of at least equal biodiversity value to cattle pastures, with a wide range of taxa more characteristic of forests also represented.

Ant species richness in oil palm plantations in Sabah was affected by their diver- sity in adjacent forest fragments and, in turn, on fragment size (Lucey et al. 2014). That investigation was predicated in the belief that knowing the properties of frag-

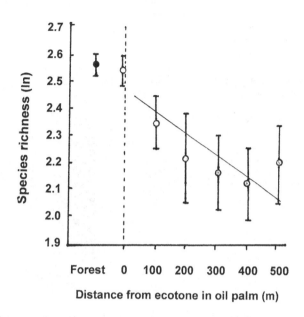

Fig. 3.3 The species richness of ants across the ecotone from forest into oil palm plantation, indicating decline in richness with distance from the forest edge (point '0') (Lucey et al. 2014) (error bars are standard error of the mean number of species sampled at each station)

ments that can support native species richness within plantations is important – not least because retaining fragments that are too small, too isolated, or too poor in quality to display clear conservation benefits (perhaps including spillover of beneficial species into plantations) might lead to economic losses and/or negligible benefits.

Ant richness was high (forest: 167 species/morphospecies; plantations 130), and the richness in plantations decreased with increasing distance from the forest edge ecotone (Fig. 3.3), with an overall decline of 22% in oil palm. Fragment size was demonstrably important, and large (> 200 hectare) fragments and continuous forest contained ant assemblages that were very distinct from those on oil palm, but those in smaller fragments were more similar - and those small fragments supported very few species that did not persist in the plantation areas. Trophic structure was also influenced by fragment size, with the number of predatory ant species also greater in plantations adjacent to large forest fragments. Larger fragments thereby harboured distinctive ant assemblages and increased diversity, from spillover, in plantation areas. Benefits to pest management in plantations are likely, with retention of larger fragments increasing wider conservation values of the matrix. In that landscape, smaller fragments might be useful stepping stones promoting connectivity, but Lucey et al. (2014) noted that the conservation priority should be on protecting the more valuable larger forest fragments.

Collectively, many aspects of crop protection and landscape ecology become increasingly relevant to conservation of insects in the remaining forest domains, themselves thereby of increasing wider conservation significance, within the universal belief that intensification and increased scale of agriculture increasingly

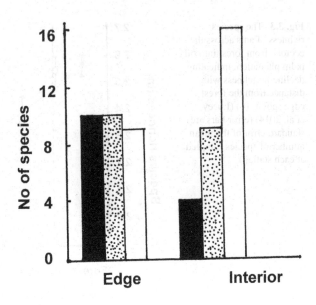

Fig. 3.4 The species richness of saproxylic beetles in relation to vertical stratification of a closed canopy deciduous forest in the Czech Republic. Note higher canopy 'preference' in the forest interior. Columns are (left to right) understorey (black), no preference (dotted), canopy (open). (Vodka and Cizek 2013)

affects biodiversity. Edge effects, spatial scale, and dispersion of natural resources and disturbances incorporate many aspects of agricultural conversion.

Presence and proximity of forest strongly influences the wellbeing and diversity of insects in adjacent agricultural systems. Many 'gradient studies' have demonstrated diminished diversity, and assemblage changes, with increasing distance into a crop from a forest edge, and the converse. Form of the crop is also influential, affecting the levels of structural contrast between crop and forest environments. Coffee plantations bordering a forest fragment in Chiapas, Mexico, contained more twig-nesting ants (from a pool of 86 species across the comparative survey) in the more shady and diverse coffee area than in an area with low canopy cover (Armbrecht and Perfecto 2003). However, the influences of the adjacent forest extended to greater distances from the forest edge in the traditional coffee habitat but dropped rapidly in the shaded regime.

Forest edges are renowned for the high insect richness that can occur, essentially reflecting the meeting of forest and outside forest faunas, with a number of 'edge specialists' augmenting these. Thus, saproxylic beetles were > 60% richer on the edge of closed forest in the Czech Republic than in the forest interior (Vodka and Cizek 2013). That study, using flight intercept traps, compared beetles in both canopy and understorey levels, each level being sampled by eight traps at the forest edge and in the forest interior. Over April–September 2006, 4739 beetles of 289 saproxylic species were trapped, with differences in richness very clear (Fig. 3.4), and the two strata giving very similar trends across treatments. The 'canopy-understorey gradient' was thus much less pronounced than the 'edge-interior gradient'.

3.4.1 Spillover

Reasons for 'spillover' of species into adjacent ecosystems from those they characteristically frequent, such as that noted above, can be difficult to clarify. Lucey et al. (2014) noted that over the distances they studied, microclimate and vegetation structure varied little in the plantations, so that local environmental conditions may not have contributed obviously to the patterns observed. The strong relationships between ant species richness in forest and forest fragment size implied that properties of the forest contributed to maintaining local diversity – with this, in turn, influencing plantation species richness from the spillover of forest species from different pool sizes colonising adjacent plantations.

Differences from an earlier study of butterflies (Lucey and Hill 2012) suggested that spillover effects can differ markedly across taxa. Butterfly richness (here, of Nymphalidae, from fruit-baited traps) decreased by 54% in oil palm plantations. Abundance was far higher in plantations than in forest, but five species dominated plantation catches and comprised nearly 70% of the 782 individuals captured. Of the total 26 species in plantations, nine were considered vagrants, because apparent lack of larval food plants would not enable them to breed there. Butterfly diversity declined with distance into the plantation. One such vagrant, *Dophla evelina*, occurred one Km into the plantation, and another three were found 700 m in. Distribution patterns of some common resident species showed either increasing (*Discophora necho*, *Amathusia phiddipus*) or decreasing (*Mycalesis horsfieldi*, *M. anapita*) abundance with distance into the plantation, and none showed such effects with distance into the forest. In all, 34 butterfly species were found in forest, and this study indicated that conversion of forest to oil palm plantation had significant negative impact on butterflies – but spillover of vagrant species into plantations played a large part in the observed decline in diversity with distance from the ecotone and reflected changed abundances of some species associated with palms or grasses as larval foods. Retention of forest patches as corridors or stepping stones within plantation areas may facilitate landscape-level butterfly conservation.

3.5 Agroforestry

One need for agroforestry is 'to conserve or create a diverse layer of multi-purpose shade trees that can be pruned rather then removed when crops mature' (Tscharntke et al. 2011). These authors pointed out that many tropical landscapes support agroforestry systems that are the closest local structural analogues to natural forest. Those systems can have substantial value for biodiversity, and their management complement that of natural forests to enhance both rural crop production and livelihoods, and conservation. Strong focus on some key crops had led to an enormous literature dealing with agroforestry, including shading agroforestry impacts, and conservation – for examples, Tscharntke et al. (2011) noted 602 studies for cocoa

(77 dealing with shading), 257 (32) for coffee arabica, and 117 (7) for coffee robusta. Cocoa and coffee are both usually grown under shaded conditions to reduce physiological stress. Protection against crop pests may also eventuate, and shaded systems for both crops can support higher biodiversity than unshaded systems, with increased shading in some cases associated with increased spillover from native forests. Reduced pesticide applications associated with increased agroforestry plantations can protect functionally important taxa such as pollinators and the natural enemies of key pests, with these activities facilitated through spillover effects.

Low plant diversity agroforests are likely to sustain only a small proportion of the native forest insects, notably those that are widespread generalists (Schulze et al. 2000). However, perceptive planning (as for urban forests) and selecting a range of suitable tree species, can do much to remedy or avoid that situation.

An important driver of agroforestry development is simply seeking to diversify farm income, and provide a long-term investment component by planting trees (commonly with projected harvest periods of 20–30 years) on land that otherwise provides little or highly unreliable economic return and as a part of mixed farming operations. Anciliary benefits may include shade and shelter for stock and crops from wind and rain.

The prospects for agroforestry to be tailored to enhance the supply of particular tree species needed by notable insects, or of variety to favour wider insect diversity may be considerable, but more practical incentives may prevail. Some forms of traditional agroforestry incorporate a mixture of woody perennial crops with annual crops to constitute an integrated cropping system that combines food production with biodiversity conservation. Wider ecological benefits can occur by those areas being refuges for local species in environments where the wider natural forest has been destroyed by wildfire. Although demonstrated for birds rather than insects, this role was evident for severely burned areas of Guatemala (Griffith 2000). Two agroforestry sites were largely protected from surrounding wildfire by cleared buffer strips and rapidly extinguishing fires that moved beyond their perimeters, leaving them as sites with relatively intact vegetation in largely devastated surrounds and, in Griffith's words, they 'may have served as a critical refuge during a habitat bottleneck for many forest species'.

References

AAS (Australian Academy of Sciences) (1981) Australia's forests. Their roles in our future. Australian Academy of Sciences, Canberra

Ahern LD, Yen AL (1977) A comparison of the invertebrate fauna under *Eucalyptus* and *Pinus* forests in the Otway ranges, Victoria. Proc Roy Soc Vic 89:127–136

Alonso-Rodriguez AM, Finegan B, Fiedler K (2017) Neotropical moth assemblages degrade due to oil palm expansion. Biodivers Conserv 26:2295–2326

Anonymous (1997) Plantations for Australia: the 2020 vision. Standing committee on forestry, Canberra

Armbrecht I, Perfecto I (2003) Litter-twig dwelling ant species richness and predation potential within a forest fragment and neighboring coffee plantations of contrasting habitat quality in Mexico. Agric Ecosyst Environ 97:107–115

Australian Forestry Standard (2003) The Australian Forestry Standard Project Office, Canberra

Australian Forestry Standard (2007) Standards Australia, Sydney

Australian Government (2007) Australia's native vegetation; a summary of Australia's major vegetation groups, 2007. Department of the Environment and Water Resources, Canberra

Berndt LA, Brockerhoff EG, Jactel H (2008) Relevance of exotic pine plantations as a surrogate habitat for ground beetles (Carabidae) where native forest is rare. Biodivers Conserv 17:1171–1185

Bonham KJ, Mesibov R, Bashford R (2002) Diversity and abundance of some ground-dwelling invertebrates in plantation vs. native forests in Tasmania, Australia. For Ecol Manag 158:237–247

Brockerhoff EG, Berndt LA, Jactel H (2005) Role of exotic pine forests in the conservation of the critically endangered ground beetle *Holcaspis brevicula* (Coleoptera: Carabidae). N Z J Ecol 29:37–43

Brockerhoff EG, Jactel H, Parrotta JA, Ferraz SFB (2013) Role of eucalypt and other planted forests in biodiversity conservation and the provision of biodiversity-related ecosystem services. For Ecol Manag 301:43–50

Carnus J-M, Parrotta J, Brockerhoff E, Arbez M, Jactel H et al (2006) Planted forests and biodiversity. J Forestr 104:65–77

Carron LT (1985) A history of forestry in Australia. Australian National University Press. Canberra

Cerda Y, Grez AA, Simonetti JA (2015) The role of the understory on the abundance, movement and survival of *Ceroglossus chilensis* in pine plantations: an experimental test. J Insect Conserv 19:119–127

Cunningham SA, Floyd RB, Weir TA (2005) Do *Eucalyptus* plantations host an insect community similar to remnant *Eucalyptus* forest? Austral Ecol 30:103–117

Doughty RW (2000) The eucalypts: a natural and commercial history of the gum tree. Johns Hopkins University Press, Baltimore

DPIE (Department of Primary Industries and Energy) (1997) Plantations for Australia: the 2020 vision. Department of Primary Industries and Energy, Canberra

Fitzherbert EB, Struebig MJ, Morel A, Danielson F, Bruhl CA, Donald PF, Phalan B (2008) How will oil palm expansion affect biodiversity? Trends Ecol Evol 23: 538–545

Foster WA, Snaddon J, Turner EC, Fayle TM, Cockerill TD et al (2011) Establishing the evidence base for maintaining biodiversity and ecosystem function in oil palm landscapes of South East Asia. Phil Trans R Soc Lond B 366:3277–3291

Frew A, Nielsen UN, Riegler M, Johnson SN (2013) Do eucalypt plantation management practices create understory reservoirs of scarab beetle pests in the soil? For Ecol Manag 306:275–280

Gentle SW (1981) The environmental issues. In: Australia's forests. Their roles in our future. Australian Academy of Sciences, Canberra, pp 76–84

Gilroy JJ, Prescott GW, Cardenas JS, Del Pliego Castaneda PG, Sanchez A et al (2015) Minimizing the biodiversity impact of Neotropical oil palm development. Glob Change Biol 21:1531–1540

Griffith DM (2000) Agroforestry: a refuge for tropical biodiversity after fire. Conserv Biol 14:325–326

Gunther MJ, New TR (2003) Exotic pine plantations in Victoria, Australia: a threat to epigaeic beetle (Coleoptera) assemblages? J Insect Conserv 7:73–84

Harris LD (1984) The fragmented forest: island biogeography and the preservation of biotic diversity. University of Chicago Press, Chicago/London

Hartley MJ (2002) Rationale and methods for conserving biodiversity in plantation forests. For Ecol Manag 155:81–95

Irwin S, Pedley SM, Coote L, Dietszch AC, Wilson MW et al (2014) The value of plantation forests for plant, invertebrate and bird diversity and the potential for cross-taxon surrogacy. Biodivers Conserv 23:697–714

Lanfranco D, Dungey HS (2001) Insect damage in *Eucalyptus*: a review of plantations in Chile. Austral Ecol 26:477–481

Lindenmayer DB, Hobbs RJ (2007) Fauna conservation in Australian plantation forests – a review. Rural Industries Research and Development Corporation publication no 05/128, Canberra

Lindenmayer DB, Cunningham RB, MacGregor C, Tribolet C, Donelly CF (2001) A prospective longitudinal study of landscape matrix effects on fauna in woodland remnants: experimental design and baseline data. Biol Conserv 101:157–169

Lindenmayer DB, Franklin JF, Fischer J (2006) General management principles and a checklist of strategies to guide forest biodiversity conservation. Biol Conserv 131:433–445

Livingston G, Jha S, Vega A, Gilbert L (2013) Conservation value and permeability of neotropical oil palm landscapes for orchid bees. PLoS One e78523:8910. https://doi.org/10.1371/journal.pone.0078523

Lucey JM, Hill JK (2012) Spillover of insects from rain forest into adjacent oil palm plantations. Biotropica 44:368–377

Lucey JM, Tawatao N, Senior MJM, Khen CV, Benedick A et al (2014) Tropical forest fragments contribute to species richness in adjacent oil palm plantations. Biol Conserv 169:268–276

Luiza-Andrade A, Brasil LS, Benone NL, Shinmano Y, Farias APJ et al (2017) Influence of oil palm monoculture on the taxonomic and functional composition of aquatic insect communities in eastern Brazilian Amazonia. Ecol Indic 82:478–483

Maeto K, Sato S, Miyata H (2002) Species diversity of longicorn beetles in humid warm-temperate forests: the impact of forest management practices on old-growth forest species in southwestern Japan. Biodivers Conserv 11:1919–1937

Makino S, Goto H, Hasegawa M, Okabe K, Tanaka H, Inoue T, Okochi I (2007) Degradation of longicorn beetle (Coleoptera, Cerambycidae, Disteniidae) fauna caused by conversion from broad-leaved to man-made conifer stands of *Cryptomeria japonica* (Taxodiaceae) in Central Japan. Ecol Res 22:372–381

Martikainen P, Siitonen J, Puntilla P, Kaila L, Rauh J (2000) Species richness of Coleoptera in mature managed and old-growth boreal forests in southern Finland. Biol Conserv 94:199–209

Nair KSS (2007) Tropical forest insect pests: ecology, impact and management. Cambridge University Press, Cambridge

Neumann FG (1979) Insect pest management in Australian radiata pine plantations. Aust For 42:30–38

O'Callaghan CJ, Irwin S, Byrne KA, O'Halloran J (2016) The role of planted forests in the provision of habitat: an Irish perspective. Biodivers Conserv 26:3103–3124

Oxbrough A, Garcia-Tejero S, Spence J, O'Halloran J (2016) Can mixed stands of native and non-native tree species enhance diversity of epigaeic arthropods in plantation forests? For Ecol Manag 367:21–29

Paine TD (2016) Insects colonizing eucalypts in California. In: Paine TD, Lieutier F (eds) Insects and diseases of Mediterranean forest ecosystems. Springer, Cham, pp 711–730

Paine TD, Millar JG, Daane KM (2010) Accumulation of pest insects on eucalyptus in California: random process or smoking gun? J Econ Entomol 103:1943–1949

Paquette A, Messier C (2010) The role of plantations in managing the world's forests in the Anthropocene. Front Ecol Environ 8:27–34

Pawson SM, Brockerhoff EG, Meenken ED, Didham RK (2008) Non-native plantation forests as alternative habitat for native forest beetles in a heavily modified landscape. Biodivers Conserv 17:1127–1148

Pedley SM, Martin RD, Oxbrough A, Irwin S, Kelly TC, O'Halloran J (2014) Commercial spruce plantations support a limited canopy fauna: evidence from a multi taxa comparison of native and plantation forests. For Ecol Manag 314:172–182

Plath M, Mody K, Potvin C, Dorn S (2011) Establishment of native tropical timber trees in monoculture and mixed-species plantations: small-scale effects on tree performance and insect herbivory. For Ecol Manag 261:741–750

Purchart I, Tuf IH, Hula V, Suchomel J (2013) Arthropod assemblages in Norway spruce monocultures during a forest cycle – a multi-taxa approach. For Ecol Manag 306:42–51

Ranius T, Johansson V, Schroeder M, Caruso A (2015) Relative importance of habitat characteristics at multiple spatial scales for wood-dependent beetles in boreal forest. Landsc Ecol 30:1931–1942

Russek LA, Mansilla C, Crespin SJ, Simonetti JA (2017) Accompanying vegetation in young *Pinus radiata* plantations enhances recolonization by *Ceroglossus chilensis* (Coleoptera: Carabidae) after clearcutting. J Insect Conserv 21:943–950

Schulze CH, Linsenmair KE, Fiedler K (2000) Understorey versus canopy patterns of vertical stratifciation and diversity among Lepidoptera in a Bornean rain forest. Plant Ecol 153:133–152

Sinclair JE, New TR (2004) Pine plantations in south eastern Australia support highly impoverished ant assemblages (Hymenoptera: Formicidae). J Insect Conserv 8:277–286

Spellerberg IF, Sawyer JWD (1996) Standards for biodiversity: a proposal based on biodiversity standards for plantation forests. Biodivers Conserv 5:447–459

Spellerberg I, Sawyer J (1997) Biological diversity in plantation forests. In: Hale P, Lamb D (eds) Conservation outside nature reserves. Centre for Conservation Biology, University of Queensland, Brisbane, pp 517–522

Spence JR, Langor DW, Hammond HEJ, Pohl GR (1997) Beetle abundance and diversity in a boreal mixed-wood forest. In: Watt AD, Stork NE, Hunter MD (eds) Forests and insects. Chapman and Hall, London, pp 287–301

Steinbauer MJ, McQuillan PB, Young CJ (2001) Life history and behavioural traits of *Mnesampela privata* that exacerbate population responses to eucalypt plantations: comparisons with Australian and outbreak species of forest geometrid from the northern hemisphere. Aust Ecol 26:525–534

Stephens SS, Wagner MR (2007) Forest plantations and biodiversity: a fresh perspective. J For 105:307–313

Strauss SY (2001) Benefits and risks of biotic exchange between *Eucalyptus* plantations and native Australian forests. Aust Ecol 26:447–457

Sweaney N, Driscoll DA, Lindenmayer DB, Porch N (2015) Plantations, not farmlands, cause biotic homogenisation of ground-active beetles in South-Eastern Australia. Biol Conserv 186:1–11

Tscharntke T, Clough Y, Bhagwat SA, Buchori D, Faust H et al (2011) Multifunctional shade-tree management in tropical agroforestry landscapes – a review. J Appl Ecol 48:619–629

Turnbull JW (1999) Eucalypt plantations. New For 17:37–52

Turner J, Gessel SP, Lambert MJ (1999) Sustainable management of native and exotic plantations in Australia. New For 18:17–32

Vangansbeke P, Blondeel H, Landuyt D, De Frenne P, Gorissen L, Verheyen K (2017) Spatially combining wood production and recreation with biodiversity conservation. Biodivers Conserv 26:3213–3239

Vodka S, Cizek L (2013) The effects of edge-interior and understorey-canopy gradients on the distribution of saproxylic beetles in a temperate lowland forest. For Ecol Manag 304:33–41

Watt AD, Stork NE, Eggleton P, Srivastava D, Bolton B et al (1997) Impact of forest loss and regeneration on insect abundance and diversity. In: Watt AD, Stork NE, Hunter MD (eds) Forests and insects. Chapman and Hall, London, pp 273–286

Yu X-D, Luo T-H, Zhou H-Z (2010) Distribution of ground-dwelling beetle assemblages (Coleoptera) across ecotones between natural oak forests and mature pine plantations in North China. J Insect Conserv 14:617–626

Chapter 4
Insects in Native and Alien Forests in Australia

Keywords Alien insects · Bark beetles · Coleoptera diversity · Endemism · Forest litter · Insect conservation priorities · Insect pest management · Pest outbreaks

4.1 Introduction: The Diversity and Ecological Roles of Australia's Forest Insects

The taxonomic and ecological variety of forest insects, and their intricate relationships to the functioning and integration of forest ecosystems, encompasses many different trophic interactions and ecological contexts. These were summarised broadly, with emphasis on the temperate region forestry of Europe and North America, by Dajoz (2000), whose wide-ranging account gives much useful background to the awareness of insect diversity and importance. Dajoz made only passing reference to Australia, and his treatment of tropical forests insects is also brief. A wealth of studies on insects in forests in other parts of the world furnishes at least an initial template for considering the ecology and vulnerability amongst Australian insects for which direct documentation is more obscure. Thus, general protestations over measures such as removing logging residues from clearing or thinning in Australia are numerous, but studies paralleling that of de Jong and Dahlberg (2017, p. 159) in Sweden are sparse.

The major categories of Australia's characteristic forest insects have aroused interest over many years, because of their taxonomic and ecological complexity and variety, and the impracticalities of advancing understanding in a generally poorly-documented fauna. Frameworks of the biology of many taxa are gathered in some classic early compendia, most notably those by Froggatt (1923, 1927) and the series of volumes by French (1891–1923). These eminent State Entomologists, for New South Wales and Victoria respectively, had wide reponsibilities to document important forest insects, and their synopses laid firm foundations for later studies. Those works confirmed that many insect groups contained significant pests of trees, with many of the species being members of complex species radiations that stll challenge definitive interpretation. Many members of those radiations appear to be localised, relatively rare, and ecological specialists that may prove vulnerable to non-specific measures instigated to suppress their 'pest relatives'.

A persistent and major dilemma, at times a conflict, in forest management involves the widely advocated tradeoff between needs for conservation and increasing risks and intensity of forest pest attacks by the measures undertaken. For bark beetles in boreal forest, prescribed burning and gap cutting affected richness and abundance, with many Scolytinae depending on the availability of dead wood within the forest (Karvemo et al. 2017). Very few of the 102 species of bark beetles in northern Europe can attack living trees, but these include several serious pest species whose dynamics have major economic implications for production forestry. Those 'primary bark beetles' are thereby of particular concern, and were significantly more abundant in burned stands than in the gap-cut stands and control (untreated) stands in northern Sweden. Species richness of Scolytinae responded similarly to both restoration treatments, but abundance of primary species and tree mortality were five times higher in burned than in gap-cut stands. Forest restoration thus increased general bark beetle richness, but burned stand assemblages differed more from reference stands than they did from gap-cut stands. The higher response by primary species may be due to fire damaging, but not killing, many trees and rendering them temporarily suitable for beetle attack. The most abundant primary bark beetles in burned areas were *Pityogenes chalcographus* and *Polygraphus polygraphis*, both known from other studies to respond rapidly and strongly to fire-stressed trees, where they may achieve outbreak status. Prescribed burning may indeed increase the numbers of trees susceptible to primary bark beetle attack, and accelerate risks of outbreaks in nearby forests. Gap-cutting, in this context, was considered to be the preferable management option.

Taxonomic inadequacy pervades insect surveys of major Australian environments, with outcomes of many expressed in terms of 'morphospecies' rather than by unambiguous binomials. It is thus difficult in many cases to prepare definitive inventories of taxa, and to cross-reference outcomes of different surveys to ensure valid interpretation and comparison, especially in the absence of carefully labelled and responsibly archived voucher specimens, and across different sampling methods and intensity. By contrast, the far stronger foundations of insect inventories in Europe and North America enable many species to be recognised and identified consistently, even by non-specialists. Commonly used 'approximations' in Australia, as in other lesser known faunas, include allocation to only higher groups (genera, rather than species, and sometimes sorting insects only to family level), with correspondingly reduced capability to interpret subtle or species-level biological information. Forest insects are no exception, and high proportions of putative species in many collections are difficult to name to species level, and the specialist help needed itself sparse or simply unavailable. Claims have been made that 'higher taxon richness' or 'selected taxon richness' may be a reliable surrogate of 'species richness' – as Oliver and Beattie (1996) implied for beetles, in which they found a significant correlation between total richness and the richness of three focal families. Likewise, Ohsawa (2010) explored surrogate values of selected beetle families of forest Coleoptera in Japan. Malaise trap catches across 52 forest stands yielded 76 beetle

Table 4.1 Numbers of species and morphospecies of the five most species-rich beetle families in inventories from forest litter in New South Wales, to indicate concordance between morphospecies and species richness estimates (Oliver and Beattie 1996)

Number in inventory				
Family	Morphospecies	Corrected morphospecies	Species	Error (%)*
Staphylinidae	91	66	74	23
Curculionidae	99	59	62	60
Pselaphidae	58	48	51	14
Scarabaeidae	23	21	22	5
Carabidae	23	19	21	10

*Error = (no. morphospecies − no. species).(no. species)$^{-1}$.100

families, collectively with 869 species. Diversity of three of the four richest families (Cerambycidae, 198 species; Curculionidae, 96; Elateridae, 50) were strongly correlated with overall Coleoptera diversity, and were suggested as valid surrogates for this. Good taxonomic information was a major advantage, with the saproxylic Cerambycidae, in particular, clearly largely dependent on forest ecosystems, and the other two families more widespread in their habits. Their correlations were valid across different forest types, including plantations and secondary forests, and different stand ages and management regimes.

Oliver and Beattie (1996) compared beetles, ants and spiders on the ground, by pitfall trapping and litter extractions in four forest types in New South Wales, and used three levels of taxonomic 'penetration' to interpret richness. These were (1) morphospecies categories made by non-specialists; (2) corrected morphospecies listings by taxonomists so that (as far as possible) a morphospecies equated to a taxonomic species; and (3) listings with specialist examination to correct the above – but this desirable progression toward achieving greater accuracy also incurred substantially greater costs. Five beetle families were relatively rich in species (Table 4.1). Amongst them, Staphylinidae and Curculionidae remain difficult to identify to species level, but the other three were regarded as potential practical surrogates and, with ants also included, might be useful in wider forest invertebrate assessments. Development of the approaches initiated by Oliver and Beattie might in due course help development of more representative monitoring of those assemblages.

A significant endemic radiation of eucalypt-associated moths exemplifies the complexity of defining either richness or distribution of individual species within a group of massive ecological and evolutionary relevance in understanding the development of the country's insect fauna. Predominant in forest and woodland environments, Oecophoridae reflects the many insect groups that remain insufficiently documented to formulate realistic conservation needs and measures other than in the most general terms. Oecophoridae is by far the richest family of Lepidoptera in Australia: Common (1994) recognised about 1850 valid species names, and estimated the total fauna to be more than 5000 species. A very high proportion of species for which any biological information is available use Myrtaceae as larval food.

Highly unusually amongst Australian Lepidoptera, larvae of many feed on dead
Eucalyptus leaves and collectively appear to be very important in breaking down
eucalypt litter on the forest floor. Dead eucalypt foliage typically has high concentrations of tannins but only low nitrogen content, so that many oecophorids have
specialised on an abundant food source that is not available to many competing
insect herbivores in other groups. The larvae of many species construct 'cases' of
foliage, in some initially on the trees from which they later drop into the litter, where
later development occurs. Some species feed on dead foliage attached to fallen
branches. Whilst some species can occur in very high numbers in litter, many of the
taxa are known from very few specimens and few localities. Common (1994) commented that their evolution in Australia has included 'an enormous multiplication of
species adapted to a diversity of habitats, especially in sclerophyll forest, woodland,
mallee ...'

A parallel habit occurs amongst a group of leaf beetles (Chrysomelidae,
Cryptocephalinae). Larvae of many construct cases, as do Oecophoridae, and also
feed on dead foliage on the ground. Many of the 500 (or so) Australian species are
poorly known, but the beetles collectively occur in many kinds of forest and woodland litter and appear to be significant contributors to breakdown of dead foliage.

Loss or erosion of either of these groups might have severe ecological consequences, as directly affecting the rates of litter breakdown and nutrient recycling.

Taxonomic problems will assuredly persist amongst Australia's forest insects,
and the approach developed by Sebek et al. (2012), of using subsets of identifiable
(or only easy-to-identify) species for monitoring wider changes in diversity, has
considerable potential amongst such inadequately characterised assemblages. In
that study, beetles were enumerated from 67 surveys in France and Belgium, collectively involving a large data set (of 42 forest plots, and 856 species). Paralleling
other studies (including Oliver and Beattie 1996, above), Sebek et al. defined three
levels of difficulty in identifying the beetles captured, as (1) least difficult – those
easy to identify by visual or other unambiguous features; (2) moderate – those
requiring detailed identification keys that are, nevertheless, available and likely to
be comprehensive and reliable; and (3) difficult – groups containing or comprising
species that can be identified only by specialists, reflecting features such as poorly
understood complexity, lack of recent revisionary studies, and the need for intricate
examination such as by genitalic dissection and comparison. The last is not unusual
amongst species radiations, in which speciation is recent and distinctions between
taxa small. Many Australian forest insects fall into this category, and some are
largely intractable for detailed identification and documentation. Substantial risks
of misidentification, and consequent spread of conveying 'misinformation' remain
and, despite the use of molecular approaches to differentiate closely-related taxa,
are likely to persist for some time.

Sebek et al. (2012) ranked their 'best' surrogates according to each of five criteria (Table 4.2), each consisting of several divisions, and recognising that the practical value of any such group depends on how simply and realistically it may be used,
as well as minimising the costs of analyses. In this case the most useful surrogates
were those needing least identification time, with the greatest proportions of

Table 4.2   Some criteria for evaluation of surrogacy in assessing diversity of saproxylic beetles in western Europe (Sebek et al. 2012)

Ecological representativeness	
Host tree groups	
	Conifer
	Deciduous
Feeding guild	
	Mycophagous
	Predator
	Saproxylophagous
	Secondary wood decayer
	Large species (> 10 mm length)
	Cavity-dwelling
Conservation interest	
	Common
	Rare
Identification costs	
	Total no. species
	Einv (1 – proportion of easy-to-
	identify species in subset
	considered)
	Psp (proportion of species to be
	identified in subset from all
	trapped species)
	Pind (proportion of individuals to
	be identified in species subset from
	all trapped individuals)
Surrogacy potential	
	R^2 (correlation between subset and
	total species richness)
Variability in surrogacy	
	With forest type and geographical
	range
	With trap-baiting
	With spatial scale

easy-to-identify taxa. In contrast, surrogates based on individual families of beetles did not predict overall richness well, in contrast to the outcome found by Oliver and Beattie. Likewise, the number of rare species was a poor surrogate for total species richness – again in contrast to some other studies in which use of red-listed species had been advocated (Muller and Gossner 2010). In European temperate forests, monitoring subsets of 'easy-to-identify' families may be sufficient to reflect changes in saproxylic assemblages (p. 96) (Muller and Gossner 2010). Likewise, focusing on the near-ground fauna was sufficient to represent general patterns of diversity.

The widespread difficulties of distinguishing insect species for comparisons of richness and 'diversity' as indicators of forest management or 'quality'(especially in the tropical regions where numerous insects are not easily diagnosed) have led to exploration of alternative approaches. One such path (Aoyagi et al. 2017) involved examining tree community composition amongst the two distinct groups of tree genera of mature or disturbed forest. This approach obviates need for specific iden- tification of the trees and, in Borneo, helped to indicate the responses of tree com- munities to logging disturbances through assessing changes amongst key resources for insects.

Native terrestrial insects living in, and dependent on, Australia's forests are diverse, largely endemic, many of them narrowly distributed and to some extent host plant specific or restricted, and cover a wide array of higher taxa (see Appendix). Distributions can be restricted by climate or floristic features, so that 'gradient dif- ferences' (p. 93) are common, both in elevation and microhabitat features such as changing vegetation or exposure.

As noted above, most species have not been studied in detail, many are unde- scribed, and many are also known from very few specimens, so that any comments on their distribution and vulnerability are highly tentative. They include a number of very diverse radiations, notably of some herbivore groups associated with plant radiations, so that suites of closely related insect species are associated with parallel suites of plant species. These are perhaps most notable and diverse on *Eucalyptus* s.l. species and *Acacia*, with approximately 700 and 1000 species, respectively, in these dominant plant genera. The significance of those insect radiations in the Australian fauna has been emphasised repeatedly (New 1983), and some recent taxonomic studies are progressively characterising their diversity more accurately. Early focus was mainly on eucalypt-defoliating taxa associated with tree health, and so of economic significance as well as ecological interest.

4.2   Major Forest Pests

The influences of forestry practice and forest condition on some key native insect pests, the most intensively investigated information source on forest insects, provide much of value for other less-heralded insect inhabitants, including informing their conservation needs and status.

As examples, studies on the population dynamics of some eucalypt psyllids (Clark and Dallwitz 1975) and eucalypt phasmids (Readshaw 1965) are early clas- sics of Australian forest entomology that – as well as providing basic information on their focal insects – have added more generally to knowledge of insect population fluctuations and their causes. This focus has ensured that phytophagous insects causing visible or tangible damage to their host trees have received the greatest attention. The major phytophagous groups involved fall into two broad functional trophic groups – as chewing defoliators (Coleoptera, Lepidoptera, Hymenoptera:

Symphyta, Phasmatodea) and sap-suckers (Hemiptera, with major radiations within several lineages).

Taxonomic uncertainty, associated with so-called 'cryptic diversity', manifests in many ways – but one foundation is that many specialised saproxylic and other forest arthropods can form very isolated populations. In the case of some saproxylic species, multiple generations may pass within a single log, in stable microhabitats and with low (if any) level of dispersal, so leading to spatially and genetically isolated lineages. Conservation strategies that seek to conserve genetic variety recognise that traditional 'morphological' taxonomy underestimates genetic diversity, and the enormous number of 'cryptic species' that are each genetically distinct. Discussed by Garrick et al. (2006, 2012), studies on genetic diversity of forest invertebrates in Tallanganda State Forest (New South Wales) sought to inform forest management and conservation through identifying areas that support evolutionarily distinct lineages of log-dependent taxa. Such relatively sedentary species, perhaps especially those associated with circumscribed resources such as decaying logs, can be excellent study organisms for investigating patterns of genetic diversity and phylogeography. Target species included a saproxylic collembolan (Neanuridae) that showed very high local endemism over scales of only around 10 Km (Garrick et al. 2004). Management strategies that focus on vertebrates and vascular plants in forest may unwittingly neglect a large proportion of the real 'biodiversity' present amongst sedentary saproxylic invertebrates – even if individual taxonomic species are acknowledged, substantial genetic variety is easily overlooked. Identification and reservation of areas in which such high variety persists can increase value and representation, as contrasts to 'whole species' focus in conservation strategy. Cases such as this neanurid demonstrate that the saproxylic habit may be a major generator of invertebrate specialisation.

That study also demonstrated the very local distribution of distinctive components of notable invertebrate lineages. Indeed, the levels of local endemism among insects in Australia's forests – often suggested or inferred to be high – have only rarely been explored comprehensively by extensive comparative surveys. One notable study involved survey of endemism amongst the flightless insects found in the upland rainforest of the Wet Tropics region of Queensland. Yeates et al. (2002) used 274 selected species of flightless insects restricted to that area to explore patterns of endemicity at two levels: (1) the 'regional endemics' confined to the Wet Tropics, and (2) the subset occurring only in a single subregion of the 14 upland (> 300 m a.s.l) subregions distinguished, as 'subregional endemics'. The insects were collected across about 350 sites over a 20-year period, using a variety of different techniques. Of 15 families included, the most diverse were Aradidae (Hemiptera, 46 morphospecies), and the beetle families Carabidae (78 morphospecies), Scarabaeidae (32 morphospecies) and Tenebrionidae (87 morphospecies). The use of flightless species in this study reflected that those distributions had not been affected by long-distance dispersal, so that a high level of habitat/site fidelity might indicate the extent of local endemism amongst these taxa. Half the species were found only in a single subregion, and the three subregions with highest species richness also supported the greatest numbers of subregional endemics (21, 17, 28), and only one subregion yielded no subregional endemic. Overall, eight subregions supported nine or more subregional endemics, and were suggested to support larger rainforest areas continually and likely to constitute refugia. Those areas, in particular, are irreplaceable if the low vagility insects are to persist, and the high numbers and restricted distributions of localised endemic species emphasise the conservation importance of the wider rainforest ecosystems of the Wet Tropics.

The two studies noted above, very different in approach, both demonstrate the fine levels of local specialisation and distinctiveness among both individual 'species' and wider assemblages. Features such as low mobility, ecological specialisation, and long-term occupancy render such taxa powerful tools in interpreting past environmental changes, with genetic analyses helping to interpret past range changes, gene flow, and divergence of populations.

4.3 Alien Insects on Native Trees

Pest insects in Australian forests include both native and alien species and, as categorised by Dajoz (2000) for Europe, these can be divided into four broad groups according to the damage they cause in forestry. Those effects are (1) deterioration in the quality of wood after the trees have been attacked, such as distortion, or holing of timber; (2) loss of production through repeated defoliation leading to reduced tree growth or distortion of normal growth patterns; (3) killing of trees from massive infestations such as sporadic or more regular 'outbreaks'; and (4) disease introductions and spread by insect vectors. Other bases for defining groups (such as by feeding guilds or host range) may be preferred by ecologists, but the above scheme illustrates major concerns over insects causing economic losses, and leading to the causative agents needing to be suppressed, eradicated or their damage prevented in order to safeguard commercial interests. Many pest species are closely related to species of conservation interest and concerns, and which are also susceptible to the measures used for pest management. Harmonising the priorities of forestry and insect conservation in Australia's forests is a continuing need and concern.

Rather few alien insects have become pests on *Eucalyptus* in Australia, and several of these are polyphagous species – indeed, all such species are generalists that attack many host taxa. Most are associated only with juvenile eucalypts. Both Paine et al. (2011) and Nahrung et al. (2016) suggested that Australia's strong biosecurity regulations may have helped to limit the numbers of introductions, together with most alien insects lacking a wide range of preadaptations to feeding on eucalypts should those insects arrive. *Heteronychus arator* (p. 143) is one of few serious insect pests and occurs mainly on newly planted plantation seedlings.

As elsewhere, most information on alien forest insects has accrued from studies on economically important invasive pests, undertaken primarily to minimise losses of wood production. More recently, those interests have spread to encompass pest and pathogen taxa that can devastate natural ecosystems, including forests. In the United Kingdom, invasive pathogens considerably outnumber invasive insects of equivalent importance for forestry, but the insect examples discussd by Freer-Smith and Webber (2017) span a variety of taxa and ecological roles, and impose a range of phytosanitary needs. Fungal pathogens, however, also pose concerns for native insects they may encounter. In Britain the pathogen *Hymenoscyphus fraxinus* (Chalara ash dieback) has affected enormous numbers of European ash trees (*Fraxinus excelsior*), a widely distributed host tree to numerous specialised insects amongst the approximately 953 species associated with it (Mitchell et al. 2014). Loss of ash trees from mixed forest stands is associated with disruptions to ecosystem functions and species declines, and has led to discussions over whether suitable 'replacement' tree species might be planted to sustain species normally dependent on ash. Mitchell et al. (2014) suggested that the oak *Quercus robur/petraea* might be a suitable surrogate, because it hosts 69% of the species normally found on *F. excelsior*. This novel practice is essentially 'off-setting' death of native trees from pathogen attack by providing alternative hosts for at least some of the insect species that might otherwise become more vulnerable. The four actions suggested by Mitchell et al. to assess the impacts of such a tree disease extend easily to insect pests, as (1) identify the ecological functions associated with the threatened tree species and assess how these might change if the tree is lost or replaced; (2) collate information on which species use the tree, how they use it, and how they interact with other species/guilds present; (3) assess suitability of alternative tree species to replace the threatened species; and (4) identify the management options that can overcome or reduce the impact of the disease (or pest). Gaining the biological information implicit in the first two of these can be complex – for the ash case in Britain, Mitchell et al. (2014) noted 239 invertebrate species present but 131 of those had unknown importance, some probably being casual incidences, and only 29 were obligate users of *Fraxinus*, with a further 24 species 'highly associated' and rarely using any other plant species. Setting priorities in this example incorporated combining the level of association with any already signalled species' conservation status – so that 'high risks' extended to 37 invertebrates that were considered likely to severely decline or be extirpated if tree dieback continued. The recommended oak replacement was the highest ranked of 22 alternative trees considered, and it was also recognised that planting mixtures of substitute species might increase the number of ash-associated taxa conserved.

This example is important also in that a number of management options available as countermeasures were assessed. They ranged from non-intervention, to promoting natural regeneration, felling and removal of *Fraxinus* without further intervention, and felling and removing trees. Attempts made to predict impacts, by canvassing expert opinion, on the priority species included four factors and scenarios (Table 4.3), but predictions from all indicated that many priority species would decline or be extirpated if *Fraxinus* was lost from an area.

Impacts of alien insects and pathogen outbreaks on forests are predicted widely to increase in frequency and intensity, aided by changing climates (Cale et al. 2016). Mortality of infested tree species inevitably imposes changes on the structure of mixed forests, with possible cascade impacts on resident species as their habitats change. Scales of impact range from large forested areas to more restricted and

Table 4.3 Ash (*Fraxinus excelsior*) dieback in the United Kingdom: management scenarios and factors for assessment of their impacts (after Mitchell et al. 2014)

Management scenarios defined for testing
1. Non-intervention: stands develop naturally
2. No felling, with natural regeneration promoted – such as by fencing
3. Felling – all trees removed in single operation; no subsequent interventions
4. Felling and replanting – as above, with gaps replanted with alternative tree and shrub species appropriate to the region, with subsequent management to develop overstorey species

Predicted impact factors for each of the above scenarios
1. Complete loss of living ash trees: ash-obligate species go extinct, highly associated species decline except for short-term increase in species associated with dead wood
2. Reduced numbers of living ash trees: obligate and highly associated species decline, except for increase in species associated with dead wood
3. Increase in dead wood: increase in species associated with dead wood
4. 'Alternative tree species': highly associated species may not decline if replacement species can be used or ground/shrub cover provides suitable alternative habitats for ash-associated species

intensively managed areas such as urban forests, in which pest or disease outbreaks can pose serious management needs (p. 212) to restore their health.

The Sirex wood wasp (*Sirex noctilio*, Siricidae) has been a significant alien pest of Australia's *Pinus radiata* plantations in the south east of the continent, and its outbreaks have been associated widely with trees that are in some way 'stressed', for example by drought or fires. The wasp established initially in Tasmania (1952) and was found in Victoria by 1961, thereafter spreading to all pine plantation areas in the adjacent states of South Australia and New South Wales by 1984. In addition to plantations, many shelter belts of pines on farms were damaged. Then innovative control by use of nematode worms injected into the trees initially led to almost total destruction of wasps, but this efficiency later declined (Collett and Elms 2009). Nevertheless, the nematode *Beddingia* (formerly *Deladenus*) *siricidicola* is still the primary and most effective biological control agent, but with contributions also from several imported wasp parasitoids. Together with changed plantation management through increased thinning and tree health, this is one of few alien forest pest management programmes in Australia for which neither pesticides nor non-target impacts from agents pose any major concerns. It is likely that continually improving stand hygiene and thinning schedules will allow this scenario to persist, by preventing *Sirex* populations from reaching damaging levels.

Many forest-dwelling insects pose concerns for suppression or conservation, with greatest concerns across these contrasting themes in Australia being for Lepidoptera, Coleoptera and Hemiptera, as perhaps the most universal and widely studied forest insect groups in which pest and conservation interests often co-occur. For the first two orders, a historical legacy of distributional and biological knowledge has accumulated from naturalists and collectors and, whilst this benefit is tiny in Australia compared with rewards from the longer and more intensive 'collecting tradition' in Europe and North America, has given a preliminary template of awareness of relative scarcity and habitat specialisation amongst species in some families. Many of the rarer or more elusive taxa can be recognised, at least tentatively, as such. However, inventories of taxa in and across forest systems remain sparse, so that the use of these orders in conservation assessments (as used in some northern hemisphere temperate region forests: Flensted et al. 2016) is rarely practicable. In Flensted et al.'s study in Denmark, nationally red-listed species of eight groups of organisms, including butterflies and saproxylic beetles, for which comprehensive national distribution record sets were available, were assessed by mapping distributions on a 10 × 10 Km grid across the entire country. The 12 forest-associated butterflies were associated with floristically-rich forest glades, and the 55 beetle species with dead wood, many in semi-open forest conditions. Both these groups, and others, were also associated with larger forest areas, in which old-growth forest in conjunction with forest continuity was an important habitat component. Lessons from this correlative survey may have wide relevance elsewhere – with the principles that continuous forests are important for native biodiversity and that specialised red-listed species may depend on mature forest environments that are not replaceable by reforestation or plantations, even if these are allowed to persist for many decades.

Concerns over alien or invasive forest insect pests extend to their potential to attack native plant species in areas they newly inhabit, or to wider pest concerns arising from spread in those areas. Thus, the eucalypts established widely as plantation crops or urban amenity trees in New Zealand can be attacked by Australian insects that may affect native flora, and may lead to other impacts. The Gum-leaf skeletoniser moth (*Uraba lugens*, below), for example, may both attack native trees in urban areas of New Zealand and also pose a health hazard from the larval urticating hairs (Withers et al. 2011), through either or both of forming self-sustaining populations and temporary spillover of larvae from primary hosts. Such examples pose occasional, but significant, needs for managing susceptible native plants that have conservation importance.

Ability to manage most key defoliating pest insects in Australian production forests is hampered by lack of knowledge of the factors that predispose development of outbreaks of taxa that normally occur only in low, non-damaging, numbers. Both in natural forests (Collett and Fagg 2010) and plantations (Ostrand et al. 2007), relatively few native insects from among the many relevant candidates may be regarded as serious, if sporadic, pests for which monitoring of incidence and abundance is important. In mixed-species eucalypt forest in eastern Victoria, different native Lepidoptera species caused severe defoliation in 2003 and 2005 (Collett and Fagg 2010). In 2003, larvae of a cup moth (*Doratifera* sp., Limacodidae) were the major influence, with a mean of 71% defoliation across 10 sites, with Gum-leaf skeletoniser (*Uraba lugens*, Nolidae) secondary. In 2005 the order was reversed, with *U. lugens* the primary agent contributing to a far lower level, of 17% defoliation, across the same sites. The two episodes were regarded as separate events, with causes of the *Doratifera* outbreak unclear, and that of *Uraba* linked more reliably to low regional rainfall. Lack of more precise predictive capability renders planning for any effective countermeasures difficult.

Clarification of the factors influencing development of damaging ('outbreak') levels for Australian forest insects is thwarted by their often seemingly sporadic incidence and rapid progress, but outbreaks appear to be rarer (or, at least, are less often reported) in natural forests than in managed regimes. However, it seems clear that individual case studies display that different insect taxa and systems respond in different ways and that any general pattern is elusive. Individual studies are also valuable in clarifying the population dynamics of the focal species, and complementary information from different species of Lepidoptera, Coleoptera and Hemiptera contribute to wider awareness.

Thus, massive infestations of psyllids, *Cardiaspina* spp., occurred over consecutive years (2009–2013) in the Cumberland Plains Woodland (near Sydney, New South Wales), itself listed as a critically endangered ecological community. The psyllid causes defoliation of its host tree *Eucalyptus moluccana* (Grey box) over several thousand hectares (Hall et al. 2015). Psyllid populations declined in early 2013, attributed to depletion of food resources and to extensive summer heat waves affecting reproduction and early stages. That study supported the more generally reported pattern of many other insect outbreaks, described commonly as 'boom and bust', but parasitoids (the major natural enemies of the psyllids) appeared not to contribute significantly to the declines.

Paropsine leaf beetles (Chrysomelidae) are major defoliators of plantation eucalypts in Australia and New Zealand, and have generally been managed by aerial applications of broad-spectrum insecticides – with almost inevitable non-target effects, not least on related non-pest paropsines occurring on the same or nearby trees: only about a dozen of the 400 or so native paropsine beetles are significant pests of *Eucalyptus*. That control option is more effective than most others suggested (Elek and Wardlaw 2013), with the current 'best strategy' for a pest management programme combining changes to produce tree stocks less susceptible to insect attack (the 'landscape option' of Elek and Wardlaw) and so reducing the frequency of pest attacks, with insecticide use if outbreaks occur and pending development of some form of specific 'attract and kill' trapping. Nevertheless, aerial spraying is a high business risk for forest managers, and any reductions of broad spectrum chemical uses are desirable – those chemicals used widely can kill both non-target terrestrial insects, including desirable natural enemies, and most are also toxic to aquatic fauna (Elek and Wardlaw 2013). 'Landscape options' are the most desirable alternatives, and further research is needed to develop plantation stands that can resist or tolerate paropsine defoliation, together with increased use of specific attraction methods based on natural chemicals such as pheromones. Silvicultural measures include retention of dead trees to foster natural enemy populations locally, and landscape options to satisfy forest certification standards in Tasmania also demand that (1) genetic engineering is not used to produce resistant trees, and (2) no alien tree species are introduced.

The major approach to control of pest insects in forests has been remedial, largely using insecticides, implemented after damage has become evident. As Showalter (1986) remarked, those measures may be too late to prevent damage, and do not change the conditions that may predispose the forest to further pest attack. The measures undertaken are often costly in both materials and labour as well as in possible hazards of chemical use, notably possible non-target effects. In contrast, preventative silvicultural options for pest management by potentially mitigating pest activities over more extensive periods and affecting the communities in which pest species operate may pose viable alternatives to short-term remedial measures. Polyculture plantations during reforestation, for instance, and other factors that influence stand structure and uniformity, plant growth and insect activity, as well as insect community structure and predisposition to pest attacks, may occur through measures such as pre-harvest thinning. A wide array of silvicultural practices, from site preparation to fate of residues after harvesting, may each contribute to effective reduction of pest insect attacks and the needs for more interventionist controls.

Silvicultural management, however, is not suitable for all insect pests of plantation forests (Carnegie et al. 2005). Extending knowledge of the use of semiochemicals from their known values in northern hemisphere forests to the southern hemisphere may have considerable promise (Nadel et al. 2012), through processes such as mating disruption by pheromones and other largely species-specific measures likely to become valued far beyond their limited uses in species detection and monitoring (Wingfield et al. 2011). They have the additional attraction of lacking most non-target impacts, and their potentials have been demonstrated by trials against *Mnesampela privata* (p. 38), *Uraba lugens* (p. 67) and other forest Lepidoptera.

The development of insect pest management (as integrated pest management, IPM) in forests essentially devolves on forest protection from economic damage, but has emerged recently as a broader ecological requirement, as reviewed for Canada by Alfaro and Langor (2016). IPM domination has gradually given way to greater focus on assuring the sustainability and health of forests at the landscape level, and considering forest ecosystems as a whole. Forests are viewed as complex ecosystems that sustain livelihoods through a variety of products and environmental services – including conservation of biodiversity – and so balancing their social, economic and ecological values. This transition is reflected strongly in forest management that seeks to emulate natural disturbance regimes, and within which native insects are components of the ecosystems that must be conserved and managed. Anticipating trends for the twenty-first century, Alfaro and Langor (2016) highlighted four areas of entomological knowledge that would benefit from greater investment toward increasing understanding (Table 4.4). These transfer easily to Australia, where there is equal need for managers to understand the variety of ecological processes operating over a variety of scales.

The frequent 'traditional' approach of evaluating insect representation simply by comparative estimates of species richness across different vegetation types or other ecosystems can easily overlook the importance of microhabitat heterogeneity in each as, perhaps, the more important driver of richness than broader-scale differences. Thus, for the 159 ground-dwelling beetle morphospecies in Yellow Box – Red Gum grassy woodland (a critically endangered ecological community in the Australian Capital Territory) pitfall-trapped at two scales of macrohabitat (vegetation types averaging 1266 m apart) and microhabitat (traps in open areas, under

Table 4.4 Four broad areas of entomological endeavour that need further effort/investment in order to help understand the entomological elements of forests (as discussed for Canada by Alfaro and Langor 2016)

1. Study of the non-outbreak phases of population cycles: understanding how and why outbreaks occur is critical for planning optimal control of key pests, and needs knowledge of ecological interactions with hosts, competitors, natural enemies, and the abiotic environment
2. Cumulative environmental impacts and restoration ecology: values of insects as indicators of multiple disturbances affecting forests; increased entomological involvement in addressing sound sustainability indicators and restoration approaches
3. Climate change influences: addressing responses of native insects, including spread to new environments and associations, possible losses from changed environments, impacts of changed population cycles on humans and wider interactions; predictions of further trends
4. Non-native species: colonisation and spread of alien species, most of them with unknown ecological impacts on plant and animal taxa in the receiving environments; resistance of forest ecosystems to establishment of non-native insects

trees and next to logs, averaging 71 m apart), no major differences in richness, abundance and evenness were found between vegetation types (Barton et al. 2009). In contrast, heterogeneity was high in species composition at logs, with the location of the logs (under trees or in the open and more generally across the woodland landscape) important for beetle diversity. Distribution of logs as a key resource across the landscape was then a significant management consideration. Incorporating scale in harmonising production forestry operations with effective conservation of resident insects is an important theme to pursue.

4.4   Development of Conservation Concern for Insects in Australia's Forests

Changes to Australia's forests, both tropical and temperate, have been varied and severe. They have also generated massive concerns over the sustainability of these key Australian environments. The importance of conserving eucalypt-dominated ecosystems, collectively encompassing many forest, woodland and more open savanna ecosystems in Australia and nearby islands, and subjected to 'new' kinds of disturbance over the 200 years or so of European settlement, is recognised widely. Norton (1997) summarised much historical background in the perspective of a wide survey of eucalypt biology (Williams and Woinarski 1997), but also reiterated the reality that invertebrates have played little part in assessments of diversity and loss.

Collectively, disturbances to Australia's natural wooded ecosystems encompass clearance and fragmentation, as well as grazing by alien stock, other alien species impacts, and changes in fire and hydrological regimes, each posing problems for interpretation beyond the most obvious visible changes in vegetation structure and composition. Norton noted five important issues for maintenance of eucalypt ecosystems that – although challenging to pursue – are likely to contribute significantly to conservation. These, listed here to indicate the wide scope of needs and responsibilities and the massive complexities involved in advancing them, are (1) developing a more integrated approach to conservation and management, such as through recognition of 'bioregions' as assessment units; (2) resolving units for conservation and management, driven by recognition of the limitations of the most commonly used ecosystems that are defined by floristics as not being reliable surrogates for habitat needs of much fauna; (3) creating a national system of protected areas, including evaluation of the biodiversity present and developing criteria for selecting new areas for reservation; (4) eliminating threatening processes, needing effective regulatory and policy controls but reflecting that 'threatening processes' are acknowledged under much Australian legislation; and (5) developing effective government policy, as long-term and landscape-based management prescriptions and incorporating precautionary approaches to management.

The continuing concerns over forest conservation in Australia have generated significant and highly polarised political debates, some from the mid-nineteenth century (Legg 2016). Campaigns for 'wildlife' conservation in Victoria's forests since the 1920s and building on campaigns mainly for other purposes from about the 1860s, in large part reflected that many forests were publicly owned and managed (as Crown Reserves) for a variety of public purposes – but were exploited predominantly by private industry that in many cases largely ignored or circumvented official restraints. Public conflicts arose from some relatively unusual sources. Legg (2016) summarised the interactions with the gold-mining lobby in Victoria over the period 1865–1907, when there were massive needs for mining timbers and shaft supports, firewood for steam engines used to pulverise quartz and pump water, as well as demands for mining town constructions and clearing for local agriculture. Unlike some mining operations elsewhere in the world, for which plantations were established for timber supply, the Victorian mining industry relied entirely on public forests.

After that period, calls for forest conservation increased, with biological focus generated through activities of groups such as the Field Naturalists' Club of Victoria (founded in 1891), which pressed hard for creation of National Parks. The values of forests in water catchments gained increased prominence from the 1920s, with recognition that urban water quality depended on the interdependence of forestry, soil, and water management.

Perhaps the first major debate in which entomological concerns were voiced directly was the Commonwealth Senate's (1976) 'Inquiry into the impacts on the Australian environment of the current woodchip industry programme'. This was motivated by widely expressed concerns over the extent and likely future expansion of forest clearing, over which minimal controls were evident, here for export of woodchips to Japan. Submissions by conservation bodies expressed wide concerns over general lack of information that would allow informed predictions of impacts – with examples from forest mammals and birds illustrating these lacunae. A submission from the Australian Entomological Society's Conservation Committee (New 1976) was augmented by those from several eminent entomologists, and these collectively documented many concerns for insects flowing directly from lack of basic knowledge. Taylor (1976, and expanded by Taylor 1983) here introduced the term

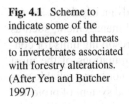

Fig. 4.1 Scheme to indicate some of the consequences and threats to invertebrates associated with forestry alterations. (After Yen and Butcher 1997)

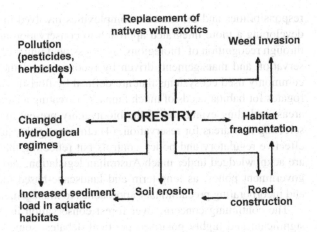

'taxonomic impediment', subsequently used widely in conservation debates involving insects in rich and poorly documented faunas, and here eliciting almost incredulous concerns from the examining Senators. For the first time, the paucity of basic taxonomic and ecological knowledge of Australia's forest insects, together with the unique nature of the endemic fauna, forcefully entered the public arena.

The Australian Entomological Society and associated submissions (discussed further by New and Yen 2012) introduced themes later endorsed by the Society in related conservation contexts. These included (1) need for a multidisciplinary review of forestry practices and impacts; (2) reservation of areas of special ecological or other scientific interest, and adequately buffering such reservoir habitats against disturbance; and (3) controls over rotation periods and coupe sizes in forestry. Specific guidelines for mosaic clearing, planned clear-felling and regeneration were also suggested, but Taylor's continued emphasis that the lack of ability to document accurately the insect fauna of forests and hence to evaluate their vulnerability and the extent of losses, and largely precluding widespread use of many insect groups in monitoring impacts of forestry practices and other land use intrusions, rendered any firm conclusions open to debate. That debate continues. In their comprehensive overview of conservation of non-marine invertebrates in Australia, Yen and Butcher (1997) noted the several earlier reviews of forestry impacts on insects – and emphasised that all of those noted the lack of adequate information on invertebrates. Some, indeed, simply drew on the scanty earlier information available, whilst others ignored invertebrates completely or confined comments to pest species. That overview also displayed the great variety of pertinent environmental changes associated with forestry (Fig. 4.1) and also commented that their overall effects on invertebrates are difficult to assess, because (1) the complexity of forestry operations that necessarily include a series of stages and processes of varying complexity and intensity; (2) that invertebrates are so diverse that it is difficult to monitor more than a few species or higher taxa at any one time; and (3) different responses by different invertebrates can flow from seasonal, geographical and habitat factors.

This important perspective contributed to the wider recognition of vulnerability of forest insects to production forestry processes.

Nevertheless, many of the insects of Australia's forests appear to be naturally rare, and have been encountered only infrequently – very little is known of the ecology and distribution of numerous species, and their very scarcity – real or not – accords them conservation interest. Validating that scarcity and any vulnerability to forest changes is a central task of assessing conservation status and needs. In essence, lack of knowledge of most of the apparently scarce (or, at least, infrequently recorded or seen) forest insects renders characterising species' conservation status highly uncertain. Invoking the 'precautionary principle' on the side of caution has both benefits (in signaling the large number of putatively needy and threatened species) and frustrations (swamping available resources for conservation and, whilst initially aiding attention to many individual species, leading also to loss of credibility if new information reveals those species to be more common than supposed, and secure).

'Sampling rare species is always challenging' (Martikainen and Kouki 2003), and the intensity of sampling or trapping needed to gain the large sample sizes of insects needed to produce reliable inventories can be difficult, or ethically and logistically prohibitive, to obtain. Much interpretation of 'rarity' almost inevitably relies on relatively small samples in which widespread common species predominate. Many of those species are ecological generalists and their incidence and abundance may not vary with forest management history, and many rarer species, even when they are present in the sampled areas, may not be detected. The samples avilable do not display realistic conservation needs and priorities as effectively as the presences of notable (for examples, scarce, threatened, near-threatened, red-listed) taxa. Targeted sampling, assessing representation from a combination of several sampling methods, may in part help to overcome this. Many of the 'high interest species' really or supposedly occur in very small numbers, and are detected largely by chance. Their presence, however, can signal heightened conservation value of the site(s) where they are found. Martikainen and Kouki (2003) suggested that threatened and near-threatened species of beetles are the most important species in formulating forest conservation actions. Some beetles have become 'flagship species' for forests in Australia (Chapter 6), but the status of many putatively threatened insect species will remain difficult to validate.

References

Alfaro RI, Langor D (2016) Changing paradigms in the management of forest insect disturbances. Canad Entomol 148:S7–S18

Aoyagi R, Imai N, Fujiki S, Sugau JB, Pereira JT, Kitayama K (2017) The mixing ratio of tree functional groups as a new index for biodiversity monitoring in Bornean production forests. For Ecol Manag 403:27–43

Barton PS, Manning AD, Gibb H, Lindenmayer DB, Cunningham SA (2009) Conserving ground-dwelling beetles in an endangered woodland community: multi-scale habitat effects on assemblage diversity. Biol Conserv 142:1701–1709

Cale JA, Klutsch JG, Erbilgin N, Negron JF, Castello JD (2016) Using structural sustainability for forest health monitoring and triage: case study of a mountain pine beetle (*Dendroctonus ponderosae*)-impacted landscape. Ecol Indic 70:451–459

Carnegie AJ, Stone C, Lawson S, Matsuki M (2005) Can we grow certified eucalypt plantations in subtropical Australia? An insect pest management perspective. N Z J For Sci 35:223–235

Clark LR, Dallwitz MJ (1975) The life system of *Cardiaspina albitextura* (Psyllidae), 1950-1974. Aust J Zool 23:523–561

Collett NG, Elms S (2009) The control of sirex wood wasp using biological control agents in Victoria, Australia. Agric For Entomol 11:283–294

Collett NG, Fagg PC (2010) Insect defoliation of mixed-species eucalypts in East Gippsland. Aust For 73:81–90

Common IFB (1994) Oecophorine genera of Australia. I. The *Wingia* group (Lepidoptera: Oecophoridae). CSIRO Publishing, Melbourne

Dajoz R (2000) Insects and forests. Intercept, Andover

de Jong J, Dahlberg A (2017) Impact on species of conservation interest of forest harvesting for bioenergy purposes. For Ecol Manag 383:37–48

Elek J, Wardlaw T (2013) Options for managing chrysomelid leaf beetles in Australian eucalypt plantations: reducing the chemical footprint. Agric For Entomol 15:351–365

Flensted KK, Bruun HH, Ejrnaes R, Askildsen A, Thomsen PF, Heilman-Clausen J (2016) Red-listed species and forest continuity – a multi-taxon approach to conservation in temperate forests. For Ecol Manag 378:144–159

Freer-Smith PH, Webber JF (2017) Tree pests and diseases; the threat to biodiversity and the delivery of ecosystem services. Biodivers Conserv 26:3167–3181

French C (1891–1923) A handbook to the destructive insects of Victoria, Vols 1–5. Government Printer, Melbourne

Froggatt WW (1923) Forest insects of Australia. Government Printer, Sydney

Froggatt WW (1927) Forest insects and timber borers. Government Printer, Sydney

Garrick RC, Sands CJ, Rowell DM, Tait N, Greenslade P, Sunnucks P (2004) Phylogeography recapitulates topography: very fine-scale local endemism of a saproxylic 'giant' springtail at Tallaganda in the Great Dividing Range of south-east Australia. Molec Ecol 13:3329–3344

Garrick RC, Sands CJ, Sunnucks P (2006) The use and application of phylogeography for invertebrate conservation research and planning. In Grove SJ, Hanula JL (eds) Insect biodiversity and dead wood. Proceedings of a symposium for the 22nd International Congress of Entomology. USDA General and Technical Report SRS-93, Ashville, NC, pp 15–22

Garrick RC, Rowell DM, Sunnucks P (2012) Phylogeography of saproxylic and forest floor invertebrates from Tallaganda, South-Eastern Australia. Insects 3:270–294

Hall AAG, Gherlenda AN, Hasegawa S, Johnson SN, Cook JM, Riegler M (2015) Anatomy of an outbreak: the biology and population dynamics of a *Cardiaspina* psyllid species in an endangered woodland ecosystem. Agric For Entomol 17:292–301

Karvemo S, Bjorkman C, Johansson T, Weslien J, Hajlten J (2017) Forest restoration as a double-edged sword: the conflict between biodiversity conservation and pest control. J Appl Ecol 54:1658–1668

Legg S (2016) Political agitation for forest conservation: Victoria, 1860-1960. Int Rev Environ Hist 2:7–33

Martikainen P, Kouki J (2003) Sampling the rarest: threatened beetles in boreal forest biodiversity inventories. Biodivers Conserv 12:1815–1831

Mitchell RJ, Beaton JK, Bellamy PE, Broome A, Chetcuti J, Eaton S et al (2014) Ash dieback in the UK: a review of the ecological and conservation implications and potential management options. Biol Conserv 175:95–109

Muller J, Gossner MM (2010) Three-dimensional partitioning of diversity informs state-wide strategies for the conservation of saproxylic beetles. Biol Conserv 143:625–633

Nadel RL, Wingfield MJ, Scholes MC, Lawson SA, Slippers B (2012) The potential for monitoring and control of insect pests in Southern Hemisphere forestry plantations using semiochemicals. Ann For Sci 69:757–767

Nahrung HE, Loch AD, Matsuki M (2016) Invasive insects in Mediterranean forest systems; Australia. In: Paine TD, Lieutier F (eds) Insects and diseases of Mediterranean forest ecosystems. Springer, Cham, pp 475–498

New TR (1976) Submission on behalf of the Australian Entomological Society to the Australian Senate Standing Committee on Social Environment Inquiry into the impact on the Australian environment of the current woodchip industry programme. Australian Senate Official Hansard, pp 3628–3636

New TR (1983) Systematics and ecology: reflections from the interface. In: Highley E, Taylor RW (eds) Australian systematic entomology; a bicentenary perspective. CSIRO Publishing, Melbourne, pp 50–79

New TR, Yen AL (2012) Insect conservation in Australia. In: New TR (ed) Insect conservation: past, present and prospects. Springer, Dordrecht, pp 193–212

Norton TW (1997) Conservation and management of eucalypt ecosystems. In: Williams JA, Woinarski JCZ (eds) Eucalypt ecology: individuals to ecosystems. Cambridge University Press, Cambridge, pp 373–401

Ohsawa M (2010) Beetle families as indicators of Coleoptera diversity in forest: a study using Malaise traps in the central mountainous region of Japan. J Insect Conserv 14:479–484

Oliver I, Beattie AJ (1996) Designing a cost-effective invertebrate survey: a test of methods for rapid assessment of biodiversity. Ecol Appl 6:594–607

Ostrand F, Elek J, Steinbauer MJ (2007) Monitoring autumn gum moth (*Mnesampela privata*): relationships between pheromone and light trap catches and oviposition in eucalypt plantations. Aust For 70:185–191

Paine TD, Steinbauer MJ, Lawson SA (2011) Native and exotic pests of *Eucalyptus*: a worldwide perspective. Annu Rev Entomol 56:181–201

Readshaw JL (1965) A theory of phasmatid outbreak release. Aust J Zool 13:475–490

Schowalter TD (1986) Ecological strategies of forest insects: the need for a community-level approach to reforestation. New For 1:57–66

Sebek P, Barnouin T, Brin A, Brustel H, Dufrene M et al (2012) A test for assessment of saproxylic beetle biodiversity using subsets of 'monitoring species'. Ecol Indic 20:304–315

Taylor RW (1976) Submission to the Australian Senate Standing Committee on Social Environment Inquiry into the impact on the Australian environment of the current woodchip industry programme. Australian Senate Official Hansard, pp 3724–3731

Taylor RW (1983) Descriptive taxonomy: past, present, and future. In: Highley E, Taylor RW (eds) Australian systematic entomology; a bicentenary perspective. CSIRO Publishing, Melbourne, pp 91–134

Williams JA, Woinarski JCZ (eds) (1997) Eucalypt ecology: individuals to ecosystems. Cambridge University Press, Cambridge

Wingfield MJ, Roux J, Wingfield BD (2011) Insect pests and pathogens of Australian acacias grown as non-natives – an experiment in biogeography with far-reaching consequences. Divers Dist 17:968–977

Withers TM, Potter KJB, Berndt LA, Forgie SA, Paynter QE, Kriticos DJ (2011) Risk posed by the invasive defoliator *Uraba lugens* to New Zealand native flora. Agric For Entomol 13:99–110

Yeates DK, Bouchard P, Monteith GB (2002) Patterns and levels of endemism in the Australian Wet Tropics rainforest: evidence from flightless insects. Invert Syst 16:605–619

Yen AL, Butcher RL (1997) An overview of the conservation of non-marine invertebrates in Australia. Environment Australia, Canberra

References

Chapter 5
Studying Insects for Conservation in Forests

Keywords Ants · Boreal forests · Bracket fungi · Canopy insects · Carabidae · Forest edges · Indicators · Insect diversity · Sampling forest insects · Saproxylic insects · Stratification

5.1 Introduction: Problems of Access and Enumeration

The massive diversity of insects in forests reflects the elaborate structure of the environment, the persistence and complexity of these botanically diverse systems, and long histories of evolutionary associations. Quantifying that diversity and the patterns of its variation, series of surveys in several parts of the world have confirmed both high insect species richness and the difficulties of ecological interpretation, not least because of taxonomic inadequacy with many species undescribed. The roles of locally trained parataxonomists for inventory surveys of forest insects (Basset et al. 2004) may be important contributions to increasing understanding of diversity and its trajectories in relation to forest condition and management. Those surveys have progressively enabled extrapolations from limited samples to incorporate the full range of host tree species present, as the template that could in due course lead to estimations of impacts from forest changes. Comparisons between intact forest and forests modified in various ways provide evidence of impacts of imposed changes, as bases for conservation concern and possible reaction by changed management or level of protection.

Each major forest type, and perhaps every forested site, may have a different complement of characteristic arthropods, with these differing with vegetation category, elevation and latitude. Kitching et al. (1993), for example, found the numbers of Collembola in Australian forests decreased northward from the cool temperate forests sampled, whilst Diptera showed the reverse abundance trend. Profiles of insect orders showed significant differences between high and low (2–6 m from the ground) canopy samples. Again, Collembola were far less numerous in the upper canopy – however, the upper canopy yielded several 'specialist' species not found near the ground so this difference was not simply attenuation from a richer near-ground fauna.

© Springer International Publishing AG, part of Springer Nature 2018   77
T. R. New, *Forests and Insect Conservation in Australia*,
https://doi.org/10.1007/978-3-319-92222-5_5

5.2   Assessing Diversity

Sampling methods and limits on extrapolation need to be considered very carefully, but evaluation of 'diversity' (as both richness and abundance) is central to such studies. Thus, sampling saproxylic beetles (p. 96) adequately may demand a variety of approaches (Okland 1996), each with both advantages and disadvantages but with the recognition that substantial sampling effort and cost may be needed to assure relatively reliable quantitative information. In general, either for inventory surveys or for comparative evaluations of abundance and richness, a primary aim is to gain as much valid and relevant information as possible over the shortest period, and with lowest costs - but without sacrificing reliability. Many studies have led to suggested suitable protocols for particular insect groups or habitats.

Tropical forest insects, reflecting their richness and abundance as well as difficulties of access and often poor local infrastructure and facilities, pose significant problems of evaluation, and have led to continuing debates over themes such as (1) the optimal taxa to study; (2) the level of taxonomic interpretation needed; (3) the relative importance of common and rare species in monitoring; and (4) the reliability of more simplistic structural analyses in reducing sampling effort. Thus, whilst rare species are often regarded as especially significant in conservation surveys, the common species present might contribute far more to the ecological responses of communities – the debate is by no means confined to forest environments, but acknowledges that common species are the predominant components of community richness and are the main 'victims' of habitat losses and other threats. Although rare species are often regarded as more susceptible to local or regional extinctions, their loss may have only minor impacts on their host communities. For fruit-feeding butterflies in Brazil, Graca et al. (2017) suggested that giving higher importance to more common species can help optimise survey returns in monitoring, but rare taxa must still be considered in wider studies on community ecology and in conservation.

A widespread investigative approach is exemplified by a study in Cameroon, where Watt et al. (1997) compared plots with the treatments of (1) intact forests without any ground disturbance, as a 'control'; (2) complete clearance with all trees and other vegetation removed using a bulldozer, with resulting soil compaction of bare ground; (3) partial mechanical clearance, in which about half the large trees and most understorey vegetation were removed by bulldozers; and (4) partial manual clearance, where ground vegetation and some small trees were cleared by hand, machete and chainsaw, and with little or no soil compaction from machinery use. As in other treatments there, including open ground left after clearing as 'farm fallow', only about half the number of species of termites and butterflies were found in treated plots compared with intact forest. Plantation forest management also had marked effects, with all the insect groups assessed being more diverse in plantations on partially cleared plots than in plantations established after complete clearance. Here, as in many other surveys undertaken with similar perspective, intensity of intervention increased loss of insect diversity.

Interpreting patterns and composition of insect richness in relation to forest management is often difficult, but some probable generalisations have emerged from northern temperate forests, in particular and have indicated some constructive management steps to promote conservation. Thus, for the beetles of boreal forests of Fennoscandia, Niemela (1997) noted that old-growth specialists tend to disappear from clear-cut areas, but local species richness may increase there from a combination of persistence of forest generalists and the arrival of many more characteristically open habitat species. Beetle species richness alone may thereby give a misleading impression of the impacts of the clearing. On larger scales, particularly, the more sensitive and specialised species decline as landscape homogenisation continues. These general trends led to three complementary recommended approaches to maintain biodiversity and continue timber harvesting there: (1) undisturbed old-growth forest must be left alone to conserve specialist species and act as source areas for those species to colonise other areas that may become suitable; (2) those reserves are highly unlikely to ever be sufficient (in size and representation) to fully serve that purpose, so that informed silvicultural practices are needed to aid conservation, using natural disturbance regimes as guidelines; and (3) restoration and promotion of natural regeneration is necessary for recovery of the associated species assemblages of the area.

Close parallels in trends were found across the Carabidae of Fennoscandia and the Nearctic region and the above trends show that simply maximising species richness and diversity alone may be inadequate to counter the compositional changes that flow from forest cutting.

As discussed later, much inference on changes of northern forest faunas has been based on studies of saproxylic beetles (p. 96), as a well-surveyed, diverse and circumscribed group that is relatively easily sampled and contains numerous species of conservation interest. Similar trends may occur amongst non-saproxylic beetles and many other taxa. Thus, Martikainen et al. (2000) surveyed beetle assemblages in the oldest age classes of managed forest and old-growth forest. Among the 553 species, both saproxylic and non-saproxylic species were more abundant in old-growth than in mature managed forests, but with differences significant only for saproxylic beetles – and most pronounced for those that are not bark beetles and their associates, and for rare saproxylic species (Fig. 5.1). Non-saproxylic beetles were relatively unaffected by forest management, so that the common emphasis on saproxylic taxa in planning conservation measures seemed well-justified.

Comparison of insect catches from 'fogging' (below) trees in primary forests and various forest treatments provide substantial information on anthropogenic disturbances, whether the catches are from high canopy (not always available after disturbance) or lower-growing understorey trees, and commonly using 'time-since-disturbance' as an index of potential recovery. As one example, beetles from 10 understorey trees in four forests (primary forest and forests left to regenerate for 5, 15 and 40 years after clearing) in Sabah (East Malaysia) may represent a more general pattern. There, Floren and Linsenmaier (2003) found that primary forest yielded no common species, and every fogging sample was largely distinctive, so that the Coleoptera assemblage present could not be distinguished

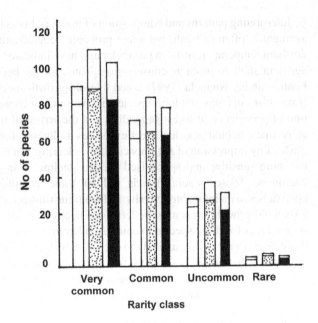

Fig. 5.1 Species richness of saproxylic beetles separated by rarity level among three forest categories in southern Finland: old-growth forest (black), mature forest (open), over-mature forest (dotted); open space at top of each bar is number of species occurring only in that forest category class. (Martikainen et al. 2006)

consistently from random assemblages. Species diversity was much lower in the disturbed forests, the proportion of frequently-occurring species was higher, and some species were collected regularly from most individual trees examined. The number of species (of Chrysomelidae and Curculionidae as major phytophagous groups: together with Staphylinidae these always comprised more than half the species found) increased with regeneration time (84, 91, 186 species with increasing time) and all were far lower than in the primary forest (which yielded 285 species).

5.2.1 Canopy Fauna

The most influential pioneering study on tropical forest insect diversity is the often-cited survey of Neotropical forest canopy beetles by Erwin (1982). In that classic paper Erwin set out, for the first time, a series of assumptions and testable hypotheses, based on collections by insecticide fogging of beetles from the canopy of 19 individual trees of *Luehea seemannii*, that could be used to estimate species diversity. They have generated considerable further investigation. These (Table 5.1) included testable evaluations of levels of host tree specificity and number of tree species present, and implied that – to some extent - every tree species may support a unique suite of arthropod species. Erwin's paper stimulated considerable debate over his initial global diversity estimate of 30 million arthropod species. The assumptions have all been queried and modified by later commentators, but remain

Table 5.1 The numbers of arthropod species in the tropics: the assumptions made in extrapolating from samples of beetles from the canopy of a single species of rain forest tree, *Luehea seemannii* in Panama (based on Erwin 1982)

1. Number of beetle species from canopy fogging of 19 individual trees	1200
2. Number of host-specific species (four trophic guilds)	163
3. One hectare of tropical forest has on average 70 genus-group trees	
Number of host specific beetles per hectare (70 × 163)	11,410
4. Plus the remaining transient species.	
Number of beetle species per hectare (11,410 + 1038 transient species)	12,448
5. Beetles constitute 40% of all arthropod species	
Number of canopy arthropod species per hectare (12,448 × 100/40)	31,120
6. The canopy fauna makes up 2/3 of the total	
Total arthropod fauna per hectare (Erwin 'added a third')	41,389
7. There are an estimated 50,000 species of tropical trees.	
Based on the above, total number of tropical arthropod species	30,000,000

one of the most stimulating and effective contributions to rationally assessing insect diversity in tropical forest canopies.

In addition to further New World tropics surveys, parallel studies in Africa, Borneo, Papua New Guinea and Queensland have increased basic documentation and understanding of tropical forest canopy insect diversity. In contrast to the considerable attention to enumerating insects of tropical forest canopy, equivalent vertical stratification studies of insects in temperate forests have received relatively little attention – but, nevertheless, are important. Some workers (such as Basset et al. 2003) have suggested that such vertical gradients may be more evident in tropical than in temperate forests, and Lowman et al. (1993) earlier suggested that peak insect diversity may be near the ground in temperate forests, in contrast to this occurring in the canopy of tropical forests. Vertical distribution surveys have understandably focused on larger and better-known insects, for which interpretations may be valid. Some major groups, despite their immense diversity and ecological importance, cannot be studied properly. For most Hymenoptera, other than some conspicuous aculeate groups, few such surveys have been attempted, and have been largely thwarted by taxonomic complexity and high proportions of singletons and other low abundance outcomes, rendering sound intepretation of samples very uncertain. From their surveys comparing canopy and understorey parasitoid Hymenoptera in North American forests, and reflecting on their roles as higher level trophic participants, Vance et al. (2007) referred to them as being 'seriously underrepresented' in canopy studies. Similarly, Stork (1988) found 437 of the 739 species of Chalcidoidea in his tropical fogging samples represented by single individuals, so any reliance on these for interpretation is questionable. Similar ambiguities occur

Table 5.2 The dead wood species of saproxylic beetles found in canopy and forest floor samples in three forest types in France (Bouget et al. 2011)

Data set	Canopy	Forest floor
Beech-fir mountain		
Forest, window traps		
No. individuals	906	2329
No. unique species	5.1(24%)	11.9(58%)
Mean species richness	12.9	21.2
Oak branches, reared		
Mean no. individuals	23.9	37.8
No. unique species	42(21%)	94(47%)
Included singletons	14	41
Mean species richness	12.8	15.1
Pine branches, reared		
Mean no. individuals	14.5	8.9
No. unique species	17(44%)	16(41%)
Included singletons	8	3
Mean species richness	2.5	2.6

for Diptera, for which different studies have reported higher or lower diversity and relative abundances in different forest strata (Maguire et al. 2014).

For the diverse saproxylic insects of northern temperate forests, Bouget et al. (2011) noted the likely vertical layers operating as environments that differ in microclimates and with differing conditions and interactions as forest floor, understorey, mid-canopy, and tree top, and also noted the paucity of studies on this diverse insect group across those levels. They set out to assess several relevant hypotheses from comparative sampling of saproxylic beetle assemblages at two levels in three temperate forests in France, namely (1) whether the bulk of diversity is indeed near the ground; (2) whether there are species characteristic of either vertical stratum; (3) whether there are canopy specialists, as implied in some other insect groups; and (4) whether canopy assemblages are nested subsets of understorey assemblages, or if their distribution between canopy and understorey is essentially equal. The trial involved three different forest systems and involved comparisons of (1) canopy and understorey flying beetles in mountain beech/fir forests; (2) canopy and ground saproxylic beetle comparison in pine dead wood; and (3) canopy and ground saproxylic beetle comparison in oak dead wood, yielding respectively 158, 201 and 39 species. Significant differences between strata occurred, and the first comparison supported assertion that the ground fauna were the more diverse (Table 5.2). A similar trend occurred for the oak wood comparison, but a substantial proportion of species (31%) were retrieved from both levels. The three comparisons thereby gave somewhat contrasting outcomes, but each stratum harboured taxa not found in the other, and canopy assemblages were not nested subsets of the richer ground fauna. Aerial dead wood appears to comprise a somewhat different substrate from ground dead wood, and its possibly distinctive roles as beetle habitats imply that it should not be removed unless it becomes necessary – as, for example, if it is affecting public safety and/or tree stability through being likely to fall.

The canopy arthropods of northern forests are clearly diverse: Thunes et al. (2003) found 510 species from *Pinus sylvestris* in Norway by canopy fogging, for example. Many were new records for the country, and some undescribed – again emphasising the needs for such surveys to advance basic faunal documentation. Species diversity was similar in 'old' (250–330 years) and 'mature' (60–120 years) trees, but the number of species new to Norway or to science was higher in the former group.

Assessment of forest insect diversity clearly necessitates multi-stratum evaluations in temperate forest as well as in the better-publicised tropical canopies, as advocated by Su and Woods (2001). Information from tropical forests of Queensland has been complemented by surveys of southern forests in both the south east and south west of Australia, confirming that those faunas are also diverse, and locally characteristic.

Arthropod diversity among canopies of temperate region eucalypt forests were investigated through a one-year comparative knock-down sampling project in Western Australia (emphasising *Eucalyptus marginata* and *E. (Corymbia) calophylla*) and eastern New South Wales (*Eucalyptus crebra, E. moluccana*), in which ten trees of each species were sampled in each of the four seasons (Majer et al. 2000). Morphospecies analysis of the captures revealed 687 species (representing 176 families) from Western Australia, and 976 species (from 173 families) from New South Wales forest. Little over half (53%) of the families were found in both regions. Collectively, the samples were dominated by Hymenoptera (450 morphospecies), Coleoptera (363), Diptera (252), Araneae (168), and Homoptera (150). All these richness estimates were regarded as conservative in relation to total diversity present, reflecting the limitations of sample numbers and lacking the numerous taxa not susceptible to knockdown techniques because they are associated with bark, flowers, fruits or wood, rather than being exposed on foliage to the insecticide used. Extrapolating from these data, Majer et al. noted that these limited samples from only four of the approximately 700 eucalypt species and at single sites could nevertheless represent about 1% of the (then) estimated 140,000 Australian insect species.

Despite frequent allusions to very high diversity of forest canopy insects, some studies have found this to be rather low. For three eucalyptus forest canopies sampled by direct examination of bagged branchlet samples, only low numbers were recorded (Ohmart et al. 1983). The sample method might itself prove less inclusive than some others, but that survey indicated that insect herbivore numbers were more similar to some of those reported from the northern hemisphere, and countered the more often cited implication that Australian canopy insects are by far the more abundant.

Whilst claims have been made that the forest canopy insect fauna is both characteristic and largely restricted to that stratum of vegetation, Kitching (2004) noted that, whether or not this is so, the extent to which canopy insects are distinct from those elsewhere in the forests has engendered considerable debate. It is also tempting to claim, on the basis of the above and related studies, that tropical forest canopies support the greatest forest insect richness, but other studies introduce doubts over this. Not least, occasional confusions in natural distributions can occur from disturbances to the forest – for example, De Vries (1988) noted that edges and gaps (p. 245), may be treated by canopy butterflies as if 'the canopy has come to the ground', with the disturbance changing light levels and distorting the usual vertical stratification. The similarity between the butterfly communities of disturbed and undisturbed habitats may then be increased and vertical stratification patterns distorted (Fermon et al. 2005). A parallel finding by Willott (1999) was that some moth species common only in the canopy of undisturbed forest may occur near the ground in recently logged areas. As Intachat and Holloway (2000) surmised, a sudden disturbance such as logging may influence their normal flight behaviour and cause them to fly nearer the ground within cleared areas – so that samples by light traps might suggest increased abundance simply because the moths then become 'trappable'. That increase may be temporary and cannot be sustained if suitable larval food plants are not available. However, current lack of knowledge largely precludes further interpretations in most tropical areas.

Willott (1999) warned, however, that such inter-site comparisons can be misleading, because differences in assemblages across primary forest understorey sites may be as great as those between understorey and logged forest sites. Generally low similarity between understorey and canopy faunas supported the occurrence of a distinct canopy fauna in three Sabah forests: that about 10% of the species were confined to primary forest was taken to imply a minimum estimate for the number of species that could be lost following logging. The high moth richness (at least 1850 species) in the area of this study (Danum) includes many for which biology is essentially unknown. Willott's surveys yielded 1238 species, with diversity across sites varying considerably.

Species diversity amongst butterflies in primary forest is often higher at sites with more open canopy – so that assemblages may respond to changes in shading when selective logging or other gap creation occurs. Hamer et al. (2003), comparing butterflies of primary forests and forests selectively logged 10–12 years previously in Borneo, noted little difference in overall diversity but substantial differences in assemblage composition. This linked with species' gap preferences – species strongly preferring shade (and often with narrower distributions) were those most strongly affected, whilst cosmopolitan species preferring open (light) regimes benefited from the logging. Comparison of butterflies in primary rainforest with an adjacent site logged only 6 years previously in Borneo (Willott et al. 1999) gave no evidence that selective logging had changed assemblage composition from locally distributed species to more widespread species: low logging intensities do not necessarily affect butterfly richness or abundance to any significant extent. More species (151) occurred in logged than in primary forest (121) and of the total 180 species, 59 were found only in logged forest and 29 only in primary forest – but almost all 'restricted' species were scarce, with few individuals recorded.

Hammond et al. (1996) assessed the beetles of different habitats in lowland forest of Sulawesi, collecting 4500 species. That survey indicated that about three times as many beetle species are ground specialists than are canopy specialists. The arthropod fauna of the soil appeared there to be at least as rich as that of the forest canopy.

Vertical distributions of arthropods in forests are determined by four main groups of factors (Basset et al. 2003), namely abiotic factors, forest physiognomy and tree architecture, resource availability, and behaviour – with many species foraging at levels at which microclimates and food are most suitable. Differences between tropical and temperate forests can broadly reflect the less pronounced vertical changes in microclimates, mainly in temperature and humidity, and which some ecologists have likened to gradients of latitudinal richness (Turner et al. 1987). Those gradients can also incorporate biological influences such as predation pressure. Insects preferring shady conditions, high humidities and lower temperatures may be found only, or predominantly, in the lower and less-exposed forest strata.

The studies pioneered by Erwin involved insecticide 'knockdown' of insects from fogging tree canopies, and catching the falling bodies on trays or in funnels near the ground. That technique has remained useful, but has been augmented by an array of other methods that have increased the breadth and comparative values of samples, based on increasingly original and innovative sampling approaches that progressively augment biological information on the taxa obtained. One informative approach utilised a trap that combines different trapping principles, so separating insects that move upward into a collecting bottle (Malaise trap) and those which drop downward into a trough of preservative (flight interception trap). The trap can be suspended in the canopy, hoisted on ropes and with paired ground level traps enabling comparison between strata. That comparison in lowland tropical rainforest in northern Queensland provided important information on the assemblage composition of beetles at the two levels (Stork and Grimbacher 2006). Paired traps at five sites and operated over four years, with 45 two-week sampling periods, yielded 29,986 beetles representing 1473 species and 77 families across the replicated segregates of canopy/ground Malaise and canopy/ground intercept traps. The two strata had similar numbers of individuals (canopy, 14,473; ground, 15,513). As is usual in large ecological samples of insect assemblages, many species were represented by single individuals or two specimens, whose omission reduced the above totals to 649 species (canopy) and 612 species (ground), of which 72% were retrieved from both levels. Many beetle species were therefore vertically widespread – but for the more abundant taxa similar proportions (24–27%) were strongly associated with one or other sample set so were considered 'specialists' to that level. Of the remaining 115 'non-specialists', 42 species were each three times more abundant in one stratum than in the other.

Knowledge of Australian tropical forest canopy invertebrates was transformed with construction of the Australian Canopy Crane in the Daintree Forest of Queensland. This provided direct access to the initially 680 trees (representing 82 species) in about 0.95 hectare of forest (Stork 2007). Each tree can be located individually for study from above, using compass direction and distance along the crane boom in relation to a detailed GIS-based map that shows the identity, basal size and height of all included trees. That access has enabled detailed studies of the canopy insect fauna – for example, survey of canopy beetles by Wardhaugh et al. (2012) in which beetles were sampled from the different microhabitats of mature leaves, new leaves, fruit, flowers and suspended dead wood from 23 canopy plant species.

Direct close-range observation and inspection of forest canopy, facilitated by use of canopy cranes, 'rafts' and other techniques allows for direct observation and more detailed small-scale investigations with minimal disturbance. Access to the forest canopy was for long the major limitation to studying their structure and biota, but a variety of techniques pioneered in the last quarter of the twentieth century (Lowman and Wittman 1996) have progressively led toward more comprehensive and penetrating studies of ecological interactions and relationships, gradually augmenting purely descriptive studies of diversity (which remain important) by functional interpretations as well as increasing recognition of the complexity and heterogeneity of forest canopy environments. Many – perhaps most – surveys of forest insects have been confined to the more easily accessible understorey levels, and any extrapolation to the canopy fauna is necessarily cautious. Thus, for Lepidoptera, biases include food plant availability (species, nutritional quality, plant defences), presence of woody vegetation, and unknown influences of stratification (Stireman et al. 2014), so that the assemblages reflect processes of both fragmentation and subsequent resource-driven invasion and colonisation.

The functional contributions of insects may differ substantially across forest strata, as indicated by representation of a key pollinator group, bees, which were compared by trapping in the canopy and near the ground of the same trees in a temperate hardwood forest in Georgia (Ulyshen et al. 2010). *Augochlora pura* (Halictidae) was by far the most abundant of the 71 species captured, and was more than 40 times more abundant in the canopy than near the ground. In general, canopy bees were richer (57 species compared with 47 species near the ground) and more abundant (6300 individuals compared with 353). However, 14 species (10 of them singletons) were found only in the lower traps, and some of these were probably restricted by being ground-nesters. Prevalence of bees in the canopy might reflect feeding on resources such as honeydew produced by hemipterans feeding on the younger foliage there, and sap, over periods when supplies of plant nectar and pollen are low (Ulyshen et al. 2010).

In addition to resource supply, the vertical distributions of insects in forests are also likely to reflect microclimates, and their interaction with resources. Thus, vertical layering of saproxylic insects might in some cases be related to the importance of sun exposure (p. 166), and so become relevant in conservation management. Vodka et al. (2009) drew the distinction that (1) if species that require sun-exposed substrates breed high in the forest/woodland canopy, they may be secure without active human intervention, whilst (2) if those species breed mainly in the understorey, more active management may be needed to conserve them. In woodlands of the Czech Republic, freshly cut oak (*Quercus robur*) 'baits' of stems, branches and twigs were exposed to ovipositing beetles in several situations – meadows, canopy shade, canopy sun, understorey shade, and understorey sun. 'Canopy' baits were suspended 17–22 m above ground, and 'understorey' baits were hung about 1 m above ground. Reared beetles comprised Cerambycidae (17 species) and Buprestidae

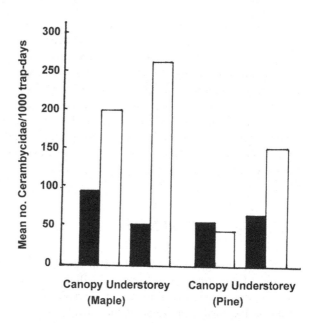

Fig. 5.2 Numbers of Cerambycidae (given as mean/1000 trap-days) in maple and pine forests in Ontario, Canada, comparing abundance in canopy and understorey by flight interception traps at two levels (Vance et al. 2003) (bottom bottle, black; top bottle, open bars)

(five species) but, despite some assemblage differences, influences of stratification seemed minor compared with those of sun or shade regimes. In general, species richness was greater near the ground, but this reflected that the more common canopy taxa also occurred in the understorey, whilst species preferring understorey were largely absent from the canopy. Several rare species were retrieved only in the understorey, but the major management recommendation for these oak woodlands was to restore open regimes together with assuring supplies of sun-exposed dead wood.

Cerambycidae of the understorey and canopy of White pine (*Pinus strobus*) and Sugar maple (*Acer saccharum*) forests in Ontario comprised a total of 28 species, six of which were captured at both levels, whilst each level had 11 species not found elsewhere (Vance et al. 2003). The pine sites had more species (22, compared with 18 from maple) but fewer individuals, and differences (Fig. 5.2) implied that the assemblages could be defined only by sampling from both strata, so increasing representation of their species constitution. The second implication was that the pine stands are important in sustaining cerambycid diversity, and should be incorporated actively into maintaining tree diversity in the maple-dominated forests.

Empirical studies on the vertical stratification of insects in forests have somewhat overshadowed that their distributions across the 'horizontal axis', from trunk to outer canopy edge may also differ. Using the rich and well-documented beetle fauna of oak trees in southern England, Stork et al. (2001) showed some differences in species incidence near and more distant from the trunk of isolated trees, as well as direction-related patterns of within-tree distribution. Their knockdown samples from 36 trees yielded 144 species of Coleoptera, and broad patterns can be summarised as (1) some species being associated more closely with the trunk; (2) others were more closely associated with the outer regions; (3) some species 'prefer' particular compass sectors of the tree; and (4) distribution patterns of some species changed during the year. Patterns are thereby complicated, but their occurrence may influence interpretation of arboreal insect trap catches in other contexts, from far less understood faunas and without any knowledge of edge effects or even of the basic biology of most species encountered. The edges of single trees may, on a different scale, parallel the forest edge effects that are more frequently considered. As an additional example from the same trapping sequence, Barnard et al. (1986) found several species of Neuroptera more abundant near the trunks than towards the edges of the crowns.

5.2.2 Litter and Soil Fauna

After clear-cutting, some form of soil preparation and surface clearance may be employed to aid tree regeneration or facilitate further plantings. Soil preparation and related measures for afforestation of arable land may also have strong impacts on the surface and near surface fauna, including ground-active insects, but those impacts have only rarely been studied directly. Effects of soil treatment of clearcut areas of a closed canopy pine forest, and undertaken preparatory to planting with pine seedlings, on carabid beetles were assessed in Poland (Sklodowski 2017). Pitfall trapping across two years and six treatments yielded 73 carabid species, very few of which had any clear indicator value for any treatment.

In general, abundance of beetles decreased with increased depth of soil preparation, with highest numbers found in untreated plots and those on which vegetation was simply slashed. More intensive disturbance, notably deeper soil preparation over larger areas, had stronger impacts – and even shallow ploughing or manual preparation constituted local disturbances likely to have some impacts. In general, low intensity preparation over limited areas of a site was recommended, and deeper ploughing over larger areas should be avoided. Sklodowski also noted the conservation benefits of leaving a few piles (of at least a few metres long) of branches as refuges in the centre of clear-cut areas and, where possible, also some tree retention on sites.

Soil preparation had little clear impact on forest floor carabids in Finland (Koivula and Niemela 2003), although some species either benefited or were disadvantaged by changes to microhabitats caused by changes to the humus layer and logging residues. One forest species, *Calathus micropterus*, was slightly less abundant in gaps with mechanical soil preparation (removing the humus layer in approximately 50 cm wide strips) than in unlogged control sites.

Litter amount and quality in forests greatly influence the abundance and diversity of invertebrates, many of which contribute to the major processes of decomposition and recycling nutrients. In pine forests, the abundance of litter invertebrates may increase with mass of leaf litter (Burghouts et al. 1992), and distinct differences in relative abundance of taxa may occur between logged and primary forests. Thus,

Table 5.3 The four provisional functional groups of termites used in interpreting responses to forest disturbance in Cameroon (slightly abbreviated from Eggleton et al. 1995)

1. Soil-feeders. Termites distributed in the soil profile, surface litter materials and/or epigeal mounds, and apparently feeding on mineral soil; some included species may be root-feeders, even without strong evidence at present.
2. Soil/water interface-feeders. Termites collected only or predominantly within soil under logs, within soil plastered onto the surface of rotting logs or within highly decayed wood.
3. Wood-feeders. Termites feeding on wood and excavating galleries in items of woody litter, which in some cases become colony centres. Category also supports termites with arboreal nests and others with subterranean nests with cultivated fungus gardens.
4. Litter foragers. Termites that forage for leaf litter and small woody litter.

beetles and cockroaches were relatively more abundant in such a comparison of litter fauna in Sabah.

Changes in litter fauna following forest maturation and disturbance are usual, but the mechanisms of loss and any subsequent recovery are often not clear. Litter beetles of wet forest (*Eucalyptus obliqua*) in Tasmania compared across young and mature stands (Baker 2006) showed that the twin assemblages differed considerably. Indicator analysis identified seven species characteristic of young forest and nine species characteristic of mature forest. The remaining 21 species did not characterise either forest stage, but some had very different abundance in the two treatments: for example all 370 individuals of *Mecyclothorax ambiguus* (Carabidae) were from mature forest, and 545/547 of an *Anabaxis* sp. (Staphylinidae) were from young forest litter. Many beetle species displayed some form of successional 'preference', and this study also showed the values of examining beetles other than the most usual focus of Carabidae alone.

Leaf litter from different tree taxa can differ considerably and (in part following Swift et al. 1979) many ecologists have emphasised the gradient between 'mor' and 'mull' humus. The former typically has low quality and initially low fragmentation rates and pH, and the latter is higher quality with high fragmentation rates and pH range. Mor litter tends to support high abundance of fungivorous mites and Collembola, whilst mull litter supports larger invertebrates capable of fragmentation, rather than grazing alone. Comparisons of the invertebrate litter communities may thereby reflect features of the litter and its origin, and the inhabitants influence rate and pathways of litter decomposition.

Ants and selected beetle families have been the most intensively studied litter insects in comparative surveys, but parallel functional groups occur also in other groups. Termites (Isoptera) in forests in southern Cameroon were distributed across four functional groups to assess changing representation in relation to forest disturbance (Eggleton et al. 1995), building on previous accounts demonstrating that forest clearance reduces termite species richness. Representation of those groups (Table 5.3) differed across the five treatments (Fig. 5.3), with the most disturbed blocks having far fewer termite species than forested blocks. Soil-feeders, for example, were almost absent from the former but the predominant termites in forest treatments. Altogether, 88 species (in 50 genera) were distinguished. Reasons for this

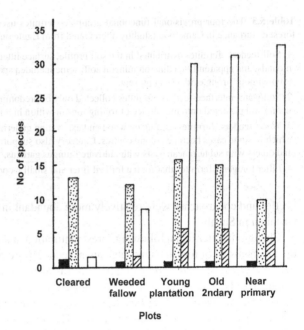

Fig. 5.3 The relative species richness of functional groups of termites from five categories of forest plots in Cameroon. Litter, black; wood, dotted; soil, open; wood/soil, diagonal hatch. (Eggleton et al. 1995)

disparate representation were not wholly clear, but the large amounts of dead wood left in Old Secondary and Young Plantation plots might enable more wood-feeding and wood-soil interface termites to occur. Eggleton et al. likened those responses to 'gaps' within which those species might occur normally. The insects of litter and dead wood can overlap somewhat (p. 130): some stag beetles, for example, occur mainly in the soil near and under fallen logs.

The leaf-litter ants of mature Atlantic Forest in Brazil are both richer than the assemblages in regenerating forests in that area, and distinct from them, as demonstrated by litter samples and bait traps in several successional stages (Silva et al. 2007). Disturbed sites had considerably lower richness, with the 89 species from primary forest samples reduced to a total of 69 species across the five other treatments sampled, and the number of 'unique' species declined from 50 to 31. An innovative suggestion from this study was that the regeneration stages form three distinct groups associated with changes in litter ant assemblages, as highly simplified, intermediate stages, and secondary forest, a gradient along which the progressive changes in assemblages could be documented.

Ground-dwelling arthropods, whether or not associated strongly with woody debris or litter, are diverse in forests. Woody debris is, however, a key influence, and is discussed in Chapter 8. Its extent may influence overall richness and abundance of many arthropod groups. Relatively little is known of many other taxa in litter, and of their relationships – if any – with dead wood and its more encapsulated and characteristic fauna. In Long-leaf pine forest communities in Florida, pitfall trap catches of arthropods (collectively of more than 932 genera) were compared from traps deployed near fallen logs (as possibly habitats or surrogate drift fences) or along

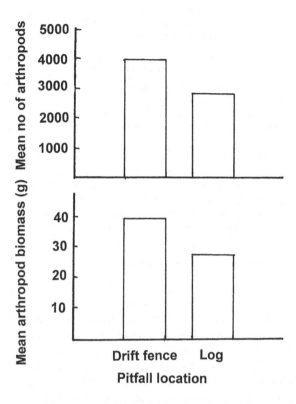

Fig. 5.4 Relative catches of arthropods in pitfall traps deployed near logs and near drift fences arthropods in pine litter in a Florida forest. (Hanula et al. 2006)

drift fences (both ca 3 m long). Catches were considerably greater in the latter (Fig. 5.4, Hanula et al. 2006). Despite lack of species-level identifications, Hanula et al. also noted that a number of taxa captured near logs were likely to be associated functionally with them. Removal of coarse woody debris (p. 152) lowered the abundance of some taxa, but it was not clear if (1) those species were affected by the physical removal of the wood or (2) whether the wood removal simply affected their ability to maintain viable populations in the forest. Similar uncertainties are not uncommon, and the wider influences of dead wood on epigaeic (but non-saproxylic) arthropods in forests remain largely unclear. The changing constitution of dead wood, with 'preconditioning' by pest beetles sometimes creating the conditions needed by numerous other arthropod species, dictates the need for parts of trees killed by bark beetles and wood-boring beetles to be left *in situ* in order to enhance conservation of forest biodiversity (Hilszczanski et al. 2016). Both standing and fallen timber may provide critical resources for a number of rare and threatened species: Hilszczanski et al. cited beetle examples of species of *Bothrideres*, *Cucujus* and *Dendrophagus*.

Table 5.4 Priorities for future research on influences of edge effects on invertebrates in forest canopy or other habitats, suggested by Foggo et al. (2001)

1. Studies of the microclimate of canopies in different vertical strata, and comparison with microclimates at different horizontal distances from the forest edge.
2. Studies on the vertical distribution of organisms in the canopy in relation to microclimate and biotic factors.
3. Studies of groups likely to be highly responsive to microclimate through indirect influence such as food availability – epiphyte-feeders, for example, might be affected by edge effects on their food.
4. Studies of 'dorsal edge effects', such as acid rain, on taxa predicted to be sensitive – such as lichens and lichen-feeding invertebrates.
5. Studies of edge effects due to gaps created by selective logging, and how these are influenced by geometry, distribution and temporal synchrony.
6. Studies of the biotic and abiotic similarity of canopies to open and gap habitats, and their importance for dispersal of open-habitat species.

5.3   Insects and Forest Edges

An 'edge' is essentially an ecological interface between adjacent habitat types, most commonly viewed as a physical boundary between vegetation structures or categories and functioning as an ecotone that reflects changes in microclimate, structure, and floristic composition. The mixing of two or more communities produces a variety of so-called 'edge effects' that can include heightened species richness at the interface. Most pertinent studies on insects involve edges produced by anthropogenic activity, such as forest clearance, which produce relatively abrupt transitions ('hard edges'). Studies on insects across more naturally developed and more gradual 'soft' edges are far fewer, especially in the tropics where, in any case, 'Studies of edge effects on invertebrates in tropical forest have been relatively scarce' (Foggo et al. 2001). Foggo et al. noted, for example, that the forest canopy abuts three major and distinct habitats – the atmosphere, the understorey, and the lateral bounding vegetation type, whether this last is natural, plantation, or agricultural or urban ground. Tropical rain forests are subject to three predominant forms of exploitation that will influence edge effects in the future, as (1) clear-felling and clearing, leading to fragmentation and increased edge lengths around the remaining forests; (2) plantation forestry; and (3) selective logging and tree extraction from areas of forest. Foggo et al. (2001) combined these to develop a suite of priorities for future research to help understand the far-reaching implications (Table 5.4) and formulate mitigation strategies for the future.

Features of forest edges, increasing in extent under much modern forestry, permeate almost all biological and microclimate gradients in addition to leading to changes among insect assemblages and affecting the 'performance' of resident taxa. Thus, the gradient in desiccation of wood (felled bolts of *Picea abies*) placed at intervals from clearcut to 50 m into forest was associated with increased attack densities of two key bark beetle species (*Hylurgops palliatus*, *H. glabratus*) towards the forest interior (Peltonen and Heliovaara 1999). In that study the forest edge effects

Fig. 5.5 Average numbers of individuals (black bars) and species (open bars) of Carabidae in traps at forest interior (30–60 m into fragment), edge (55 m inside forest at edge and 15 m into clearcut) and in clearcut areas (30–60 m into clearcut). (Heliola et al. 2001)

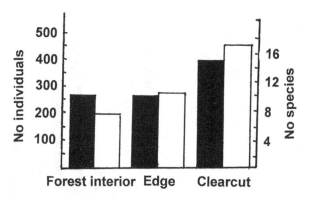

on a key resource substrate affected colonisation and breeding success by the beetles. Such influences lead to changes in insect incidence in natural and managed forests, but are likely to be highly specific.

Numerous 'gradient studies' have examined the changes in insect assemblage composition across the transition zones of forest edges. Ground beetles, Carabidae, sampled by pitfall traps at regular intervals from the forest interior to equivalent distances into the outside matrix, have been appraised in relation to distance from forest edge, the degree of contrast between the forest and adjacent area, and (where relevant) the area of the forest fragment. Inferences from a study comparing pitfall trap catches of Carabidae along transects crossing the zone between boreal forest and clearcut areas are likely to have much wider relevance. That study, in Finland and involving 'hard edges' as abrupt transitions – the most common pattern resulting from forestry activity – included 32 ground beetle species (Heliola et al. 2001). The regimes differed markedly in vegetation, and sampled sites spanned from 60 m into each of the forest and clearcut areas. The major findings (Fig. 5.5) were (1) the edge region assemblage more clearly resembled that of boreal forest interior than that of clearcuts; (2) there were no 'edge specialists', but the edge zone was not avoided actively; (3) open habitat species only occasionally penetrated far into the forest interior (as demonstrated well by *Pterostichus adstrictus*); and (4) some forest species were less abundant in clearcuts.

The lack of decrease near the edge was paralleled in Canada (Spence et al. 1996), where some old-growth forest specialist taxa in forest fragments were progressively exposed to more generalist and open-habitat carabids in clearcut and regenerating areas. The dramatic impact of the forest edge is illustrated by distributions of two species with abrupt changes along the forest-clearing gradient (Fig. 5.6). However, the forest assemblages in residual stands included open-habitat species. Old-growth forest specialists are especially vulnerable to forest clearance, but details of responses to disturbance are usually unclear, especially in relation to the history of plots studied.

Chronosequence surveys, comparing the fauna of forest stands of different ages or recovery periods, or under different management regimes have also been informative. Although the forest floor carabid fauna of *Eucalyptus obliqua* forests in

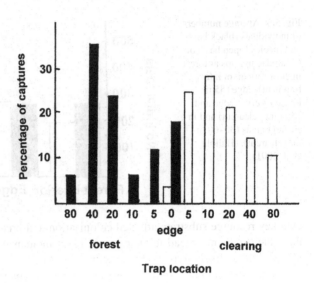

Fig. 5.6 Edge effects on Carabidae: the captures of *Synuchus impunctatus* (n = 29, open bars) and *Scaphinotus marginatus* (n = 17, black bars) at different distances (m) across the forest-cleared edge in Canada. (Spence et al. 1996)

Tasmania is relatively small (18 species sampled), a chronosequence pitfall trap survey implied that it was not changed significantly by forest management (Michaels and McQuillan 1995). Across the 14 sites sampled, four of them old-growth, the others regenerated over varying periods following a single cycle of clearfellng, burning and aerial seeding, differences occurred in relation to litter depth and age of the regeneration, but not to the proportion of bare ground or soil pH. The habitat mosaic that resulted from commercial forestry could support the carabids by a combination of survival in fire-free refuges and recolonisation of the disturbed areas. Dispersal potential was clear from the predominance of winged species in the pioneer communities, but forest fragmentation could lead to reduced carabid richness if remnant old-growth patches were too small to sustain populations of their specialist species, and too separated to allow dispersal between patches. Thus, Michaels and McQuillan (1995) found that such specialists (in their study exemplified by *Stichonotus piceus*, *Pterocyrtus* sp., *Chylnus ater* and *Notonomus politulus*) may become vulnerable through essentially irreversible habitat changes, with long-term increased fragmentation of old-growth forest progressively threatening more generalist species as population segregates are increasingly isolated.

In the relict temperate forests of Fray Jorge, Chile, the structure of such beetle assemblages was affected by fragmentation (Barbosa and Marquet 2002). Forest interior had the greatest beetle abundance and, in contrast to several other studies, there was no trend of increase in richness toward the edge. That study compared beetles of three fragments of different sizes (0.3, 2, 14.2 hectares) within the Fray Jorge National Park and, agreeing with expectation, beetle richness and abundance were lowest in the smallest patch.

Axes for comparative gradient surveys include space, as above, time, and successional stage. Thus boreal forest successional surveys summarised by Niemela (1997) revealed that specialist old-growth forest carabids were a relatively small proportion of the total assemblage and that the great majority of species occurred in early regeneration (where trees were cut <20 years previously). Niemela showed that the invertebrates of boreal forest had suffered greatest impacts at either the climax stage (with old forest decreased directly by clear-cutting) or the early regenerative stages, in which burning had ceased because of effective preventative control, and lack of burned substrates had led to declines of many species. The faunas of the two stages differed markedly and, as Niemela commented in a sentiment relevant to many other forests throughout the world 'Unfortunately, there is no general set of prescriptions for the ecologically sustainable use of such a complex and extensive biome as the boreal forest'.

Chronosequence surveys of forest insects, with 'time since clearing' or 'time since [other, specified] disturbance' have provided much information on assemblage resilience and vulnerability, in addition to changes of individual species. Thus, in southern Finland, beetles (sampled by window traps) were appraised on 40 sites logged 1–16 years previously, with 20 of those sites burned after logging (Toivanen and Kotiaho 2007). Some taxa continuing to thrive in conventionally managed forests, most notably some rare saproxylic and red-listed beetles, strongly benefited from the combined retention and burning regimes. Such disturbance-adapted species may depend on such measures for conservation, with a continually renewed supply of dead wood needed. Both richness and abundance of saproxylic beetles increased with burning, but that effect depended also on retention trees being present in clear-cut stands. Particularly high numbers of some rare species occurred on burned sites 3–8 years old, and subsequent declines implied that the quality of the burned wood then decreased.

In Black spruce (*Picea mariana*) forests in eastern Canada, surveys of ground-active Carabidae by pitfall traps across 31 stands representing an age gradient from 0–340 years constitute one of the most complete appraisals of faunal change across a full natural forest succession (Paquin 2008). Thirty-eight species were trapped. The forest sites were allocated among four age classes (0–2, 21–58, 70–170, 177–340 years), in sequence 'burned', 'regenerating', 'mature' and 'old-growth', across which the beetle assemblages differed considerably. Whilst richness was rather similar – in the above sequence, trapping yielded 19, 18, 23, and 19 species - each age class had species that were significantly associated with it alone, and only one species (*Pterostichus punctatissimus*) occurred in all stages. The chronosequence thus identified species of carabids that were likely to be indicators of particular forest age groups during succession.

Implications of such changes are exemplified by studies of the ground beetles in plantings of conifers on agricultural lands, as afforestation (Butterfield et al. 1995). The rarer species characteristic of the moorland and grassland habitats replaced by forest can be retained only by retaining the original habitats within the forest plantings. Although subsequent rotational management indeed re-creates open habitats, these are insufficient for those beetles to persist. In this study, in northern England, the newly created closed canopy conifer plantations supported relatively few carabid species, but the eventual mosaic – ranging from clear-felled areas to more mature forest – may indeed support higher diversity than the areas without afforestation. Assemblages on forest sites were distinct from those of open sites, but the main absentees from the forest were the more poorly dispersing open habitat taxa.

In northern Spain, some forests include a considerable proportion (53%) non-native trees, alien conifers or eucalypts, providing a relevant scenario to compare the Carabidae of different native and non-native forests and detect any species characteristic of each (Martinez et al. 2009). Distribution of assemblages overlapped considerably, and no specific assemblages were distinguished among the 27 species retrieved by pitfall trap surveys. The conifer forest fauna was not poorer in species than that of deciduous forest, and most species were forest generalist taxa. The similarity might represent lack of habitats suitable for specialised species, wider homogenisation of the forest ground features, and poor habitat quality in the pine plantations (Martinez et al. 2009).

5.4   Some Key Groups and Concerns

Carabid beetles are among the few groups of insects, delimited either by ecology or taxonomy, in forests that have proved especially informative in understanding wider responses to changes and subsequent conservation needs. The lessons from these focal groups or guilds, largely from northern hemisphere studies, have direct relevance in Australia, both as exemplars and for the ecological information they convey. They also help to display some of the practical problems of transferring such lessons to the relatively poorly documented Australian fauna.

5.4.1   Saproxylic Beetles

The term 'saproxylic' refers to species that depend on dying or dead wood. Those insects are ecologically diverse, reflecting the variety of wood decay stages and substrates available to beetles and others. 'A dying or dead tree is an arboreal megalopolis' (Bouget et al. 2005), in which insects feed directly on the wood, with different decay stages forming well-defined successional habitats exploited by a parallel succession of insect taxa. Many insects feed only on wood of particular tree species, limited decay stages, and particular tissues, so that their roles can become difficult to define. In attempting to standardise the varied descriptors applied to saproxylic insects, both structural and physiological parameters become involved for species that feed directly on wood. In addition, a wide range of associated food resources and microhabitats lead to further complexity, and species can be categorised according to feeding regime (such as wood-associated fungi or sap, or as commensals or predators) or by microhabitat (using, and perhaps restricted to, fungal sporophores, sap flows, tree holes, galleries of previous occupants, and so on). Understanding the interplay of the numerous species involved is a clear need for conservation, and the scheme advanced by Bouget et al. (2005) helps considerably by demonstrating the above diversity.

As a major ecological component of forest life, saproxylic insects associated with old trees are one of the most endangered invertebrate groups (Speight 1989, McLean and Speight 1993). Their importance and diversity has led to their elevation to become one of the most significant suites of taxa in assessing forestry impacts, with informed calls for their conservation far exceeding those for most other forest insect taxa. Some are highly specific in habit, and are restricted to particular tree taxa and/or decay stages, although wider generalistions are incomplete. Species of Cerambycidae with larval development in dead wood material are often much less host specific than those utilising living or recently-dead trees, possibly reflecting that the tree's defensive chemicals are not as important or become redundant in dead wood (Kletecka 1996).

The need to develop use of saproxylic insects in evaluating forest conservation studies in Europe, flowing from some 1988 recommendations of the Council of Europe, publicised and elevated the importance of considering dead wood in forest management, together with old 'habitat trees' (Harmon 2001). It also stimulated development of 'rapid habitat assessment' for these insects, most successfully in France (Bouget et al. 2014), where a wide range of ecological attributes for beetle diversity were distinguished. A few features (notably the density of cavity-bearing or fungus–infested trees and snags, and the extent of local forest openness) seemed to have most influence, but were inconsistent over different forest types. No single variable clearly explained differences in species richness in either deciduous or coniferous forests. Key habitat factors were inconsistent both for rare species, and for all species combined. Large logs and large snags were clearly important, but development of such 'indirect biodiversity indicators' that can be used easily by non-specialists under field conditions still needs further attention. Likewise, in montane beech (*Fagus sylvatica*) forests in central Europe, large amounts of dead wood are particularly important for maintaining high saproxylic beetle diversity (Lachat et al. 2012), but site microclimate (temperature) was also influential: in general, warmer sites (mean annual temperature 8.4 °C) supported more indicator species, from IndVal analyses (74 species) than cooler sites (mean annual temperature 6.8 °C), with 28 species. The temperature effect, however, was overridden by wood supply as the most important single factor.

Studies on saproxylic beetles, a diverse and almost universal and ecologically and economically important functional group across the world's forests, have done much to display the subtle ecological differences and variety amongst a trophic guild. They have also increased understanding of how 'indicators' (p. 113) may be defined and employed to monitor environmental changes, and to help interpret changes in forest insect assemblages. They also have direct relevance in relation to forest management, notably in relation to retention or wider disposition of their key resource, dead wood. As discussed later (Chapter 8), intensity of forest management is linked strongly with the dead wood resources left as habitat for native insects (Grove 2001). Ecologically informed appraisals of saproxylic beetle diversity have been adopted widely as important and feasible tools for monitoring forest conservation measures, predominantly using some form of flight interception traps to do so. However, cautions are needed as trap replication and other sampling effort variables can lead to considerable differences in outcomes: the large sample sizes extending over several years and desirable for reliable interpretation of assemblages are not always available. In Europe, increasing trap numbers and trapping periods can have dramatic effects. At the plot-sample level, addition of a further trap (increasing from one to two traps/plot) gave an average increase of 50% in the number of species detected (Parmain et al. 2013), with changes most marked for rare species – for which a 75% increase occured. Sampling rare saproxylic species is especially challenging.

The numerous detailed studies on the Fennoscandian saproxylic beetles have emphasised the many different ecological variables that affect the insects, and provide many principles for discussion and comparative investigation in other contexts. Thus, for a spruce forest fauna, Okland et al. (1996) allocated the influencing variables amongst five categories, namely variables associated with (1) forms of decaying wood, (2) wood-decaying fungi, (3) the level of disturbance, (4) the landscape context, and (5) the vegetational structure. All the saproxylic species sampled were represented in variable richness in groups of (1) obligate saproxylic beetles in general, and obligate saproxylic beetles found in decaying wood of (2) *Picea*, (3) *Populus*, (4) *Betula*, (5) in deeply decayed wood, (6) in wood-inhabiting fungi, or (7) species present in the Norway or Sweden Red Lists. In particular, the features 'density of dead tree parts', 'number of large diameter trees (> 40 cm at breast height)' and 'number of polypore fungi' were commonly associated with increased richness and abundance. That detail of information directly benefits conservation by forest management, simply by acknowledgement that mature forests with high levels of wood-frequenting fungi and dead wood are widespread key correlates of beetle richness, and provide the initial data for evaluating threshold levels of those resources. Thus, extensive clear-cutting and thinning are both likely to be detrimental, with adverse effects most pronounced for species with low dispersal capability and for which connectivity across habitat units may become evident.

Association of key resource variables, as above, with forest categories adds considerable conservation credibility to comparative studies across different forest treatments. However, whilst it is acknowledged widely that many saproxylic species are 'threatened', many of those species have not been studied in detail, so that measures needed for species-level conservation may be unclear (Martikainen et al. 2000), together with reasons for specific vulnerability. Understanding of individual species remains an expedient key to wider appraisals, as Speight (1989) demonstrated by selection of 33 European saproxylic insect species for his pioneering synthesis. Those species were selected expediently by a series of criteria (Table 5.5), but a wider variety of around 200 species was later adopted to increase coverage of the review by including further taxa that also fulfilled the listed criteria.

Monitoring of selected threatened species may lead to wider values in conservation assessment. The endangered European *Aegosoma scabricorne* (Cerambycidae), one of the largest saproxylic beetles in the region, is associated primarily with large declining or freshly dead trees. Its incidence and abundance can be monitored easily

Table 5.5 The practical criteria employed to select saproxylic insect species to be used as indicators of site quality in European forests (Speight 1989)	
	1. Species associated with the dominant, long-lived climax tree species of the forests.
	2. Dependent on dead wood of senescent and dead trees for their habitats.
	3. Regarded as being currently excessively localised in their European distribution.
	4. Of moderate to large size.
	5. Relatively easy to find.
	6. Relatively easy to identify.

by the characteristic exit holes in wood, demonstrating that the beetle occurs in a wide range of broadleaved tree hosts. This implied that it could be an 'umbrella species' whose presence might also protect other saproxylic species (Foit et al. 2016).

The saproxylic beetles of boreal forest, predominantly those associated with pines or aspen, have thus received more detailed study than those of any other forest type, and the influences of modern forestry and landscape changes from forest management have been investigated in considerable detail. The twin groups of studies encompass single species autecological studies and richness/diversity studies, both in relation to forest state and management regimes. Emphasis on saproxylic taxa, many of them scarce, reflects the importance of dead wood as a wider resource for insects (Speight 1989), and also the richness of the associated beetle assemblages. Those assemblages contain a high proportion of the beetles signaled as threatened ('red-listed') in northern Europe, for example. The study environments also enable comparisons of faunas across areas with different known history and periods of succession and defined management, and undisturbed forests as a 'baseline' for unaffected diversity. The ecological variety of forest beetles is itself diverse, with saproxylic species comprising the two major guilds of 'bark beetles' and 'others' (Simila et al. 2002).

Managed pine forest sites in Finland generally yielded fewer species, and semi-natural sites (including recently burned sites) were important for rare threatened and near-threatened species (Simila et al. 2002). Early succession habitats, and dead wood in open sunny locations were important for those species. The relative representations of 'rare' species along successional gradients in both managed and unmanaged forests clearly demonstrate this importance (Fig. 5.7), although numbers of individuals of most species were very low, and endorse the need for younger successional stands to be incorporated in forest management. The early stages of forest succession thus had considerable conservation value because they supported beetle assemblages that were very different from those of closed forest. A mosaic of different-aged forest types, and structural variations maintained by imposed

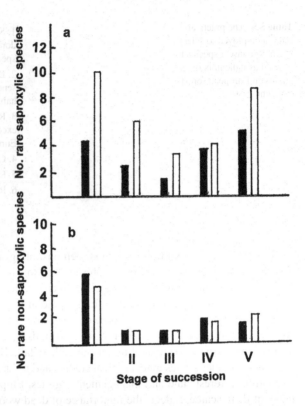

Fig. 5.7 The richness of (**a**) rare saproxylic beetle species and (**b**) rare non-saproxylic beetle species in five successional stages (I–V) of managed (black) and non-managed (open) forests. (Simila et al. 2002)

disturbances appeared necessary for successful conservation of all or most of the species present.

The richness of dead aspen as a beetle habitat was illustrated by more than 10% of the entire Norwegian beetle species being obtained from only 40 dead trees, with those captures including 18 red-listed saproxylic species, mostly in very small numbers (Sverdrup-Thygeson and Ims 2002), with one such species (*Scaphisoma boreale*, Scaphidiidae) comprising 77 of the 122 individuals. Level of sun exposure, from 'low' in closed forest to 'high' in clear-cut areas, and whether in logs or snags also linked with occurrence of red-listed species (Fig. 5.8), demonstrating that sun exposure influences beetle assemblages, and endorsing needs for retention trees in open areas (p. 200). Snags yielded more species, including red-listed species, than logs – as Jonsell et al. (1998) reported also from Sweden. Both snags and logs contribute to local saproxylic beetle assemblages, and retaining aspen trees in otherwise clear-cut areas will retain sun-exposed dying wood in the landscape.

Sunlight also helps to determine saproxylic beetle richness on oak trees (*Quercus robur*) in spruce plantations in Sweden (Widerberg et al. 2012), where oaks in more open areas attracted more oak-associated species then those shaded by plantation trees. Not least because of increased exposure, clearing around retained oaks could benefit local saproxylic beetle richness. That survey, using window traps, encompassed 226 beetle species, including 18 red-listed species (again all in very low

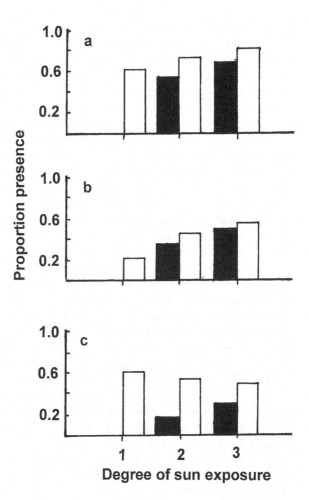

Fig. 5.8 The relative incidence of red-listed saproxylic beetle species in clearcuts in relation to substrate type (snags, open bars; or downed logs, black bars), and exposure to sun (given as three categories from low to high). One species of the 18 taxa involved, *Scaphisoma boreale*, was by far the most common; (**a**) all species; (**b**) *S. boreale*; (**c**) the remaining species. (Sverdrup-Thygeson and Ims 2002)

numbers, with a maximum of 10 for *Xyleborus monographus*, and nine represented by singletons). Nevertheless, standing oaks in spruce plantations can sustain diverse beetle assemblages, and larger groups of trees generally associated with higher beetle richness – as long as some dead branches of sufficient size were available on the oaks. Accessibility of any such trees is clearly important, with beetle species richness highest in stands with large free-standing oak trees, and nearby forest growth detrimental to many beetle species (Ranius and Jansson 2000). Grazed stands, possibly because of greater insolation, contained more saproxylic species and endorsed earlier inferences that sun-exposed oaks are preferred. Conservation of rarer saproxylic beetles can include management measures such as (if necessary) resumption of grazing, and cutting of nearby shrubs and trees – with the caveat that dramatic microclimate changes may hasten the death of old oak trees.

The major focus on boreal forest saproxylic beetles does not mask that the beetles are also a predominant forest insect group elsewhere. Those of the Mediterranean region of Europe, for example, are diverse. Surveys from forest near Salamanca, Spain, yielded 2563 saproxylic beetle species across 44 families (Ramilo et al. 2017), and the appearance of some was strongly seasonal. The highest temperatures (> 40° C) in August-early September limited activity of many species and suggested that more studies on seasonal patterns of beetle richness might be useful in predicting impacts of future climate changes.

Many large groups of saproxylic beetles in Australia are still taxonomically intractable to non-specialists, and a study of wider beetle assemblages in lowland tropical rainforest in the Daintree region of Queensland illustrates some of the more general difficulties of applying such information locally. Grove (2002b) could only obtain tentative species-level identifications for about a third of taxa in the more tractable families of the 51 families represented in his surveys, leading him to comment that sampling saproxylic insects using ground-based flight interception traps 'presents numerous practical difficulties, and cannot be recommended as a method for monitoring the impacts of forest management in the tropics'. The key questions Grove raised, in addition to beetle recognition, were (1) the wider applicability of his findings beyond any individual study area; (2) recognising the importance of larger-diameter trees, and determining the threshold sizes that should be left, together with (3) the quantity of such trees needed to sustain the saproxylic biota and how that might be sustained in management prescriptions. However, it was also clear from the Daintree study that beetle abundance and richness found in old-growth forests was not maintained in logged and regrowth forest (Grove 2002c), with the last having considerably fewer detritivores and predators.

Whatever part of the world they inhabit, many saproxylic and other arboreal beetles (and other insects) may have far more specialised and restricted ecological needs than might be apparent initially. At one level, the endangered European cerambycid *Ropalus ungaricus* is monophagous on living Sycamore (*Acer pseudoplatanus*) trees, but more detailed investigation showed that its survival depended on supply of sun-exposed damaged and declining trees in montane forests (Kasak and Foit 2018). The beetles were more abundant on sun-exposed trees in the Czech Republic, and the warmer south- and east-facing quarters were preferred. Another European beetle, *Boros schneideri* (Boridae), became extinct in Western Europe because of extensive logging (Gutowski et al. 2014), and (although the species has a broad host range) larvae live under bark of standing dead trees. In Poland, they were most common on pine trees. In contrast to *R. ungaricus*, *B. schneideri* largely avoided trees in sunny locations and was more characteristic of mesic to wet coniferous forests. The key need for conserving *Boros* was considered to be assured access to large diameter dead trees with intact bark, and assuring that resource may need very strict protection of selected forest stands.

5.4.1.1   Fungi

The large fruiting bodies of wood-decaying fungi comprise a well-defined discrete substrate for saproxylic insects, and have been used in studies to assess effects of forest management on insect colonising ability, and the dynamics and abundance of the fungi. Reduced availability of dead wood leads to corresponding reduction of associated specialised insect resources such as basidiocarp fungi, so that habitat loss and fragmentation can affect beetles and others associated with those resources. Surveys of dead bracket fungi, *Fomes fomentarius*, in Norway – a single basidiocarp of this fungus may support successive beetle generations over the years before it disintegrates – showed that all the more abundant beetle species were less common when distance to other infested trees increased (Rukke 2000), reflecting reduced migration success between trees. Reduced habitat size was also associated with fewer beetles, so that conservation of larger forest fragments is desirable to counter those effects.

The assemblage of saproxylic beetles associated with the wood-decaying fungus *Fomitopsis pinicola* in Norway helped to elucidate whether the presence of this conspicuous fungus is a valid 'continuity indicator', as sometimes advocated (Sverdrup-Thygeson 2001), and whether its presence might predict (1) richness of saproxylic beetles and/or (2) the incidence patterns of red-listed species. Beetles were sampled by flight interception traps mounted on fruiting bodies of *Fomitopsis*, and by rearing them from dead fungal fruiting bodies. Fifty-seven traps yielded 109 saproxylic beetle species, many of them collected on single or few occasions. The reared material increased the total to 113 species, including seven red-listed species. However, no significant associations between the numbers of fungi and beetle richness and abundance of red-listed species were evident – and variations across red-listed species confirmed the widely recognised reality that it may be unwise to presume that 'group responses' occur, rather than examining the more individualist responses by each species. In this particular study, Sverdrup-Thygeson considered that any generalisations from the use of potential fungal indicator species 'must be made with great care'. Congruence of responses across different species and different higher taxa (de Andrade et al. 2014) cannot be assumed. Fungus beetles have also been considered more widely as indicators (p. 113) because of their predictable assemblage changes in relation to forest management.

Comparisons of the insects associated with *F. pinicola* in natural old-growth forest, and in fungi exposed at distances up to 1610 m from the forest edge into an intensively managed forest in Sweden (Jonsson and Nordlander 2006) showed that some species did not decline in colonisation rate with distance. Some, indeed, colonised fungi in managed forest at rates sufficient to maintain viable populations. Exceptions occurred. Colonisation by the fungivorous *Cis quadridens* (Ciidae) and the predatory fly *Medetera apicalis* (Dolichopodidae) declined with distance from old-growth forest, even from about 100 m from the reserve. Both these species were known previously to be less common in managed forests, apparently reflecting poor colonisation ability, and similar weak dispersal may occur in other saproxylic species and be the primary cause of their scarcity in managed forests.

The fruiting bodies of *Fomes fomentarius* are a specific habitat for the tenebrionid *Bolitophagus reticulatus*, a common and widespread beetle in Scandinavian forests and for which associations with the fungi have been explored in some detail (Jonsell et al. 2003). Contradicting earlier assertions that the beetle disperses only poorly, the newer information implied that adults disperse effectively to occur throughout the entire forest landscape – including managed forests with only low densities of the fungus. This beetle was shown, by baiting trials and mark-release-recapture surveys, to be capable of detecting restricted breeding sites within managed landscapes in Sweden. More generally, the changed implications flowing from this study suggest that the assumed poor dispersal powers of some other saproxylic insects might also be misleading, with at least some species dispersing more effectively than popularly presumed, as an important consideration in assessing their vulnerability across patchy landscapes (Jonsell et al. 2003).

Forest fragmentation affects the insect community associated with the old-growth forest fungus *Fomitoposis rosea* in Finland (Komonen et al. 2000). That community includes many old-growth forest specialists, and the food chains in which these participate become increasingly truncated over time following the isolation of fungus-carrying forest fragments. *F. rosea* has itself declined greatly from forest loss, but the fruiting bodies can persist for several years. Insect associates were compared across fungi in old-growth (control) forests, and two categories of fragments, isolated respectively for 2–7 years and 12–32 years, from each of which fruiting bodies were collected and emerging insects assessed over about 14 months. The 33 species retrieved were dominated by Coleoptera (19 species), followed by Hymenoptera (6), Diptera (5) and Lepidoptera (3). Most were not common, and the two most abundant captured species were the fungus-feeding moth *Agnathosia mendicella* (Tineidae) and its parasitoid fly *Elfia cingulata* (Tachinidae). However, the last was absent from the long-term fragments, so that the food chain pattern changed across treatments (Fig. 5.9), in ways attributed by Komonen et al. to fragmentation. It appeared that the small isolated fragments were associated with species loss, and that changes occurred over a relatively short period.

5.4.2 Ants

Widespread advocacy for ants as indicators flows from a combination of their high diversity and abundance, ubiquity in many natural and disturbed terrestrial environments over much of the world, that they can be sampled and identified – in most cases at least to genus level - easily, as well as them being ecologically varied and their broad functional roles reasonably well understood in relation to their responses to many environmental changes. They have been proposed often as suitable taxa for investigating impacts of forest disturbances.

In French Guyana, ants of small (<10 hectares) monoculture plantations of acacia, rubber, cocoa and pine trees within rainforest, together with those from undisturbed forests and treefall gaps (p. 245) were compared (Groc et al. 2017), and

Fig. 5.9 Changes in the structure of food chains involving the fungus *Fomitopsis rosea*, the moth *Agnathosia mendicella* and its parasitoid *Elfia cingulata* in control forest areas and in fragments isolated for shorter (2–7 years) and longer (12–32 years) periods (black: 3 trophic levels; dotted, 2 trophic levels; open, 1 trophic level; diagonal hatch, all extinct. (Komonen et al. 2000)

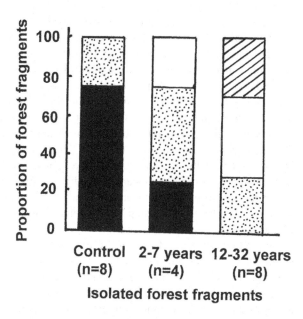

showed (1) decreased species richness from the forest to the plantations, but also (2) that such small embedded stands may limit biodiversity loss. Only 36 of the 124 forest ant species were not recovered from the plantations. However, ant community composition differed across the four plantation species, with greatest changes for acacia and rubber, far less changes for pines, and the cocoa fauna was intermediate. Characteristics of the litter appeared to influence extent of change, and was generally more homogeneous in plantations than in natural forest.

Ant assemblages along a chronosequence representing forest recovery in Brazil and divided among epigaeic, hypogaeic and arboreal ant components collectively included 77 species, representing 36 genera and nine subfamilies (Schmidt et al. 2013). Epigaeic species richness was considerably the greatest. Species composition changed along the gradient of recovery, with increasing vegetation cover, and with increasing age of the forest remnants, similarity between ant assemblages of forest and pasture habitats decreased, in large part reflecting the differing responses of open habitat and forest specialist ants to increasing cover.

The values of ants as bioindicators in Ponderosa pine (*Pinus ponderosa*) forests in Arizona implied that functional roles, rather than richness or abundance, were useful in demonstrating changing forest states (Stephens and Wagner 2006). Different functional groups predominated under different levels of disturbance. That survey used pitfall traps to compare replicated stands that were (1) unmanaged, (2) thinned, (3) thinned and burned; and (4) subject to wildfires, collectively from a pool of 20 ant species across 10 genera, and from samples totalling more than 18,000 individuals. Richness does not necessarily equate to the functional roles that can increase the values of purported indicators. In that study, opportunistic ant taxa dominated the assemblage of unmanaged stands, as they did also in thinned stands.

In contrast, thinned stands that were also treated by prescribed burning had greater relative abundance of generalists, and coarse woody debris specialists were also reduced. Greatest changes in functional group composition occurred on the wildfire sites, on which generalists were significantly predominant and apparently suppressed opportunists – perhaps through direct competition for resources. As in many parallel analyses investigating 'ants as indicators', functional group interpretations provide information not readily apparent from richness/abundance data alone, but each functional group is typified through particular taxa, whose recognition and changes across treatments is central to interpretation. Reinforcing sentiments expressed repeatedly in similar contexts, Stephens and Wagner (2006) concluded that 'there are no perfect indicators', leading to questioning of the values of ant functional groups in such comparisons. Many correlations have endorsed indicator values of ant functional groups, with suitable reservations related to contexts (Andersen 1991, 1997).

Ground-dwelling ants living in leaf litter or soil have been the most commonly studied assemblages, but the broader 'three-dimensional' structure of forests led Schmidt et al. (2013) to investigate responses to forest recovery by comparing ants in three vertical strata (trees, in/on the litter, soil) In addition, ants were compared across pasture and seven remnant forest patches with different recovery times, treating this as a chronosequence analogue for forest recovery, in Minas Gerais, Brazil, a region in which considerable loss and fragmentation of the Atlantic Forest has occurred. Composition of assemblages was the only measure that changed along the recovery gradient, and was most evident in epigaeic and arboreal ants. Hypogaeic ant composition did not change among the forest remnants. However, ant richness and evenness lacked any relationship with forest recovery, this being attributed tentatively to the short distances between fragments, so that rapid recolonisation was likely to occur after disturbance. In addition, the comparison was based on only rather short trapping periods.

References

Andersen AN (1991) Responses of ground-foraging ant assemblages to three experimental fire regimes in a savanna forest of tropical Australia. Biotropica 23:575–585

Andersen AN (1997) Using ants as bioindicators: multiscale issues in ant community ecology. Conserv Ecol [online]1 (8)

Baker SC (2006) A comparison of litter beetle assemblages (Coleoptera) in mature and recently clearfelled Eucalyptus obliqua forest. Aust J Entomol 45:130–136

Barbosa A, Marquet PA (2002) Effects of forest fragmentation on the beetle assemblages at the relict forest of Fray Jorge, Chile. Oecologia 132:296–306

Barnard PC, Brooks SJ, Stork NE (1986) The seasonality and distribution of Neuroptera, Raphidioptera and Mecoptera on oaks in Richmond Park, surrey, as revealed by insecticide knock-down sampling. J Nat Hist 20:1321–1331

Basset Y, Novotny V, Miller SE, Kitching RL (eds) (2003) Arthropods of tropical forest. Cambridge University Press, Cambridge

Basset Y, Novotny V, Miller SE, Weiblens GD, Missa O, Stewart AJA (2004) Conservation and biological monitoring of tropical forests: the role of parataxonomists. J Appl Ecol 41:163–174

Bouget C, Brustel H, Nageleisen L-M (2005) Nomenclature of wood-inhabiting groups in forest entomology: synthesis and semantic adjustments. C.R. Biologies 328:936–948 (in French, translated abstract title as quoted)

Bouget C, Brin A, Brustel H (2011) Exploring the "last biotic frontier": are temperate forest canopies special for saproxylic beetles? For Ecol Manag 261:211–220

Bouget C, Larrieu L, Brin A (2014) Key features for saproxylic beetle diversity derived from rapid habitat assessment in temperate forests. Ecol Indic 36:656–664

Burghouts T, Ernsting G, Korthals G, de Vries T (1992) Litterfall, leaf litter decomposition and litter invertebrates in primary and selectively logged dipterocarp forest in Sabah, Malaysia. Phil Trans R Soc Lond B 335:407–416

Butterfield J, Luff ML, Baines M, Eyre MD (1995) Carabid beetle communities as indicators of conservation potential in upland forests. For Ecol Manag 79:63–77

De Vries PJ (1988) Stratification of fruit-feeding nymphalid butterflies in a Costa Rican rain forest. J Res Lepidopt 26:98–108

Eggleton P, Bignell DE, Sands WA, Waite B, Wood TG, Lawton JH (1995) The species richness of termites (Isoptera) under differing levels of forest disturbance in the Mbalmayo Forest reserve, southern Cameroon. J Trop Ecol 11:85–98

Erwin TL (1982) Tropical forests: their richness in Coleoptera and other arthropod species. Coleopt Bull 36:74–75

Fermon H, Waltert M, Vane-Wright RI, Muhlenberg M (2005) Forest use and vertical stratification in fruit-feeding butterflies of Sulawesi, Indonesia: impacts for conservation. Biodivers Conserv 14:333–350

Floren A, Linsenmaier KE (2003) How do beetle assemblages respond to anthropogenic disturbance?. In Basset Y, Novotny V, Miller SE, Kitching RL (eds) Arthropods of tropical forest. Cambridge University Press, Cambridge, pp 190–197

Foggo A, Ozanne CMP, Speight MR, Hambler C (2001) Edge effects and tropical forest canopy invertebrates. Plant Ecol 153:347–359

Foit J, Kasak J, Nevoral J (2016) Habitat requirements for the endangered longhorn beetle *Aegosoma scabricorne* (Coleoptera: Cerambycidae), a possible umbrella species for saproxylic beetles in European lowland forests. J Insect Conserv 20:837–844

Graca MB, Souza JLP, Franklin E, Morais JW, Pequeno PACL (2017) Sampling effort and common species: optimizing surveys of understorey fruit-feeding butterflies in the Central Amazon. Ecol Indic 73:181–188

Groc S, Delabie JHC, Fernandez F, Petitclere F, Corbara B et al (2017) Litter-dwelling ants as bioindicators to gauge the sustainability of small arboreal monocultures embedded in the Amazonian rainforest. Ecol Indic 82:43–49

Grove SJ (2001) Extent and composition of dead wood in Australian lowland tropical rainforest with different management histories. For Ecol Manag 154:35–53

Grove SJ (2002b) Tree basal area and dead wood as surrogate indicators of saproxylic insect faunal integrity: a case study from the Australian lowland tropics. Ecol Indic 1:171–188

Grove SJ (2002c) The influence of forest management history on the integrity of the saproxylic beetle fauna in an Australian lowland tropical rainforest. Biol Conserv 104:149–171

Gutowski JM, Sucko K, Zub K, Bohdan A (2014) Habitat preferences of *Boros schneideri* (Coleoptera: Boridae) in the natural tree stands of the Bialowieza Forest. J Insect Sci 14(276):2014. https://doi.org/10.1093/jisesa/ieu138

Hamer KC, Hill JK, Benedick S, Mustaffa N, Sgerratt TN, Maryatis M, Chey VK (2003) Ecology of butterflies in natural and selectively logged forests of northern Borneo: the importance of habitat heterogeneity. J Appl Ecol 40:150–162

Hammond PM, Kitching RL, Stork NE (1996) The composition and richness of the tree crown Coleoptera assemblage in an Australian subtropical forest. Ecotropica 2:99–108

Hanula Jl, Horn S, Wade DD (2006) The role of dead wood in maintaining arthropod diversity on the forest floor. In Grove SJ, Hanula JL (eds) Insect biodiversity and dead wood. Proceedings of a symposium for the 22nd international congress of entomology. USDA General and Technical Report SRS-93, Ashville, NC, pp 57–66

Harmon ME (2001) Moving towards a new paradigm for woody detritus management. Ecol Bull 49:269–278

Heliola J, Koivula M, Niemela J (2001) Distribution of carabid beetles (Coleoptera, Carabidae) across a boreal forest-clearcut ecotone. Conserv Biol 15:370–377

Hilszczanski J, Jaworski T, Plewa R, Horak J (2016) Tree species and position matter: the role of pests for survival of other insects. Agric For Entomol 18:340–348

Jonsell M, Weslien J, Ehnstrom B (1998) Substrate requirements of red-listed saproxylic invertebrates in Sweden. Biodivers Conserv 7:749–764

Jonsell M, Schroeder M, Larsson T (2003) The saproxylic beetle *Bolitophagus reticulatus*: its frequency in managed forests, attraction to volatiles and flight period. Ecography 26:421–428

Jonsson M, Nordlander G (2006) Insect colonisation of fruiting bodies of the wood-decaying fungus *Fomitopsis pinicola* at different distances from an old-growth forest. Biodivers Conserv 15:295–309

Kasak J, Foit J (2018) Shortage of declining and sun-exposed trees in European mountain forests limits saproxylic beetles; a case study on the endangered longhorn beetle *Ropalus ungaricus* (Coleoptera: Cerambycidae). J Insect Conserv. https://doi.org/10.1007/s10841-018-0050-3

Kitching RL (2004) Invertebrate conservation and the conservation of forests. In Lunney D (ed) Conservation of Australia's forest fauna. (2nd edn). Royal Zoological Society of New South Wales, Mosman, Sydney, pp 115–126

Kitching RL, Bergelson JM, Lowman MD, McIntyre S, Carruthers G (1993) The biodiversity of arthropods from Australian rainforest canopies: general introduction, methods, sites and ordinal results. Aust J Ecol 18:181–191

Kletecka Z (1996) The xylophagous beetles (Insecta, Coleoptera) community and its succession on Scotch elm (*Ulmus glabra*) branches. Biologia 51:143–152

Koivula M, Niemela J (2003) Gap felling as a forest harvesting method in boreal forests: responses of carabid beetles (Coleoptera, Carabidae). Ecography 26:179–187

Komonen A, Pentilla R, Lindgren M, Hanski I (2000) Forest fragmentation truncates a food chain based on an old-growth forest bracket fungus. Oikos 90:119–126

Lachat T, Wermelinger B, Gossner MM, Bussler H, Isacsson G, Muller J (2012) Saproxylic beetles as indicator species for dead-wood amount and temperature in European beech forests. Ecol Indic 23:323–331

Lowman MD, Wittman PK (1996) Forest canopies: methods, hypotheses and future directions. Annu Rev Ecol Syst 27:55–81

Lowman MD, Taylor P, Block N (1993) Vertical stratification of small mammals and insects in the canopy of a temperate deciduous forest: a reversal of tropical forest distribution. Selbyana 14:25

Maguire DY, Robert K, Brochu K, Larivee M, Buddle CM, Wheeler TA (2014) Vertical stratification of beetles (Coleoptera) and flies (Diptera) in temperate forest canopies. Environ Entomol 43:9–17

Majer JD, Recher HF, Ganesh S (2000) Diversity patterns of eucalypt canopy arthropods in eastern and western Australia. Ecol Entomol 25:295–306

Martikainen P, Siitonen J, Puntilla P, Kaila L, Rauh J (2000) Species richness of Coleoptera in mature managed and old-growth boreal forests in southern Finland. Biol Conserv 94:199–209

Martikainen P, Kouki J, Heikalla O (2006) Effects of green tree retention and subsequent prescribed burning on the crown damage caused by the pine shoot beetles (*Tomicus* spp.) in pine-dominated timber harvest areas. Ecography 29:659–670

Martinez A, Iturrondobeitia JC, Goldarazena A (2009) Effects of some ecological variables on carabid communities in native and non-native forests in the Ibaizabal basin (Basque Country: Spain). Ann For Sci 66:304 (11pp)

McLean IFG, Speight MCD (1993) Saproxylic invertebrates: the European context. In Kirby KJ, Drake CM (eds) Dead wood matters: the ecology and conservation of saproxylic invertebrates in Britain. English Nature, Peterborough, pp 21–32

Michaels KF, McQuillan PB (1995) Impact of commercial forest management on geophilous carabid beetles (Coleoptera: Carabidae) in tall, wet *Eucalyptus obliqua* forest in southern Tasmania. Aust J Ecol 20:316–323

Niemela J (1997) Invertebrates and boreal forest management. Conserv Biol 11:601–610

Ohmart CP, Stewart LG, Thomas JR (1983) Phytophagous insect communities in the canopies of three *Eucalyptus* forest types in South-Eastern Australia. Aust J Ecol 8:395–403

Okland B (1996) A comparison of three methods of trapping saproxylic beetles. Eur J Entomol 93:195–209

Okland B, Bakke A, Hagvar S, Kvamme T (1996) What factors influence the diversity of saproxylic beetles? A multiscaled study from a spruce forest in southern Norway. Biodivers Conserv 5:75–100

Paquin P (2008) Carabid beetle (Coleoptera: Carabidae) diversity in the black spruce succession of eastern Canada. Biol Conserv 141:261–275

Parmain G, Dufrene M, Brin A, Bouget C (2013) Influence of sampling effort on saproxylic beetle diversity assessment: implications for insect monitoring studies in European temperate forests. Agric For Entomol 15:135–145

Peltonen M, Heliovaara K (1999) Attack density and breeding success of bark beetles (Coleoptera, Scolytidae) at different distances from forest-clearcut edge. Agric For Entomol 1:237–242

Ramilo P, Galante E, Mico E (2017) Intra-annual patterns of saproxylic beetle assemblages inhabiting Mediterranean oak forest. J Insect Conserv 21:607–620

Ranius T, Jansson N (2000) The influence of forest regrowth, original canopy cover and tree size on saproxylic beetles associated with old oaks. Biol Conserv 95:85–94

Rukke BA (2000) Effects of habitat fragmentation: increased isolation and reduced habitat size reduce the incidence of dead wood fungi beetles in a fragmented forest landscape. Ecography 23:492–502

Schmidt FA, Ribas CR, Schoereder JH (2013) How predictable is the response of ant assemblages to natural forest recovery? Implications for their use as bioindicators. Ecol Indic 24:158–166

Silva RR, Feitosa RSM, Eberhardt F (2007) Reduced ant diversity along a habitat regeneration gradient in the southern Brazilian Atlantic Forest. For Ecol Manag 240:61–69

Simila M, Kouki J, Martikainen P, Uotila A (2002) Conservation of beetles in boreal pine forests: the effects of forest age and naturalness on species assemblages. Biol Conserv 106:19–27

Sklodowski J (2017) Manual soil preparation and piles of branches can support ground beetles (Coleoptera, Carabidae) better than four different mechanical soil treatments in a clear-cut area of a closed-canopy pine forest in northern Poland. Scand J For Res 32:123–133

Speight MCD (1989) Saproxylic insects and their conservation. Nature and Environment Series no. 42, Council of Europe, Strasbourg

Spence JR, Langor D, Niemela J, Carcamo H, Currie C (1996) Northern forestry and carabids: the case for concern about old-growth species. Ann Zool Fenn 33:173–184

Stephens SS, Wagner MR (2006) Using ground foraging ant (Hymenoptera: Formicidae) functional groups as bioindicators of forest health in northern Arizona ponderosa pine forests. Environ Entomol 35:937–949

Stireman JO III, Devlin H, Doyle AL (2014) Habitat fragmentation, tree diversity, and plant invasion interact to structure forest caterpillar communities. Oecologia 176:207–224

Stork NE (1988) Insect diversity: facts, fiction and speculation. Biol J Linn Soc 35:321–337

Stork NE (2007) Australian tropical forest canopy crane: new tools for new frontiers. Austr Ecol 32:4–9

Stork NE, Grimbacher PS (2006) Beetle assemblages from an Australian tropical rainforest show that the canopy and the ground strata contribute equally to biodiversity. Proc R Soc B 273:1969–1975

Stork NE, Hammond PM, Russell BL, Hadwen WL (2001) The spatial distribution of beetles
    within the canopies of oak trees in Richmond Park, U.K. Ecol Entomol 26:302–311

Su JC, Woods SA (2001) Importance of sampling along a vertical gradient to compare the insect
    fauna in managed forests. Environ Entomol 30:400–408

Sverdrup-Thygeson A (2001) Can 'continuity indicator species' predict species richness or red-
    listed species of saproxylic beetles? Biodivers Conserv 10:815–832

Sverdrup-Thygeson A, Ims RA (2002) The effect of forest clearcutting in Norway on the commu-
    nity of saproxylic beetles on aspen. Biol Conserv 106:347–357

Swift MJ, Heal OW, Anderson JM (1979) Decomposition in terrestrial ecosystems. Blackwell,
    Oxford

Thunes KH, Skartveit J, Gjerde I (2003) The canopy arthropods of old and mature pines (*Pinus
    sylvestris*) in Norway. Ecography 36:490–502

Toivanen T, Kotiaho JS (2007) Burning of logged sites to protect beetles in managed boreal forests.
    Conserv Biol 21:1562–1572

Turner JRG, Gatehouse CM, Corey CA (1987) Does solar energy control organic diversity?
    Butterflies, moths and the British climate. Oikos 48:195–205

Ulyshen MD, Soon V, Hanula JL (2010) On the vertical distribution of bees in a temperate decidu-
    ous forest. Insect Conserv Divers 3:222–228

Vance CC, Kirby KR, Malcolm JR, Smith SM (2003) Community composition of longhorned
    beetles (Coleoptera: Cerambycidae) in the canopy and understorey of sugar maple and white
    pine stands in south-Central Ontario. Environ Entomol 32:1066–1974

Vance CC, Smith SM, Malcolm JR, Huber J, Bellocq MI (2007) Differences between forest type
    and vertical strata in the diversity and composition of hymenpteran [*sic*] families and mymarid
    genera in northeastern temperate forests. Environ Entomol 36:1073–1083

Vodka S, Konvicka M, Cizek L (2009) Habitat preferences of oak-feeding xylophagous beetles
    in a temperate woodland: implications for forest history and management. J Insect Conserv
    13:553–562

Wardhaugh CW, Edwards W, Stork NE (2012) Variation in beetle community structure across five
    microhabitats in Australian tropical rainforest trees. Insect Conserv Divers 6:463–472

Widerberg MK, Ranius T, Drobyshev I, Nilsson U, Lindbladh M (2012) Increased openness
    around retained oaks increases species richness of saproxylic beetles. Biodivers Conserv
    21:3035–3059

Willott SJ (1999) The effects of selective logging on the distribution of moths in a Bornean rainfor-
    est. Phil Trans R Soc Lond B 354:1783–1790

Willott SJ, Lim DC, Compton SG, Sutton SL (1999) Effects of selective logging on the butterflies
    of a Bornean rainforest. Conserv Biol 14:1055–1065

Chapter 6
Insect Flagships and Indicators in Forests

Keywords Carabidae · Focal species · Insect functional groups · Insect monitoring · Lucanidae · Nymphalidae · Oil palm · Red-listed species · Saproxylic insects · Scarabaeoidea

6.1 Introduction: Conservation and Flagship Insect Species in Forests

In general, very few tropical or subtropical forest insects have become individual conservation flagships. Almost all these are members of 'popular' insect groups attractive to hobbyist naturalists as well as to scientists and, most notably, are Lepidoptera and Coleoptera that have such wider appeal. Some are simply scarce, but associated with particular sites or plant species in forests. The spectacular noctuoid moth *Phyllodes imperialis smithersi*, for example, is associated with understorey vegetation of subtropical rainforest in south –eastern Queensland and northern New South Wales, where it is rare. It is confined to notophyll vine forest, where larvae feed on an uncommon endemic vine, *Carronia multisepalea*. For long recognised as a distinctive form and noted for its conservation significance, this subspecies has been designated only recently (Sands 2012), and is listed under both Commonwealth and State conservation legislations. Perhaps the best-known insect flagships are birdwing butterflies, in Australia notably the Richmond Birdwing, *Ornithoptera richmondia*, for which forest clearing in far northern New South Wales and southern Queensland was a major contributor to decline up to the middle of the last century (Sands and New 2013). However, more widespread publicity has attended the fate of a non-Australian relative, the world's largest butterfly, Queen Alexandra's Birdwing (*Ornithoptera alexandrae*) from Papua New Guinea, a flagship species of global conservation interest. The two major centres for this forest-dependent butterfly in Papua New Guinea have been altered by logging of primary forest (Managalas Plateau) and establishment and expansion of oil-palm plantations (Popondetta Plains), respectively (Mitchell et al. 2016). Indeed, the vulnerability of *O. alexandrae* has helped to draw attention to the massive consequences of oil-palm plantation expansions in the south east Asia region (p. 47), with one recent outcome in Papua New Guinea being increasing moves to establish new plantations only on already cleared land, and not to clear additional mature forest for this purpose. High global demand for palm oil, notably in the biofuel and food industries, has led to massive areas of *Elaeis guineensis* plantations. Foster et al. (2011) reported global coverage of 14.5 million hectares, with the largest producers being Malaysia and Indonesia which by 2006 accounted for 56.1% (then 6.7 million hectares) of global oil-palm cultivated area (FAO 2006). Demand for palm oil seems likely to escalate deforestation rates across the region, despite imposed formal strictures and, in Koh's (2008) words, 'will lead to a net reduction in biodiversity' as forests disappear.

The twin needs to arrest this are (1) effective reservation/preservation of remaining undisturbed large tracts of forest and of smaller patches, including remnants to promote effective connectivity, together with buffering those areas from adjacent disturbance; and (2) improved management regimes to reconcile the needs of effective conservation and oil-palm production. In Sabah, the percentage cover of old-growth forest and a plantation estate was an effective predictor of butterfly species richness (Koh 2008). Survey-based models from that study implied that (1) the natural forest surrounding an oil-palm estate is important to butterflies within the estate area, and (2) old-growth forests are more important than secondary forests to those butterflies – presumed to reflect the higher diversity of larval host plants in old forest. Oil-palm trees are usually replaced over 25–30 year rotations, leading to even-aged stands of uniformly spaced trees with sparse undergrowth, and amongst which microclimate fluctuations are greater than in primary forests. Plantations of that nature are biologically more akin to agricultural monoculture crops.

Single species' responses have occasionally been used to predict the consequences of alternative forest management methods and to compare their relative costs and benefits in relation to that species' possible fate. Some such species may be of conservation concern and can become flagship species: others are more properly 'indicators', as not being regarded as threatened, but the differences of emphasis can become blurred. Thus the extinction risk for the saproxylic beetle *Diacanthous undulatus* (Elateridae), a predatory species occurring in the dead wood of both conifers and broad-leaved trees in Sweden and found as metapopulations in a variety of forest management regimes, was estimated by simulation models in the two major alternative forest management strategies of extending rotation periods or setting aside younger forest stands (Ranius et al. 2016). All conservation effort levels modelled positively affected the beetle's persistence in comparison with no conservation effort (Fig. 6.1), when considered over a 200 year period projection. During the first century, setting aside older stands was an effective move to decrease extinction risk, but subsequently extinction risk diminished when at least some young forest was set aside. Old stands are expensive to protect, and the areas set aside are likely to remain small – for the same budget, far greater areas of young forest can be reserved and these in due course increase beetle habitat. Prolonged rotations, although effective, are less cost-efficient than set aside options, and the most cost-effective strategy to conserve *D. undulatus* is to set aside a mixture of old and young forests. The balance, reflecting societal values, involves reduced extinction rates (benefit) and lost harvest volume (cost), and can vary widely under different circumstances.

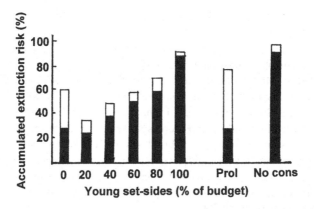

Fig. 6.1 A scheme of predicted extinction risk for the beetle *Diacanthous undulatus* in a managed forest landscape during a 200 year period, under seven scenarios with the same conservation budget and one with no conservation effort; black is century one (years 0–100) and white is century two (years 101–200); extinction risk is shown against percentage of the budget allocated to setting aside the youngest stands whilst the rest is spent on older stands; 'Prol' is a scenario with the same conservation budget but with rotation period of randomly chosen stands prolonged by 20 years; in all scenarios, some trees were retained in stands at clear-cutting. (Ranius et al. 2016)

6.2   Conservation and Indicator Taxa

Whilst single species can be useful as a 'proxy' for wider changes in forest environments, as above, most studies of 'indicators' have focused on groups of species delimited as taxa or ecologically functional groupings. Indicator taxa, those used as surrogates for assessments of wider diversity or to appraise the impacts of environmental changes, must preferably be detectable, measurable, and identifiable by nonspecialists – demands that, together with effective and logistically expedient sampling, effectively restrict the portfolio of taxa that can be used reliably. Plant features, such as changes in vegetation composition or structure, may be far more amenable to appraisal than small or obscure insects, notwithstanding wide advocacy for functionally significant insect groups, for which local knowledge of diversity and ecology may be relatively high. Those groups include ground beetles (Carabidae) and hoverflies (Syrphidae), both appraised for their changes in United Kingdom conifer plantations (Humphrey et al. 1999), using the local richness pools of 42 species (Carabidae) and 54 species (Syrphidae). In general, conifer plantations supported equivalent richness and abundance to the assemblages characteristic of semi-natural woodlands. Distinctivenes of assemblages, not unexpectedly, differed at different latitudes within the United Kingdom, with northern sites and the most southerly sites distinctive. Vertical stand structure was correlated with richness and diversity of both groups.

The suite of 'Pan-European Indicators for Sustainable Forest Management' set out by MCPFE (2003) includes those to meet the objective of 'Maintenance, Conservation and Appropriate Enhancement of Biological Diversity in Forest

Ecosystems', for which indicator criteria include the number of threatened forest species in relation to the total number of forest species. The number of red-listed species thus assumes further importance in this context but, in order to meet this aim, relevant indicators must be available (Brin et al. 2009). The indicator criteria recommended by MCPFE (2003) include the 'volume of standing and lying dead wood', so that richness and individual species' threat status of saproxylic beetles are a pertinent theme. Studies in southern France (in which 240 saproxylic beetle species were captured by window traps across 40 stands of Maritime pine [*Pinus pinaster*] plantations, in which woody debris features were also recorded) emphasised the importance of stand-level indicators, because these relate to operational scales used by forest managers. That beetle richness was correlated with quantity of dead wood and provided a basis for developing indices to suggest impacts of alternative forestry practices (Brin et al. 2009).

The presumed parallels between 'indicator species' and 'red-listed species' flow simply from the common use of the number of listed species present as a measure ('indicator') of a site's conservation value. Higher numbers of red-listed species (that status according them formally recognised conservation importance and, perhaps, priority) equate to high conservation value, and reduced richness or abundance of these is taken to reflect lesser or lost habitat suitability. That correlation needs careful evaluation (McGeoch 1998) and, more widely, many insect groups have been promoted as valuable indicators of forest condition and quality with little objective justification beyond apparent differences in incidence across treatments such as management regimes. Their validity as indicators has in many cases not been established critically. Insect groups selected must be sampleable easily and identifiable unambiguously and consistently, and the habitat units and community contexts against which changes are assessed defined clearly.

Very broadly, about seven functional categories of indicator species associated with forest condition can be recognised (Lindenmayer et al. 2000), and these comprise three major classes of indicators as defined by McGeoch (1998). Determining which of the needs displayed in Table 6.1 is a prerequisite for selecting the focal group(s) to be monitored in any individual programme clearly influences that decision – and that selection is then predicated in the need for indicators to have a functional role in sustaining forest biodiversity or assuring sustainable management, rather than being simply descriptive (Pearce and Venier 2006).

This variety reflects the need for any use of 'indicator species' to specify very clearly what meaning is relevant, or intended. Functionally, types 1, 2 and 4 in Table 6.1 have been regarded as indicators of biological diversity, and the others (types 3, 5–7) as indicators of abiotic conditions and/or changes in ecological processes. Use of indicators to monitor wider forest diversity can thus involve two broad approaches – sometimes referred to as 'species indicators' (or direct indicators) and 'environmental indicators' (or 'indirect' or 'habitat surrogate' indicators). The first describes a component of biodiversity to devise measurements such as species richness of a given taxon group, or abundance of single species, that can be correlated with total species richness. The second, more complex in concept, refers to features such as structure that affect drivers of diversity – Brin et al. (2009) cited

Table 6.1 Categories and definitions of 'indicators' applied to terrestrial insects (as discussed by McGeoch 1998)

Environmental indicator

A species or group of species that responds predictably to environmental disturbance or a change in environmental state, with change readily observable and quantifiable. Categories include (1) sentinels (species introduced deliberately as 'early warning' for disturbance); (2) detectors (species occurring naturally and which may show a measurable response to change); (3) exploiters (species whose presence indicates probability of disturbance or pollution); (4) accumulators (species that accumulate or take up chemicals in measurable quantities); and (5) bioassay organisms (used as laboratory tools to detect presence and/or concentration of pollutants, or rank toxicity of pollutants).

Ecological indicator

Species or groups of species that are used to demonstrate effects of environmental changes on biotic systems, and are thus indicators of the condition of a particular habitat, community or ecosystem. They are sensitive to identified environmental stress, and demonstrate the wider impacts of those stresses on biota, so their response represents the wider response of other coexisting taxa.

Biodiversity indicator

A group of taxa, or selected species from several groups, or functional group, whose diversity reflects some aspect of diversity of other taxa in a habitat or set of habitats being compared. Essentially a 'surrogate' of wider biodiversity, and used also to monitor changes in biodiversity.

Table 6.2 Four approaches advocated to enhance biodiversity conservation in forests, flowing from the limited knowledge of both indicator species and structure-based indicators of condition (Lindenmayer et al. 2000)

1. Establish biodiversity priority areas (such as reserves) managed primarily for the conservation of biological diversity.

2. Within production forests, apply structure-based indicators including structural complexity, connectivity and heterogeneity.

3. Using multiple conservation strategies at multiple spatial scales, spread out risk in wood production forests.

4. Adopt an adaptive management approach to test the validity of structure-based indices of biological diversity by treating management practices as experiments.

canopy openness influencing lower level microclimates that in turn influence understorey vegetation. Volume of dead wood is a commonly considered environmental indicator for saproxylic insects in forest.

For insects in forests, the most frequent intentions from purported indicators appear to be for type 1 (using a species as a surrogate for other species) and type 7 (species used to monitor condition and recovery from disturbance). Any use of indicators, however, remains 'appealing and potentially important … because of the impossibility of monitoring everything' (Lindenmayer et al. 2000). Many taxa claimed to be indicators, in forests or elsewhere, have only unclear specific roles which, in many studies, are not defined. The widespread lack of information on indicators led to listing of four urgent action needs, feasible without full information but collectively enhancing chances for conservation of forest biodiversity. Table 6.2 (after Lindenmayer et al. 2000) lists these needs, and the authors also noted the urgent need to test and (where possible) validate the values of indicators. At present, structure-based indicators can be used more reliably as potential indicators of forest biodiversity; these include features such as stand complexity and structure, floristic composition, heterogeneity and connectivity, all of which can be measured objectively by relatively straightforward means (Lindenmayer et al. 2001).

The dichotomy of 'effectiveness monitoring' and 'validation monitoring' designated by Pearce and Venier (2006) may facilitate a more adequate approach to forest management. They differ in emphasis. Effectiveness monitoring directly monitors the species' populations at each management unit level or treatment. This becomes labour-intensive – for carabids, for example, this approach requires wide-scale pitfall trapping to measure the effects of the management actions on the species. Validation monitoring, in contrast, tests hypotheses about the disturbance imposed, so assessing the sustainability of any given management practice in terms such as resemblance to patterns from natural disturbances – so that the indicators rate the sustainability of the different forestry practices. Pearce and Venier regarded the latter as more likely to lead to achieving forest management goals.

In their broad overview of the values of different arthropod groups as bioindicators in forest management, Maleque et al. (2009) noted the widespread trend that conversion of forest to plantations influences many taxa – but also that monitoring arthropods could help to design management plans to ensure sustainable forests. Within any taxonomic group, varied responses are inevitable, but Maleque et al. noted contexts in which ants, Lepidoptera, various Coleoptera (Carabidae, Cerambycidae, Scarabaeoidea), hoverflies, or parasitoid wasps had provided useful information related to forest change. The last of these groups, the diversity of parasitoid Hymenoptera, reflects their enormous taxonomic and ecological variety in most local faunas and that this diversity sometimes changes markedly across forest treatments. Thus, Maeto et al. (2009) found that monitoring parasitoid diversity may help to indicate recovery stages in plantations of *Acacia mangium* in Indonesia, by changes of assemblages amongst young and mature plantations, as well as between younger and older secondary forest.

Writing on saproxylic insects in Canada, Langor et al. (2006) suggested that they are 'perhaps the best equivalent to "a canary in the coal mine" we have for evaluating forest management', with convincing documentation of the effects of forestry practices and recognition that many species have only low dispersal powers and depend almost wholly on local conditions. Notwithstanding that, as Grove (2006) emphasised, the ecology of most saproxylic species is poorly known and may never be known in detail, more focus on the biology of selected 'key species' (such as those with especially important ecological roles or those of greatest conservation interest) is warranted in a constructive research agenda toward facilitating conservation (Table 6.3).

Within the complex saproxylic insect assemblages of northern forests, in particular, characteristics of dead wood also affect associated parasitoid assemblages (Hilszczanski et al. 2005). Thus, dead wood features and the forest type in Swedish boreal spruce forest affected parasitoid assemblages, which were compared across the different management treatments of clearcuts, mature managed forest and

Table 6.3 Summary of the major themes of a research agenda for advancing knowledge of insects and dead wood (from Grove 2006)

Ecologically orientated research: small scale
Developing standard techniques for sampling saproxylic insects.
Autecological research on key saproxylic species.
Understanding relationships between saproxylic insects and fungi.
Understanding relationships between saproxylic insects and decomposition.
Investigating value of nectar sources for saproxylic insects.
Investigating value of burned wood for saproxylic insects.
Investigating the value of dead wood for non-saproxylic insects.
Ecologically orientated research: medium scale
Investigating extent of overlap in saproxylic insect faunas of logs, trees and litter.
Investigating role of fire in shaping saproxylic insect faunas.
Ecologically orientated research: large scale
Investigate extent to which ecological traits among related saproxylic insect species are conserved across the globe – to what extent can a species' ecology be deduced from that of related species studied elsewhere?
Management orientated research: small scale
Investigating value of artificially killed wood for saproxylic insects.
Investigating value of relocating saproxylic insects or the dead wood on which they depend.
Management orientated research: medium scale
Investigating dispersal and colonisation abilities of saproxylic insects.
Developing appropriate criteria for the identification of important sites for saproxylic insects.
Investigating the feasibility of using structural surrogates for saproxylic insect faunas.
Assessing the impact of fuelwood and firewood harvesting on saproxylic insects.
Assessing the impact of feral saproxylic insects and other non-native species on native saproxylic insects.
Management orientated research: large scale
Developing suitable landscape-scale predictive models.
Developing techniques for maintaining or re-creating forest landscapes suitable for saproxylic insects.

old-growth forest, by collections made from eclector traps in each regime. At each site, catches from replicated blocks of logs (length 4 m, all obtained from the same felling operation) were compared across five treatments (burned, inoculated with either of two fungi, placed in natural shade, untreated control) and each set was supplemented by a snag (a tree cut to about 3 m high). Parasitoid complement of snags differed most from that of control logs, reflecting the preference or specificity of species associated with Cerambycidae and other wood borers largely specific to snags. Examples of inter-treatment outcomes (Fig. 6.2) demonstrated that parasitoid assemblages can be influenced strongly by forest management practices and their impacts on host availability and abundance. Creation of snags, for example, could improve host supply and so habitat availability for the wasps, and selective logging might have lower impacts than clear-felling. Hilszczanski et al. concluded that parasitoids may indeed be more sensitive to changes in forest ecosystems than are

Fig. 6.2 The mean abundance per site of major ecological categories and selected species of parasitoid wasps from logs in three forest types (clearcut: open; old managed forest: dotted; old-growth forest: black) under different management regimes (a, burned; b, control; c, *Fomitopsis pinicola*; d, natural shade; e, *Resinicium bicolor*; f, snag). (Hilszczanski et al. 2005)

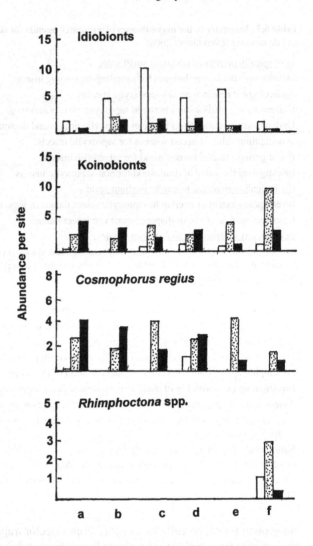

insects at lower trophic levels and so might have values as indicators of early change. Most studies on impacts of forest management on insect assemblages have indeed focused on taxa at lower trophic levels, and comparative surveys on carnivores are relatively rare, as Hilszczanski et al. pointed out. Knowledge of the habitat specificity and variety and identity of hosts that each species can attack is generally highly incomplete: in this limited duration study, most of the 24 species of Ichneumonoidea known (or believed) to be associated with saproxylic hosts were apparently restricted to hosts that depend on early decay stages of wood, but abundances of all the more common species were affected by forest type, log treatment, or both.

Conservation of predators of saproxylic taxa has also been addressed. Predatory beetles in Sweden were surveyed to compare their incidence in forestry treatments to appraise substrate and prey species preferences (Johansson et al. 2007), and motivated in part by development of viable pest management approaches necessitating conservation of the natural enemies of potential pest taxa. The ten predatory beetle species differed in abundance across treatments, with eight species more common in mature managed and/or old growth forests than in clearcuts. Two species, in contrast, occurred almost wholly in clearcut areas, so are presumed to be adapted to severe disturbances such as storms and fire. The persistence of old growth forest stands appears necessary to conserve the majority of the species, with both substrate and stand condition contributing to their wellbeing.

Most studies on indicator insect species, in forests or elsewhere, have presumed that simultaneous cross-samples in various management stages or regimes, compared with some 'undisturbed' or baseline taken as the template from which changes can be appraised, can be translated reliably to reflect a temporal gradient of change as succession or other modification or development occurs. Spatial comparison thereby purports a chronosequence, with the wide presumption that the baseline is the most desirable condition and deviations from it are damaging.

One approach is exemplified by appraisal as indicators of the insects associated with two species of dead wood-frequenting polyporous fungi in Sweden, based on laboratory rearing from collections of fungal fruiting bodies at 25 sites. The sites represented three management categories that reflected the amount of dead wood present (Jonsell and Nordlander 2002). These were (1) 'long continuity sites', with high amounts of dead wood at the time of survey and presumed to have been available continuously over the previous century; (2) 'short continuity sites', also with large amounts of dead wood but those substrates presumed scarce a century ago; and (3) 'no continuity sites', as conventionally managed forest with only low amounts of dead wood. Within each of these, species were sought that were either absent from, or less frequent in, a more intensively managed category. Species restricted to the first category may indicate sites with the most complete (unmodified) fauna, and so of high conservation value. Species more common in long and short continuity sites than in no continuity sites may be candidates for red-listing, as having declined, but not as indicators. Species of similar occurrence across the three categories were sensitive to current forestry practices. In this study, both polypores are large (reaching ca 20 cm radius), persist for more than a decade, and occur only when dead wood is plentiful. *Fomitopsis pinicola* is most common on spruce (*Picea abies*) but occurs on almost all tree species present. *Fomes fomentarius* occurs only on deciduous trees and most frequently on birch (*Betula* spp.). Few beetle species differed across management categories. The three species found only on long continuity sites were all scarce, and no species was more frequent at more intensively managed sites. Few species thus fulfilled the criteria for indicator status, but one was the tenebrionid *Oplocephala haemorrhoidalis* (from *F. fomentarius*) which occurred at most long continuity sites, but at only one short continuity site and was absent from no continuity areas. Also noted was an anobiid, *Dorcatoma minor*, but difficulties of accurate field recognition and that it is known only from a very limited area preclude confirmation – in a situation paralleled in many similar studies across the world: 'indicators' must be sufficiently recognisable, present and abundant for any response to change to be measurable. Other, related, studies in Sweden have demonstrated that insects living in dead wood substrates are very specialised in their microhabitat needs, so that different species largely occur independently and cannot be used (as sometimes presumed) to indicate the presence of each other.

Lambeck's (1997) system of 'focal species' (broadly, indicators) grouped according to microhabitat categories, with the presence in each group of those species most sensitive to forest management may help definition of those sites with the most complete fauna. However, because of the enormous and subtle differences and varieties of forested lands, management 'shortcuts' such as indicator species may have only very limited wider values (Lindenmayer et al. 2006) in biodiversity conservation. Whilst a multi-taxon approach to using insects in monitoring impacts, or as indicators of disturbance, in forests is advocated widely, the variety of resource use patterns and scales/patterns of incidence links with substantial heterogeneity in both space and time. The claim that plants may provide the best indicators of disturbance in tropical forest, not least because they are not subject to the enormous vagaries of sampling methods that affect interpretaions for almost all insect groups, was confirmed for the Brazilian Amazon (de Andrade et al. 2014).

Selection of indicator groups is often largely pragmatic rather than ecologically optimal. Costs of reliable sampling are a major limitation to exploring diverse insect groups as indicators, and the comparative rewards from many different groups of animals and plants have been investigated. Thus, Juutinen and Monkkonen (2004) recommended birds (30 species) and vascular plants (103 species) over groups such as beetles (435 species) in Finland, on grounds of cost. Their modeled inventory costings for these groups showed massive differences between beetles (Eu 34,479) and either birds (Eu 2691) or vascular plants (Eu 3868), with the other group examined (wood-inhabiting fungi, 64 species) slightly more than the last two, at Eu 5718. In terms of proven indicator values, no single group was unambiguously better than any other – but beetles were good indicators relative to the others – if inventory costs were not an issue. Juutinen and Monkkonen considered it worthwhile to pursue this, despite expense, as an inventory of beetles when the need is to complement a large conservation network in which less species-rich taxa are already covered. However, their major inference was that, if all species are to be conserved, no surrogate for complete inventory is wholly satisfactory. Forest site selection based on any indicator group or other taxon suite is likely to result in some loss of wider diversity. Wider biodiversity monitoring, often depending on a small number of indicator groups, almost inevitably fails to incorporate the full complexity of the ecosystems, and new approaches to help overcome this are suggested from time to time (Puumalainen et al. 2003). Their salutary warning, that the complexity of biodiversity almost certainly precludes any 'perfect data' for forest biodiversity assessment and monitoring in Europe, necessitates a pragmatic – rather than idealistic – approach to any selection of indicators or parallel focal groups, each of which may have only very limited potential to contribute to wider or more comprehensive practice.

Use of common taxa as indicators helped to overcome the reality that rare and threatened species are often too rare to appraise in any strict quantitative manner: numbers sufficient for convincing analysis are usually not available. If common species respond to the environmental changes being monitored, Matveinen-Huju et al. (2006) suggested that the response may be even more pronounced in related vulnerable taxa. Difficulties of applying the concepts of 'indicator species' and 'umbrella species' validly for assessing forest health and biodiversity are considerable, with recognition that either needs experimental verification. Suggested candidates for either use 'have been reasonable guesses' (Simberloff 1999), that need major efforts to transform well-meant ideas into effective practical tools.

Insect indicators of forest condition are still largely, as Grove (2002a) noted, 'the domain of entomologists and academics, rather than forestry practitioners'. However, unlike Australia where – despite information from surveys such as that of beetles in tropical forests by Grove (2002b, c) – basic information on saproxylic beetles and other potential indicator taxa is fragmentary, the greater knowledge of these insects in Europe gives them a realistic role in monitoring the impacts of forest management and identifying forest sites for conservation priority. Elsewhere, the chances of adopting saproxylic insects as good indicators of forest management or site condition are likely to remain slim (Grove 2002a).

Lists of threatened saproxylic beetles have become an important tool for indicating changes in assemblage composition in northern boreal forests but, as Martikainen et al. (2000) noted, parallel quantitative studies on non-saproxylic beetles have been relatively sparse. Ground beetles (Carabidae) and spiders are both attractive as indicators of forest management, and are perhaps more reliable in monitoring local disturbances than for landscape-level impacts (Pearce and Venier 2006). More generally, whether carabid beetles truly act as indicators for wellbeing of intact forests and their inhabitants needs further investigation (Niemela et al. 2007) to determine if, and how, the most sensitive species can survive as circumstances change.

In parallel with 'red-listing' for according priority amongst species, but emphasising distribution patterns, Sebek et al. (2012) also applied a scale of geographical rarity (using a scheme devised by Bouget et al. 2010), in which the species were categorised into four levels: (1) common and widely distributed; (2) widely distributed but not abundant, or only locally abundant; (3) not abundant and with localised distribution; and (4) very rare species known from fewer than five localities or in a single region. As elsewhere in Western Europe, information is broadly sufficient for such allocations to be made with confidence. Ecological information is also usually available, so that species could be allocated features such as host tree preference, trophic guild, body size class, and other traits, from which any 'subset' of taxa could reflect a wide range of forest components. Sebek et al. also recorded species as being listed on (1) the German List of Monitoring Species and (2) the European Red List of Saproxylic Beetles (Nieto and Alexander 2010), as further acknowledgement of conservation significance. For the red-listed species of boreal forest in Finland, a series of forest 'attributes' were each used to describe the primary requirements of the species, as far as possible. Table 6.4 (Tikkanen et al. 2006) emphasises that those species are collectively highly variable in their habitat needs and resource requirements and restrictions. The attributes described forest structure in stands and trees within stands, as the seemingly most significant key patterns that could also be related to forest management practices. Many species were not strictly confined to

Table 6.4 Attributes used to describe the ecological requirements of forest species, as used for boreal forests by Tikkanen et al. (2006)

Attribute	Category
Successional stage	Early successional, Mid-succesional, Old-growth, Indifferent
Fire	Obligate fire specialist, Prefers burned forests, Indifferent
Microclimate	Needs shady climate, Favours sun-exposed sites, Indifferent
Tree species	Specific (host genera named), Deciduous generalists, Coniferous generalists, No association
Decay stage of dead wood	1 – recent dead tree, decay not started; 2 – early decay stage, knife penetrates 1-2 cm, bark relatively intact; 3 – middle decay stage, knife penetrates 3-5 cm, bark mostly lost; 4 – late decay stage, whole knife penetrates easily into wood, bark lost; 5 – very soft trunk wholly covered by epiphytes, knife penetrates easily into wood; kelo – wood dry and hard, knife penetrates < 2 cm
Position of dead wood	Standing, Fallen, Hollow, Branch, Indifferent
Diameter of dead wood	<10 cm, >10 cm, >30 cm, Indifferent

any single category within an attribute. Tikkanen et al.'s survey included several major taxa, including plants and fungi, but Coleoptera (145 species) was by far the most diverse insect order. A high proportion of beetles (79%) were dependent on dead wood, with many occurring only in relatively fresh material (decay stage 2). About a third of the species showed no strong preference for standing or fallen wood, and branches and hollowed trees contained less than 3% of beetle species.

Focus on individual saproxylic species of conservation interest as indicators is sometimes valuable, and feasible and a number of probing studies on significant European species, in particular, illustrate the values of outcomes. Some are exemplified briefly below.

An informative study of the European Great capricorn beetle (*Cerambyx cerdo*, Cerambycidae) showed that measures to conserve this endangered species might benefit the associated assemblage of species on oak trees, which included several other red-listed species (Buse et al. 2007a, b). *C. cerdo* is an easily recognised beetle of readily acknowledged conservation significance, for example as listed under the European Union Habitats Directive, and infested trees are also recognised easily. It thereby combines 'flagship' and 'umbrella' values in central Europe, and its decline is attributed to reduction of old oak trees in open and semi-open landscapes because of changed forest and agricultural management. It was regarded as an 'ecosystem engineer', because larval feeding activities create large galleries and emergence holes (up to about 20 mm diameter) in dead wood. These constitute microhabitats for other species to inhabit, and some other beetle species were found only on trees with *C. cerdo* present. *C. cerdo* can use the same individual trees for up to several decades, and the galleries created may persist for longer than the individual capricorn population. Buse et al. surmised that dead oak trees colonised by this beetle may affect the saproxylic assemblages also over several decades. Because species richness was associated more strongly with intensity of colonisation by *C. cerdo* than with trunk diameter or exposure to sun, it seemed that the colonisation itself was the major cause of increased assemblage richness, which occurred across several feeding guilds. That process in part reflects the dispersal capability of the beetles and, as Ranius (2006) noted, it is important (but difficult) to study the dispersal of rare species in planning conservation management, especially in relation to patterns of fragmentation of their key resources. Those patterns and 'preferences' may be very subtle. Considerable information is available on microhabitat needs of *C. cerdo*, and that knowledge facilitates monitoring its distribution, both regionally and within individual old oak trees. The exit holes in trunks can be mapped, and their incidence and abundance is associated positively with trunk diameter and the openness (exposure) of the tree, and negatively with height above ground (Albert et al. 2012). Most of the beetles develop near the ground, with about a third of holes found at two sites in the Czech Republic less than 2 m above ground level. Open-grown trees are needed, and management that prevents canopy closure is crucial for the beetle's conservation (Albert et al. 2012). Sun-exposed trunks are particularly attractive for oviposition.

The translocation of 10 adults of *C. cerdo* led to one of the largest known populations of the beetle in the western Czech Republic (Drag and Cizek 2015), this appearing at a site where the species was believed to have become extinct. Beetles were detected first in 1995, and were found in >30 trees by 2011, following the translocation in 1987. This was claimed to be the first successful documented re-introduction of a saproxylic beetle, and its validity was clear from genetic information confirming the origin of the population. Translocation of *C. cerdo*, as supplementation of a small recipient population by relocating animals from a more secure source population has also been suggested within a broader set of actions proposed in Poland by Iwona et al. (2017). Monitoring the outcomes of any such exercise may be difficult.

However, the population structure of such species is usually unknown. The European helmet beetle, *Osmoderma eremita* (Scarabaeidae), may exhibit a metapopulation structure, with each tree exploited as a separate habitat patch that can potentially support a local independent population over the years that the tree remains suitable. *O. eremita* is amongst the highest priority species on the European Union Habitats Directive, and Ranius' (2002) advocacy included that it is a valuable umbrella species whose conservation will also conserve many other co-habiting species associated with trunk hollows in old trees. Survey of hollow oaks in Sweden showed that the simple presence of *O. eremita* was a better predictor of saproxylic beetle richness than most physical characters of the trees measured (Ranius 2002), and the values of those hollows increased through the extensive larval feeding by *Osmoderma*. Logistically, this species is one of the best-studied beetle species found in such habitats. It is also easier to appraise, monitor and protect than any other species of that assemblage in northern Europe. However, again as Ranius (2002) commented, some other beetles present may be even more susceptible to habitat fragmentation – and may still go extinct even when *Osmoderma* is present.

Both the above species are large and easy to identify by non-specialists, so that information is reliable and the problems that arise from taxonomically confused species groups are obviated. Several other notable European saproxylic beetles have also been appraised in some detail, and collectively demonstrate the varied ecology and levels of taxonomic knowledge that can occur within the diverse assemblages present.

Taxonomic uncertainties are exemplified by species of the cerambycid genus *Morimus*. The wingless *M. funereus* is found in dead wood of old-growth deciduous broadleaved forest and, although it is presumed to have declined, inadequate taxonomy has thwarted attempts to confirm, and gain consensus on, its status. Genetic information (Solano et al. 2013) led to suggestion that inadequacies of conservation focus on this individual taxon should be overcome by extending protected status to the entire variable species (*M. asper*), of which *M. funereus* is sometimes considered a component, with consequent protection of the primary habitat where the dubious taxon occurs.

Pytho kolwensis (Pythidae) has declined dramatically in Fennoscandia, and has vanished from most of its historical localities. Larvae live in fallen large branches of Norway spruce (*Picea abies*), and take 3–4 years to reach maturity in logs used previously by primary colonists such as some bark beetles (Siitonen and Saaristo 2000). Most records are from old-growth spruce mires, and some indication of *Pytho*'s key habitat requirements were clarified from surveys of 145 apparently suitable trees, from which 87% of the 55 trees occupied by larvae showed the character combination of: breast height trunk diameter at least 20 cm, bark cover at least 50%, penetration (measured by using a knife blade) at most 1.5 cm, mycelial cover at most 75%, and trunks without continuous contact with the ground. Historically, suitable spruce mires occurred along streams and as ecotones with open mires. Siitonen and Saaristo suggested that those circumstances could allow the beetle to perhaps move more easily between suitable trunks even when its populations were small. Subsequent draining and clear-cutting have markedly reduced those opportunities and so increased risks of extinction of small isolated populations. *P. kolwensis* may not now be able to withstand further tree removals, and conservation may rely heavily on continuity of currently occupied sites, especially those with high conservation values.

Cucujo haematodes (Cucujidae) is critically endangered, and one of the three most endangered saproxylic beetles in European forests. Its declines have been related to forest management, particularly in the western parts of its range (Horak et al. 2012), over the last few centuries. *C. haematodes* is regarded as a relict species that depends on older 'primeval' forests, and the distribution plotted by Horak et al. (Fig. 6.3) displays how a previously widely distributed beetle now has a highly disjunct relictual range in Europe.

The related *Cucujus cinnaberinus* is also of considerable conservation concern in Europe, and demonstrates how more detailed survey can change perspective of a species' needs. It appears to have declined strongly in the northern and southern parts of its range, but to be relatively well represented in central Europe (Horak et al. 2010). Unexpectedly, the anticipated stronghold host of old-growth forest

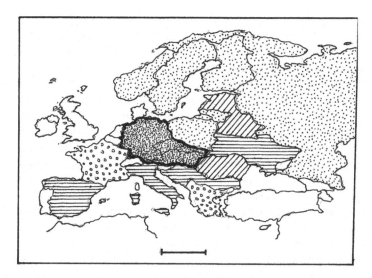

Fig. 6.3 The distribution and status of *Cucujus haematodes* in European countries, based on numbers of records reported since 2000: with numbers of sites indicated: >50 sites, darkest; 5–20 sites, fine dots; <5 sites, diagonal hatch; on records for 1950–1999, <5 sites, horizontal hatch; possible occurrence, circles; scale line 1000 Km. (Horak et al. 2012)

remnants were secondary to the beetle's incidence in riparian stands and abandoned plantation stands – mainly of fast-growing poplar hybrids established on flood plains to control soil erosion and provide timber. Many unharvested trees are dead or dying, so large concentrations of dead wood are available. Horak et al. supposed that the dying poplar plantations might simulate the later stages of pioneer forest succession, and that the beetle requires such open habitats rather than closed forest as presumed previously. Intensive forest management may threaten *C. cinnaberinus*, but the non-replacement of the dying poplar stands may lead to its decline in the near future. Management to conserve the beetle would entail planting of open canopy stands of trees such as poplars and willows. Introductions of favoured host tree species for notable saproxylic beetles has been suggested in other cases – use of the non-native red oak (*Quercus rubra*) for *Cerambyx cerdo* (p. 122) in Poland, for example. *C. cerdo* is associated strictly with mature oaks, but no evidence was found that this species was a suitable host for the beetle (Oleksa and Klejdysz 2017), and maintaining openness and high density of native old oaks was recommended for conservation.

The Rosalia longicorn, *Rosalia alpina* (Cerambycidae), is a notable flagship species that has also declined significantly across much of its European range. That decline in part reflects declines of traditional forestry practices, and that closed forest appears to be less suitable to the beetle than more open forests that are managed by means such as 'shredding' (pruning by removal of side branches, leading to more open canopy conditions), and moderate cattle grazing that prevents encroachment of woody vegetation. Closed forest, with shadowing of the substrate, may facilitate the beetle's persistence (Russo et al. 2011) and partial restoration of traditional management – at least on selected sites – may help to preserve suitable breeding habitat. However, even for such spectacular and relatively well known insects, the full ecological range and capability may not have been defined. Management of small remnant patches of old beech forest is a key to conservation of *R. alpina* in the Czech Republic (Drag et al. 2011), where mark-release-recapture studies revealed the beetle's ability to disperse between hilltop patches up to 1.6 Km apart. Beetles moved frequently over shorter distances between trees and other coarse woody debris within a patch, and most individuals remained within a few hundred metres of their emergence sites. *R. alpina* was regarded as a highly mobile species and, if sufficient suitable habitat is conserved, can track and exploit dead beech – Drag et al. suggested that practical measures needed to increase chances of population survival include (1) stopping logging and removal of dead wood from old beech remnant patches; (2) increasing the area of semi-open beech woodlands, especially on hill tops and slopes; and (3) restoring active management by measures such as pollarding and coppicing to create trees attractive for colonisation.

R. *alpina* was recently (2011) recorded from dead ash trees (*Fraxinus excelsior*) in Poland, for example, as a previously unrecorded substrate in the country (Michalcewicz and Ciath 2012) and which may have importance as roadside trees to promote connectivity between beetle populations.

The above-mentioned species collectively emphasise that dispersal of saproxylic beetles is a key aspect of considering managed resource supply (Ranius 2006). It was, for example, a component of habitat modelling for *Cerambyx cerdo* (Buse et al. 2007a, b) in conjunction with harvest levels and the needs for sufficient suitable trees to be left for the beetle.

Coleoptera and Lepidoptera are by far the most frequently studied insect candidates as 'surrogates' or 'indicators' of forest change and management. Using butterflies as a possible wider index for insect diversity in Costa Rica, for example, implied that the fauna of isolated 20–30 hectare fragments was highly depauperate in relation to a large forest patch (27 hectares) only 0.5–1 Km away. Butterflies were sampled by fruit-baited traps that may select actively for the more strongly flying species, and Daily and Ehrlich (1995) obtained about 47 species of Nymphalidae amongst which many of the Satyrinae (total 16 species) were mostly characteristic of open areas or forest edges, and had grass-feeding larvae. A more limited spectrum of the family are probably better indicators of forest interior conditions, and it seemed that large forest fragments are necessary for conservation of all the region's species. As in many parallel contexts, small patches become highly degraded within the changed landscape matrix.

Rather than changes in real or relative abundance or incidence of species, claims have also been advanced that the changed behaviour of a species attributed to forest change can also indicate habitat condition. Foraging behaviour of the ant *Paraponera clavata* differed in undisturbed primary forest and regenerating secondary forest in Costa Rica (McGee and Eaton 2014), and foraging appeared to be far less successful in the secondary forest. These omnivorous ants normally need a variety of foods, including insect prey and nectar, supplies of which are altered by changes to the forest, so that foraging distances increased and food resources decreased in regenerating forest. In primary forest, *P. clavata* can ascend to the canopy to forage, an option not available in regenerating forest.

6.2.1   Dung Beetles

Since dung beetles (Scarabaeoidea) were advocated as a useful indicator group for forest changes (Halffter and Favila 1993), they have been surveyed in many parts of the world and the criteria for their suitability applied to various other insect indicator groups. Many species show local biotope specificity. Scarabaeinae are regarded as very suitable as indicators in tropical rain forests. Reasons for this (Halffter and Favila 1993) span logistics (the beetles are easily sampled by standard methods involving baited pitfall traps), taxonomic background (a reasonably diverse and well-defined group that is relatively well-known at species level in many parts of the world), and ecological knowledge (as a well-defined functional guild, for which good foundation biological understanding exists) including responses to change. Taxonomic composition of assemblages within forest differs from that in transformed areas, with changes clearly reflecting anthropogenic changes. That collective awareness enables meaningful comparisons of species richness and compositional changes as well as their trophic diversity, including generalists and specialists, and both coprophagous and necrophagous species. Dung beetles and carrion beetles predominantly utilise resources provided by mammals, or decaying fruit or fungi.

In short, dung beetles are highly sensitive to forest conversion, to a large extent through the indirect impacts on supply of mammalian faeces. Dung beetles can respond rapidly to changes in forests, and their assemblages are affected also by fragment patch size, patch isolation, quality of the embedding matrix, and local features such as elevation, insolation and slope. Their responses are strongly context-influenced, but three main traits are linked with their sensitivity, namely body size, nesting strategy, and activity period, all discussed and reviewed by Nichols et al. (2013).

From surveys in Chiapas, Mexico, the assemblages were related only weakly to extent of patch isolation, whilst forest patch size and landscape-level forest cover were positively related to beetle richness and abundance (Sanchez-de-Jesus et al. 2016), with increased heterogeneity of the matrix surrounding the patches deemed a priority for conservation. More widely, higher proportion of open spaces, including cattle pastures or annual crops, are associated with dung beetle declines, and a number of rare species can be lost entirely from such changes: Sanchez-de-Jesus et al. considered dung beetles to be very sensitive to forest spatial changes. The above conservation measures (1) increase suitable habitat and (2) reduce edge effects so (3) promote connectivity.

Plantations and secondary forests in the neotropics may support only very impoverished dung beetle assemblages, with reduction in the average body size of the persistent species (Gardner et al. 2008) and many of the beetles clearly associated only with primary forests in Brazil. The widespread perception that regenerating forests can largely offset losses of species after primary forest removal was thus queried – with loss of the larger beetle species likely to support key processes such as dung burial and secondary seed dispersal. Protection of areas of primary forest is still a conservation need.

Changes in dung beetle assemblages in forest collectively reflect structural change (vegetation cover, soil type and condition) so indicate functional integrity in relation to parameters that can be measured. Studies in south east Asia have documented that forest clearance can reduce dung beetle richness, and habitat fragmentation decrease both abundance and richness. In the Danum Valley, Sabah, dung beetles were compared (by baited pitfall traps and flight interception traps) across primary, plantation and logged forests (Davis et al. 2001). Diversity was lower in logged than in primary forest, but overall lowest in plantation forest – whether of cocoa, *Acacia* or mahogany, with the two plantation sites differing substantially. Collectively, 86 beetle species were represented amongst the 32,279 identified individuals. Greater resemblance between assemblages in the primary and logged forest implied that some elements of the former persisted with disturbance, unlike in the more dramatically changed plantation forests.

In South America several studies of dung beetle assemblages in native and modified forests have widely supported inferences from the Sabah study. Thus, in Brazil pitfall traps with a variety of baits were used to compare dung beetles across five regimes along a gradient of disturbance (native forest, early secondary successions [5 years], mature secondary succession [15 years], cocoa plantations, cleared pasture), and demonstrated a gradient of species richness among the 112 species encountered, namely 86, 33, 69, 33, 38, respectively (Cajaiba et al., 2017). That survey implied that (1) the habitat regimes differed in richness and taxonomic composition of beetles caught, with each regime yielding exclusive and indicator species; (2) disturbed habitats supported far fewer species than native forest or mature secondary succession; and (3) forest generalist species with more preserved habitat requirements may disappear after habitat destruction or reduction. Loss of forest specialist species has been regarded as a useful ecological indicator for extent of land-use changes.

Surveys in Peru (Horgan 2005) showed that deforestation led to reduced dung decomposition by beetles, related mostly to reduced assemblage biomass with individual dung pats. Additionally, dung in forests may remain attractive to beetles for longer than dung in open sites, where it is exposed to rapid desiccation.

In Brazilian Atlantic forest, conversion to sugar cane production and cattle pasture has led to extensive forest fragmentation (Filgueiras et al. 2015). Dung beetles surveyed by pitfall traps in forest interior, at forest edges, in sugarcane plantations and in pastures around forest patches gave richness in each of 20, 20, 12, and 11 species within a pool of 45 species – also a substantial lessening in highly disturbed post-forest regimes, but with those areas supporting taxa not found in the forest. The persistence of native dung beetles in anthropogenic forest landscapes depends on 'core areas', but edge-affected areas and intermediate habitats contribute to overall diversity. The primary forest assemblages in Filgueiras et al.'s study were biased toward large-bodied species, but the form and supply of suitable dung, and the characteristics of the soil strongly influence the assemblage that can be present.

Dung beetles depend closely on resources provided by other species – notably vertebrates and with mammals predominant – and whose fate as forests disappear is of high conservation concern. The lower profile of the beetles has nevertheless generated studies, such as those cited above, on assemblages of the beetles in forests and resource-deprived altered areas that help to elucidate landscape influences. In southern Mexico, 33 dung beetle species were trapped in a comparative survey across a tract of continuous forest, forest fragments and a mosaic 'habitat island' of forest and tree crops (Estrada and Coates-Estrada 2002). No significant differences in richness occurred across these three treatments, but the largest proportion of beetles (56% of the 7332 beetles trapped) came from continuous forest, substantially more than from either the mosaic island (29%) or the forest fragments (15%). The greatest richness of mammals (27 species) was also in continuous forest, this reducing to 18 species in each of the other treatments. The single most abundant beetle species (*Canthon femoralis*) accounted for 90% of all individuals captured, and 12 species occurred in only one of the treatments. In this study, it seemed that replacement of the forest by arboreal crops (the authors noted cacao, coffee, citrus and bananas among these) led to declining numbers, mirrored by *C. femoralis* which depends on inner forest mammals – notably howler monkeys, *Alouatta palliata*. Guild analysis, displaying the relative abundance of dung rollers and dung buriers and of diurnal or nocturnal species, implied that such variety could be catered for by manipulations of vegetation that might also promote connectivity between more continuous forested areas, with additional benefits to soil quality and water retention.

In Australia's Wet Tropics (northern Queensland), about half the forest has been cleared for agricutural production, so that remnants of rich tropical forests now occur only in largely agricultural landscapes. In one of relatively few studies of impacts on native invertebrates, dung beetle assemblages were shown to differ significantly in rainforest and the cleared pastoral landscape (Kenyon et al. 2016). Of the 27 species captured in baited pitfall traps, 22 occurred in forest, and nine in pastures. Substantial differences in abundance were also found. Within the total of 5400 individuals in forest and 84 in pasture, individual species abundance differences included (rainforest, pasture) *Amphistomus* no. 5 (2446, 8), *Temnoplectron politulum* (667, 1), *Amphistomus* no. 3 (601, 0), and *Onthophagus millamilla* (517, 0), with the converse balance unusual (*Onthophagus capella*: 0, 42). Notwithstanding such major differences in abundance, only four of the 27 species occurred in both land use regimes. Separate appraisal of dung-tunnelers (17 species) and dung-rollers (10 species) showed that richness, abundance and biomass of both guilds were significantly higher in rainforest than in pasture, but body size was either significantly lower (tunnelers) or higher (rollers) in forest than in pasture plots, with implications for functional differences to occur.

Dung beetles sampled by flight intercept traps, baited pitfall traps and light traps across an open forest-rainforest ecotone in Queensland, comprised 27 species (Hill

1996). The faunas of the two forest types were largely distinct, with the rainforest fauna considerably the richer. Most rainforest species were relatively scarce at the ecotone sites, and greater numbers at 25 and 50 metres into the forest suggested that they 'prefer' the forest interior. In this case, assemblages associated with different vegetation form may be separated by only a few tens of metres.

6.2.2 Stag Beetles

Lucanidae, especially the larger species, have largely overcome the 'popularity impediment' so daunting to promoting insects for conservation, because of their appeal to collectors and consequent commercial values. The counter to their prominence as flagships for conservation among saproxylic insects is that many of the rarer species are sought by unscrupulous traders and appear on the 'black market' despite formal declarations of their protected status. The largest individuals of rare stag beetles can command very high prices, and the active breaking up of their dead wood habitats in seeking them is a serious threat to some species (New 2005, 2010). Lucanids in several parts of the world have become of conservation concern, and some publicised in sufficient detail to be pertinent flagships for the wider needs of forest-based saproxylic insects. In Britain, 'the stag beetle' (*Lucanus cervus*) is one such species, with conservation recommendations including the local supply of dead wood ('loggeries', which can be established in home gardens or larger open areas) as breeding substrate.

Perhaps the best-known such species in Australia is the tropical King stag beetle, *Phalacrognathus muelleri*, for long considered extremely elusive and much desired by collectors. However, recent assiduous collecting and searches in north east Queensland have shown it to be more widespread, and also amenable to captive breeding – under conditions in which the natural generation time of up to about four years can be shortened considerably. It is now of much lower conservation concern than it was a few decades ago (Hannay and de Keyser 2017), but because of increased knowledge rather than any additional forest conservation measures.

Australian stag beetles have received considerable conservation attention, especially in Tasmanian forests (Chapter 8). Production forests are an important invertebrate habitat in the State, and the initial schedules to Tasmania's Threatened Species Protection Act listed 79 forest-dependent invertebrates with at least part of their restricted ranges encompassed by those forests. These included several Lucanidae for which conservation needs span management in both production forests and reserved habitats. Their habitat requirements thus need to be defined clearly, as Meggs and Munks (2003) clarified for the endangered endemic and flightless Broad-toothed stag beetle, *Lissotes latidens*, for which further uncontrolled forest clearing and conversion to plantations may be a serious threat.

The key lucanid species are all very scarce and one practical advantage of surveying for some rare stag beetles (and some other larger Coleoptera) is that their dead remains are persistent and can sometimes be recognised to species level.

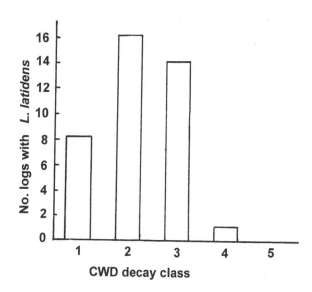

Fig. 6.4 Occurrence of *Lissotes latidens* in logs at different decay stages (1, least decayed to 5, most decayed). (Meggs and Munks 2003)

Despite the labour-intensive nature of direct searches in forest ground litter and under logs, searches for living beetles can be supplemented from any dead body parts recovered as reliable records of relatively recent incidences, and from which 'sums of parts' can approximate numbers of individual beetles – although those parts may have accumulated over at least several generations. Dead remains may far outnumber living beetles in surveys. Outcomes from surveys of two threatened species of *Hoplogonus* in Tasmania (p. 164), included discovery of 94 dead individuals but no living representatives of *H. bornemisszai*, and 13 dead individuals and one live *H. vanderschoori* (Munks et al. 2004). However, the scarcity – even of corpses – of these beetles, with discovery rates also likely to be reduced by their cryptic habits, renders 'apparent absence' from seemingly suitable sites difficult to confirm, and development of conservation management uncertain. From combined records of living and dead individuals, *Lissotes latidens* (above) was absent from dry forest corridors and occurred both in the transition zone between dry and wet forest and in patches of wetter forest along drainage lines, or in riparian corridors (Meggs and Munks 2003), in which further vegetational details help to determine possible suitable habitats. However, *L. latidens* occurred only in low densities, estimated at only 30 beetles/ha^{-1}, and mostly in areas with >10% ground cover of coarse woody debris. Of the 54 adults found by Meggs and Munks, 46 were in soil under fallen dead wood, mostly under logs with a moderate level of decay (Fig. 6.4). Past selective logging seems not to have had major impacts, but the remaining *Lissotes* populations are severely fragmented and retention and connectivity of remaining wet forest appears necessary for the beetle's survival throughout its range.

The threatened stag beetles of Tasmania's forests have received considerable attention, and several endemic species of *Hoplogonus* and *Lissotes* have been listed under the Tasmanian Threatened Species Protection Act 1995. Subsequently-detected population declines across restricted distribution areas have been attributed

Table 6.5 Species richness and abundance of Lucanidae in wet sclerophyll (Picton Forest) and dry sclerophyll (Weilangta Forest) of different regeneration ages in southern Tasmania (abbreviated from Michaels and Bornemissza 1999)

Regeneration age (years)				
	1–3	9	20–25	Old-growth
Picton Forest				
No. species	5	3	2	4
No. individuals	116	55	8	10
Percentage distribution of species				
Lissotes cancroides(122)	57	36	9	5
L. curvicornis(37)	21	69	0	10
L. politus(9)	100	0	0	0
L. rodwayi(19)	41	37	7	15
L. subcaeruleus(1)	100	0	0	0
Ceratognathus niger(1)	0	0	0	100
Weilangta Forest				
No. species	3	2	3	3
No. individuals	15	6	2	14
Percentage distribution of species				
Lissotes sp. n.(19)	41	32	0	27
L. obtusatus(13)	69	17	9	6
L. latidens(5)	55	0	27	18

firmly to forestry activities, with numbers declining along a chronosequence of forest regeneration age in both wet and dry sclerophyll forests (Table 6.5) and demonstrated need to conserve the beetles in production forest areas.

The largest endemic stag beetle in Tasmania, Simson's stag beetle (*Hoplogonus simsoni*), occurs mostly in forest leaf litter – although it was formerly believed to be log-dwelling, larvae are now considered to be edaphic in a narrow range in north eastern Tasmania. Meggs et al. (2003) increased the number of known localities, and suspected that a metapopulation structure might prevail. *H. simsoni* occurs mainly in wet eucalypt forests, in which the following correlated factors were associated with higher densities: elevation below 300 m, slope < 5°, a deep (at least 1–3 cm) litter layer, and a well-developed upper understorey layer of tall shrubs. *H. simsoni* apparently requires a relatively cool, moist and stable microclimate, implying needs for wet forest with lack of recent disturbances such as wildfires (Meggs et al. 2003).

Much of the wet forest and dryer mixed eucalypt forest within the beetle's range has been harvested by clear-felling, followed by high intensity burns to promote regeneration, and aerial sowing, with modeled subsequent harvesting at about 90 year intervals. No living *H. simsoni* were found in regenerating coupes, but low numbers of body remains (above) occurred in some areas adjacent to larger areas of unlogged forest in which beetles were more common. The preferred habitat is represented poorly in the State's reserve system, with about 19% (1700 hectares) of that within the beetle's range reserved. In contrast, 43% of the habitat type is in state forest, subject to forestry practices, and much of it with potential to be converted to pine plantations. 'Off-reserve' conservation strategies for *H. simsoni* and to ensure connectivity between source populations and isolated habitat patches centre on the need to retain contiguous areas of undisturbed wet forest throughout the full range of the beetle.

The distribution of *H. simsoni* overlaps with that of *H. bornemisszai*, but the latter does not overlap with that of *H. vanderschoori* (Munks et al. 2004). Again, only small proportions of potential habitat for these species (*H. bornemisszai*, 20%; *H. vanderschoori*, 30%) occur in reserves or areas under some form of conservation covenant – and those areas are in particular need of enlightened management. Of the 'unreserved habitats' from features defined by Munks et al. (2004), 54% (for *H. bornemisszai*) and 37% (*H. vanderschoori*) had been identified as with potential for wood production – thus, 520 hectares of potential habitat for *H. vanderschoori* had already been converted to hardwood plantations and another 210 hectares identified as suitable for later conversion. As in parallel cases elsewhere, Munks et al. affirmed that quantity and extent of habitat were critical for conservation, no further conversion of forest for plantations or agriculture could be condoned, and a moratorium should be placed on clear-fell, burn and sow silviculture pending a full review of the practice and its impacts.

A continuous supply of suitable habitat for these stag beetles can probably be assured only in old-growth forests (Michaels and Bornemissza 1999), and replacement of old forest by plantations after clear-felling is likely to cause their decline. The principles for conservation, in common with those for other saproxylic insects, include (1) continual supply of wood in all decay stages; (2) the persistence of undisturbed source areas (old-growth forests); and (3) connectivity between sites sufficient to allow beetles to move between them. Because of floristic differences between dry scerophyll and wet sclerophyll forests in Tasmania, clearfell logging could have greater impact on the beetles in dry forests there.

One problem with surveys for rare, living log-dwelling beetles is simply that direct searches involve breaking up of individual logs, effectively habitat destruction, that might increase level of threat, as Meggs and Taylor (1999) noted for the Mt. Mangana stag beetle (*Lissotes manalcas*) in Tasmania. At present, it seems that the beetle can tolerate both wildfire and the regeneration burning that follows clearfelling – the rotting logs appear to counter desiccation as well as protect beetles from fire impacts. Dwelling in the interior of decaying logs with fungal rot, and the older logs with well-developed moss covering on the wet forest floor, *L. manalcas* appears to be reasonably protected, but its needs across forests of different types need further definition. Meggs and Taylor feared, however, that plantation development in native forests where the species occurs could lead to extirpations, because rotting logs would largely be eliminated. They speculated that it might persist in windrows, but this was unlikely to extend beyond the second rotation of plantation trees.

The Tasmanian stag beetle surveys exemplify the widespread difficulties of determining absences of such localised low density and rare species, and of estimating actual population sizes in localities where they are confirmed to occur. The values of 'presence only' approaches include overcoming (at least in part) the strong limitations of distribution assessments imposed by needs for reliable 'absence data'. Reported absences, even if reliable – and there is commonly some doubt – are laborious to validate locally and almost impossible to confirm across larger areas that include apparently suitable conditions. Those values were explored for the threatened *Rosalia alpina* (Cerambycidae, p. 189) in Italy (Bosso et al. 2013). As an important generalisation, they suggested that models calculated for ecological specialist taxa (so including most insect species of individual conservation concern) may be more accurate than those for generalists, because of the restricted range of ecological conditions that are relevant. In that way, a smaller set of information may cover more of the species' ecological needs. Their model derived for *R. alpina* detected a set of environmental variables that (on a broad scale) helped to explain its non-random geographical distribution, and enabled discovery of previously unknown presences and the levels of fragmentation that could isolate populations. In addition, they suggested where creation or restoration of connectivity through new plantings could most constructively be undertaken so that landscape measures might complement more local efforts to sustain the beetle's requirements.

Concerns over conservation needs of some flagship Lucanidae have stimulated studies on other species, including close relatives of threatened taxa, and wider augmentation of relevant biological knowledge. Thus, clarification of the ecological needs and larval habits of the North American *Lucanus elaphus* by Ulyshen et al. (2017) demonstrated the relatively wide spectrum of decaying hardwood logs acceptable for breeding sites, and that the continuous supply of such logs is critical to sustaining the beetle.

6.2.3   Butterflies

Butterflies have major advantages in assessing environmental impacts involving vegetation changes, as a predominant herbivore group but also by being taxonomically tractable, diurnal, reasonably diverse, and surveyable by relatively simple and well-tried methods such as transect walks, timed observations, or fruit baits (the latter mainly for Nymphalidae).

However, whilst there is wide inference that that forest changes can have major harmful impacts on butterflies, methodological ambiguities can thwart comparisons between different studies. For example, Koh (2007) reviewed 20 studies on the impacts of land-use changes on forest butterflies in south-east Asia, all involving comparisons of butterfly species richness and/or abundance between forests and some modification from disturbance, such as clearing or selective logging. Seven of those studies found higher butterfly richness/diversity in the undisturbed or least disturbed treatments compared, nine studies found the converse (as the most commonly inferred likely outcome), three studies showed little difference across treatments, and the remaining one indicated a strong seasonal influence on logging impacts. Koh urged the need to determine the mechanisms that can influence the responses of butterflies to land-use changes: the clear lack of consensus amongst the studies then available implied considerable complexity in trying to predict the ecological features of individual species within each assemblage. Fruit-feeding butterflies in monoculture oil palm plantations and polyculture plantations (oil palm with interspersed other crops such as bananas, coconuts or sugarcane) in Malaysia differed little in species richness (Asmah et al. 2017). The same three species dominated both assemblages, in generally similar proportions (*Elymnias hypermnestra* [monoculture 54.75%, polyculture 55.58% of total butterflies], *Amathusia phidippus* [30.84%, 28.14%], *Mycalesis visala* [8.94%, 12.56%]) and, despite the increased heterogeneity in polyculture plantations, this did not clearly enhance diversity of this selected butterfly group.

In accordance with an earlier study on butterfly extinctions in Singapore (Koh et al. 2004), where almost all the original forest cover has been lost, those butterflies most likely to go extinct were forest species with greatest host plant specificity, with probability of loss potentially increased for species with small geographical ranges. Perhaps only with that level of understanding can optimal conservation strategies be developed. More generally, reports that butterfly richness and abundance are less in plantations than in natural forest (oil palm in New Britain, Papua New Guinea: Miller et al. 2011), and also with forest conversion to natural shrubland (China: Li et al. 2011), exemplify the wider trends associated with forest losses.

Fragment size and vegetation influence the richness of butterflies present, and even very small isolated fragments in Sabah rainforest contribute to regional diversity (Benedick et al. 2006). Those forests, however, were relatively poor in species, and dissimilar to intact forest butterfly assemblages – significantly, no endemic species were found in remnants less than about 4000 hectares, although even the smallest remnants surveyed (one being only 120 hectares) supported species with restricted geographical distributions. Such remnants are highly susceptible to loss, as they are often disregarded and remain unprotected. Their importance for insects may be far greater than suspected initially.

References

Albert J, Platek M, Cizek L (2012) Vertical stratification and microhabitat selection by the Great Capricorn Beetle (*Cerambyx cerdo*) (Coleoptera: Cerambycidae) in open-grown, veteran oaks. Eur J Entomol 109:553–559

Asmah S, Ghazali A, Syafiq M, Yahya MS, Peng TL et al (2017) Effects of polyculture and monoculture farming in oil palm smallholdings on tropical fruit-feeding butterfly diversity. Agric For Entomol 19:70–80

Benedick S, Hill JK, Mustaffa N, Chey VK, Maryati M et al (2006) Impacts of rain forest fragmentation on butterflies in northern Borneo: species richness, turnover and the value of small fragments. J Appl Ecol 43:967–977

Bosso L, Rebelo H, Garonna AP, Russo D (2013) Modelling geographic distribution and detecting conservation gaps in Italy for the threatened beetle *Rosalia alpina*. J Nat Conserv 21:72–80

Bouget C, Brustel H, Zagatti P, Noblecourt T (2010) The French information system on saproxylic beetle ecology (FRISBEE): an ecological and taxonomical database to help with assessment of forest conservation status. http://frisbee.nogent.cemagref.fr/index.php/en/

Brin A, Brustel H, Jactel H (2009) Species variables or environmental variables as indicators of forest biodiversity: a case study using saproxylic beetles in Maritime pine plantations. Ann For Sci 66/306:1–11

Buse J, Ranius T, Assmann T (2007a) An endangered longhorn beetle associated with old oaks and its possible role as an ecosystem engineer. Conserv Biol 22:329–337

Buse J, Schroder B, Assmann T (2007b) Modelling habitat and spatial distribution of an endangered longhorn beetle – a case study for saproxylic insect conservation. Biol Conserv 137:372–381

Cajaiba RL, Perico E, Dalzochio MS, da Silva WB, Bastos R, Cabral JA, Santos M (2017) Does the composition of Scarabaeidae (Coleoptera) communities reflect the extent of land use changes in the Brazilian Amazon? Ecol Indic 74:285–294

Daily GC, Ehrlich PR (1995) Preservation of biodiversity in small rainforest patches: rapid evaluations using butterfly trapping. Biodivers Conserv 4:35–55

Davis AJ, Holloway JD, Huijbregts H, Krikken J, Kirk-Spriggs AJH, Sutton SL (2001) Dung beetles as indicators of change in the forests of northern Borneo. J Appl Ecol 38:593–616

de Andrade RB, Barlow J, Louzada J, Mestre L, Silveira L, Vaz-de-Mello FZ, Cochrane MA (2014) Biotic congruence in humid tropical forests: a multi-taxa examination of spatial distributions and responses to forest disturbance. Ecol Indic 36:572–581

Drag L, Cizek L (2015) Successful reintroduction of an endangered veteran tree specialist: conservation and genetics of the Great Capricorn beetle (Cerambyx cerdo). Conserv Genet 16:267–276

Drag L, Hauck D, Pokluda P, Zimmerman K, Cizek L (2011) Demography and dispersal biology of a threatened saproxylic beetle: a mark-recapture study of the Rosalia longicorn (Rosalia alpina). PLoS One 696:e21345. https://doi.org/10.1371/journal.pone.oo21345

Estrada A, Coates-Estrada R (2002) Dung beetles in continuous forest, forest fragments and in an agricultural mosaic habitat island at Los Tuxtlas. Mexico. Biodiv Conserv 11:1913–1918

FAO (2006) Global forest resources assessment 2005: progress towards sustainable forest management. Forestry paper 147. United Nations Food and Agriculture Organisation, Rome

Filgueiras BKC, Tabarelli M, Leal IR, Vaz-de-Mello FZ, Iannuzzi L (2015) Dung beetle persistence in human-modified landscapes: combining indicator species with anthropogenic land use and fragmentation-related effects. Ecol Indic 5:65–73

Foster WA, Snaddon Jl, Turner EC, Fayle TM, Cockerill TD et al. (2011) establishing the evidence base for maintaining biodiversity and ecosystem function in oil palm landscapes of South East Asia. Phil Trans R Soc Lond B 366: 3277–3291

Gardner TA, Hernandez MIM, Barlow J, Peres CA (2008) Understanding the biodiversity consequences of habitat change: the value of secondary and plantation forest for neotropical dung beetles. J Appl Ecol 45:883–893

Grove SJ (2002a) Saproxylic insect ecology and the sustainable management of forests. Annu Rev Ecol Syst 33:1–23

Grove SJ (2002b) Tree basal area and dead wood as surrogate indicators of saproxylic insect faunal integrity: a case study from the Australian lowland tropics. Ecol Indic 1:171–188

Grove SJ (2002c) The influence of forest management history on the integrity of the saproxylic beetle fauna in an Australian lowland tropical rainforest. Biol Conserv 104:149–171

Grove SJ (2006) A research agenda for insects and dead wood. In Grove SJ, Hanula JL (eds) Insect biodiversity and dead wood. Proceedings of a symposium for the 22nd international congress of entomology. USDA General and Technical Report SRS-93, Ashville, NC, pp 98–108

Halffter G, Favila ME (1993) The Scarabaeinae (Insecta: Coleoptera) an animal group for analysing, inventorying and monitoring biodiversity in tropical rainforest and modified landcapes. Biol Int 27:15–21

Hannay R, de Keyser R (2017) A guide to stag beetles of Australia. CSIRO Publishing, Clayton South

Hill CJ (1996) Habitat specificity and food preference of an assemblage of tropical Australian dung beetles. J Trop Ecol 12:449–460

Hilszczanski J, Gibb H, Hjalten J, Atlegrim O, Johansson T, Pettersson EB, Ball JP, Danell K (2005) Parasitoids (Hymenoptera, Ichneumomoidea) of saproxylic beetles are affected by forest successional stage and dead wood characteristics in boreal spruce forest. Biol Conserv 126:456–464

Horak J, Vavrova E, Chobot K (2010) Habitat preferences influencing populations, distribution and conservation of the endangered saproxylic beetle *Cucujus cinnaberinus* (Coleoptera: Cucujidae) at the landscape level. Eur J Entomol 107:81–88

Horak J, Chobot K, Horakova J (2012) Hanging on by the tips of the tarsi: a review of the plight of the critically endangered saproxylic beetle in European forests. J Nat Conserv 20:101–108

Horgan FG (2005) Effects of deforestation on diversity, biomass and function of dung beetles on the eastern slopes of the Peruvian Andes. For Ecol Manag 216:117–133

Humphrey JW, Hawes C, Peace AJ, Ferris-Kaan R, Jukes MR (1999) Relationships between insect diversity and habitat characteristics in plantation forests. For Ecol Manag 113:11–21

Iwona M, Marek P, Katarzyna W, Edwards B, Julia S (2017) Use of a genetically informed population viability analysis to evaluate management options for Polish populations of endangered beetle *Cerambyx cerdo* L. (1758) (Coleoptera, Cerambycidae). J Insect Conserv 22:69. https://doi.org/10.1007/s10841-017-0039-3

Johansson T, Gibb H, Hjalten J, Pettersson RB, Hilszczanski J, Alinvi O, Ball JP, Danell K (2007) The effects of substrate manipulations and forest management on predators of saproxylic beetles. For Ecol Manag 242:518–529

Jonsell M, Nordlander G (2002) Insects in polypore fungi as indicator species: a comparison between forest sites differing in amounts and continuity of dead wood. For Ecol Manag 157:101–118

Juutinen A, Monkkonen M (2004) Testing alternative indicators for biodiversity conservation in old-growth boreal forests: ecology and economics. Ecol Econ 50:35–48

Kenyon TM, Mayfield MM, Monteith GB, Menendez R (2016) The effects of land use change on native dung beetle diversity and function in Australia's wet tropics. Aust Ecol 41:797–808

Koh LP (2007) Impacts of land use change on South-east Asian forest butterflies: a review. J Appl Ecol 44:703–713

Koh LP (2008) Can oil palm plantations be made more hospitable for forest butterflies and birds? J Appl Ecol 45:1002–1009

Koh LP, Sodhi NS, Brook BW (2004) Ecological correlates of extinction proneness in tropical butterflies. Conserv Biol 18:1571–1578

Lambeck RJ (1997) Focal species: a multi-species umbrella for nature conservation. Conserv Biol 11:849–856

Langor DW, Spence JR, Hammond HEJ, Jacobs J, Cobb TP (2006) Maintaining saproxylic insects in Canada's extensively managed boreal forests; a review. In Grove SJ, Hanula JL (eds) Insect biodiversity and dead wood. Proceedings of a symposium for the 22nd international congress of entomology. USDA General and Technical Report SRS-93, Ashville, NC, pp 83–108

Li X-s, Luo Y-q, Yuan S-y, Zhang Y-l, Settele J (2011) Forest management and its impact on present and potential future Chinese insect biodiversity – a butterfly case study from Gansu Province. J Nat Conserv 19:285–295

Lindenmayer DB, Margules CR, Botkin DB (2000) Indicators of biodiversity for ecologically sustainable forest management. Conserv Biol 14:941–950

Lindenmayer DB, Cunningham RB, MacGregor C, Tribolet C, Donelly CF (2001) A prospective longitudinal study of landscape matrix effects on fauna in woodland remnants: experimental design and baseline data. Biol Conserv 101:157–169

Lindenmayer DB, Franklin JF, Fischer J (2006) General management principles and a checklist of strategies to guide forest biodiversity conservation. Biol Conserv 131:433–445

Maeto K, Noerdjito WA, Belokobylskij SA, Fukuyama K (2009) Recovery of species diversity and composition of braconid parasitic wasps after reforestation of degraded grasslands in lowland East Kalimantan. J Insect Conserv 13:245–257

Maleque MA, Maeto K, Ishii HT (2009) Arthropods as bioindicators of sustainable forest management, with a focus on plantation forests. Appl Entomol Zool 44:1–11

Martikainen P, Siitonen J, Puntilla P, Kaila L, Rauh J (2000) Species richness of Coleoptera in mature managed and old-growth boreal forests in southern Finland. Biol Conserv 94:199–209

Matveinen-Huju K, Niemela J, Rita H, O'Hara RB (2006) Retention-tree groups in clear-cuts: do they constitute 'life-boats' for spiders and carabids? For Ecol Manag 230:119–135

McGee KM, Eaton W (2014) The effects of conversion of a primary to a secondary tropical lowland forest on bullet ant (*Paraponera clavata*) foraging behavior in Costa Rica: a possible indicator of ecosystem condition. J Insect Behav 27:206–216

McGeoch MA (1998) The selection, testing and application of terrestrial insects as bioindicators. Biol Rev 73:181–201

MCPFE (Ministerial Conference on the Protection of Forests in Europe) (2003) Improved Pan-European indicators for sustainable forest management. MCPFE Liaison Unit, Vienna

Meggs JM, Munks SA (2003) Distribution, habitat characteristics and conservation requirements of a forest-dependent threatened invertebrate *Lissotes latidens* (Coleoptera: Lucanidae). J Insect Conserv 7:137–152

Meggs JM, Taylor RW (1999) Distribution and conservation status of the Mt Mangana stag beetle, *Lissotes manalcas* (Coleoptera: Lucanidae). Pap Proc R Soc Tasm 133:23–28

Meggs JM, Munks SA, Corkrey R (2003) The distribution and habitat characteristics of a threatened lucanid beetle *Hoplogonus simsoni* in north-East Tasmania. Pac Conserv Biol 9:172–186

Michaels K, Bornemissza G (1999) Effects of clearfell harvesting on lucanid beetles (Coleoptera: Lucanidae) in wet and dry sclerophyll forest in Tasmania. J Insect Conserv 3:85–95

Michalcewicz J, Ciath M (2012) Rosalia longicorn *Rosalia alpina* (L.) (Coleoptera: Cerambycidae) uses roadside European ash trees *Fraxinus excelsior* L. – an unexpected habitat of an endangered species. Pol J Entomol 81:49–56

Miller DG, Lane J, Senock R (2011) Butterflies as potential bioindicators of primary rain forest and oil palm plantation habitats on New Britain, Papua New Guinea. Pac Conserv Biol 17:149–159

Mitchell DK, Dewhurst CF, Tennent WJ, Page WW (2016) Queen Alexandra's birdwing butterfly, *Ornithoptera alexandrae* (Rothschild 1907): a review and conservation proposals. Southdene Sdn Bhd, Kuala Lumpur

Munks SA, Richards K, Meggs J, Wapstra M, Corkrey E (2004) Distribution, habitat and conservation of two threatened stag beetles, *Hoplogonus bornemisszai* and *H. vanderschoori* (Coleoptera: Lucanidae) in North-Eastern Tasmania. Aust Zool 32:586–596

New TR (2005) 'Inordinate fondness': a threat to beetles in south-East Asia? J Insect Conserv 9:147–150

New TR (2010) Beetles in conservation. Wiley-Blackwell, Oxford

Nichols E, Uriarte M, Bunker DE, Favila ME, Slade EMet al. (2013) Trait-dependent response of dung beetle populations to tropical forest conversion at local and regional scales. Ecology 94:180–189

Niemela J, Koivula M, Kotze DJ (2007) The effects of forestry on carabid beetles (Coleoptera: Carabidae) in boreal forests. J Insect Conserv 11:5–18

Nieto A, Alexander KNA (2010) European red list of Saproxylic beetles. Publications office of the European Union, Luxembourg

Oleksa A, Klejdysz T (2017) Could the vulnerable great Capricorn beetle benefit from the introduction of the non-native red oak? J Insect Conserv 21:319–329

Pearce JL, Venier LA (2006) The use of ground beetles (Coleoptera: Carabidae) and spiders (Araneae) as bioindicators of sustainable forest management: a review. Ecol Indic 6:780–793

Puumalainen J, Kennedy P, Folving S (2003) Monitoring forest biodiversity: a European perspective with reference to temperate and boreal forest zone. J Environ Manag 67:5–14

Ranius T (2002) *Osmoderma eremita* as an indicator of species richness of beetles in tree hollows. Biodivers Conserv 11:931–941

Ranius T (2006) Measuring the dispersal of saproxylic insects; a key characteristic for their conservation. Popul Ecol 48:177–188

Ranius T, Korusuo A, Roberge J-M, Juutinen A, Monkkonen M, Schroeder M (2016) Cost-efficient strategies to preserve dead wood-dependent species in a managed forest landscape. Biol Conserv 204:197–204

Russo D, Cistrone L, Garonna AP (2011) Habitat selection by the highly endangered long-horned beetle *Rosalia alpina* in southern Europe: a multiple spatial scale assessment. J Insect Conserv 15:685–693

Sanchez-de-Jesus HA, Arroyo-Rodriguez V, Andresen E, Escobar F (2016) Forest loss and matrix composition are the major drivers shaping dung beetle assemblages in a fragmented rainforest. Landsc Ecol 31:843–854

Sands DPA (2012) Review of Australian *Phyllodes imperialis* Druce (Lepidoptera: Erebidae) with description of a new subspacies from subtropical Australia. Aust Entomol 39:281–292

Sands DPA, New TR (2013) Conservation of the Richmond birdwing butterfly in Australia. Springer, Dordrecht

Sebek P, Barnouin T, Brin A, Brustel H, Dufrene M et al (2012) A test for assessment of saproxylic beetle biodiversity using subsets of 'monitoring species'. Ecol Indic 20:304–315

Siitonen J, Saaristo L (2000) Habitat requirements and conservation of *Pytho colwensis*, a beetle species of old-growth boreal forest. Biol Conserv 94:211–220

Simberloff D (1999) The role of science in the preservation of forest biodiversity. For Ecol Manag 115:101–111

Solano E, Mancini E, Ciucci P, Mason F, Audisio P, Antonini G (2013) The EU protected taxon *Morimus funereus* Mulsant, 1862 (Coleoptera: Cerambycidae) and its western Palaearctic allies: systematics and conservation outcomes. Conserv Genet 14:683–694

Tikkanen O-P, Martikainen P, Hyvarinen E, Junninen K, Kouki J (2006) Red-listed boreal forest species of Finland: associations with forest structure, tree species, and decaying wood. Ann Zool Fennici 43:373–383

Ulyshen MD, Zachos LG, Stireman JO, Sheehan RN, Garrick RC (2017) Insights into the ecology, genetics and distribution of *Lucanus elaphus* Fabricius (Coleoptera: Lucanidae), North America's giant stag beetle. Insect Conserv Divers 10:331–340

Chapter 7
Conservation Versus Pest Suppression: Finding the Balance

Keywords Alien species · Forest dieback · Forest Lepidoptera · Outbreaks · Pest management · *Pinus radiata* · Pollinators

7.1 Introduction: Key Concerns and Resources

Exploitation of forests for economic gain leads to measures to improve and increase amounts and quality of the timber and timber products extracted, whether these are from native or alien tree species. In plantation or more natural conditions, those measures include countering the impacts of pests and diseases, with control or management of these a continuing, and sometimes predictable and substantial, need in forest management. That suppression, often equated with causing mortality of pest insects, must take place in milieux in which native insects occur and be of conservation interest either as diverse communities of localised taxa characteristic of particular floristic regimes or – more rarely – as individual species signaled as of concern.

A more widespread general concern is simply that broad-based pest management, for example by aerial applications of non-specific pesticides being broadcast across areas whose resident insects are poorly documented and in which the vulnerabilty of many native taxa (some perhaps related closely to the target pest and occurring with them) is not known. Non-target impacts of pest management campaigns to control alien pest species (such as through release of alien classical biological control agents), and many of those pests attacking commercially desirable tree species, are a major concern. Through 'spillover' or unanticipated non-specificity, possibility of movements of such agents to native environments and species can only rarely be discounted, and continue to cause concerns, despite increasingly rigorous screening of all potential agents before they can be approved for release.

Harmonising pest management and insect conservation poses problems in many contexts in which human economic priorities, essentially protecting supplies of food or commodities such as timber, can come into conflict with conservation ideals. Many concerns are couched in only rather general terms but situations such as (1) spread of alien species into native forests in their receiving environments and

© Springer International Publishing AG, part of Springer Nature 2018
T. R. New, *Forests and Insect Conservation in Australia*,
https://doi.org/10.1007/978-3-319-92222-5_7

novel competitive or aggressive interactions with native insects there (New 2015), and (2) use of non-specific chemical or biological pesticides that may drift or be carried into contact with native species or which may affect such species co-occurring with the target pest, each raise possibility of ethical and practical conflicts. Neumann and Marks (1990) emphasised the importance of cultural controls for insect pests in Australian pine plantations, but also noted that exotic conifers 'will continue to be prime targets for accidental introduced pests and diseases'.

Forest management affects risk of damage from pests in many ways, not least by influencing the conditions under which potential pests can thrive. Studies on beech trees (*Fagus sylvatica*) in Europe and levels of infestation by the Beech scale insect (*Cryptococcus fagisuga*, whose presence is linked strongly with fungal necrosis of the trees) implied stronger dependence on tree canopy structure and tree age than on management intensity (Kohler et al. 2015). Implications for management included that probability of scale infestations might be reduced by reducing gap size in the forests.

In addition to direct control measures to counter forest pest infestations, measures to predict and prevent spread of forest pests may also occur. The predicted advancing range edge of the introduced Gypsy moth (*Lymantria dispar*, Lymantriidae), a serious tree defoliator in North American forests, into the Cumberland Plateau region of Kentucky where it could threaten ecologically sensitive species in local oak forests, could enforce need for management that would influence some notable local arthropods (Rieske and Buss 2011). Non-target effects of methods used widely to control *L. dispar* were assessed by aerial spraying applications of diflubenzuron (non-specific, as a chitin-inhibitor, and with many records of non-target impacts on canopy and litter arthropods) and the biopesticide *Bacillus thuringiensis* var. *kurstaki* (specific to Lepidoptera, with prior records of non-target impacts on native Lepidoptera and their natural enemies), and noting responses of litter and ground-dwelling arthropods, as well as the arthropod communities in nearby water bodies. Short-term impacts were reported in most of the terrestrial arthropod groups sampled by pitfall trapping, in either the application year of the following one.

7.2 Alien Species

In principle, each alien tree species used for forestry plantations, in Australia or elsewhere, has potential to be attacked by insects from the tree's area of origin or previous introduced range, and/or by native insects that extend or switch their host range to incorporate the novel resource. In parallel, the former, with consumers introduced with the tree or arriving subsequently, have potential to move to native tree hosts and become invasive. Either direction of this 'two-way traffic' creates novel ecological interactions, some of which are likely to be of ecological or conservation concern. Some native Australian Lepidoptera, for example, have become sporadic pests of *Pinus radiata* plantations, and can undergo periodic damaging outbreaks there, perhaps in response to superabundant food supplies in monoculture even-aged plantations. Some outbreaks are associated with poor tree condition induced by weather conditions and/or silvicultural practices, and generally subside rapidly. Indeed, *P. radiata* plantations in Australia support a variety of native insect herbivores. However, those pest moth species are a small proportion of those recorded from *P.radiata* in Australia – by 1993, about 70 species had been reported (Britton and New 2004), and included some notable host transfers such as among the regionally endemic Anthelidae, as an isolated local lineage from which seven species had been found on pines. The main pest species included representatives of Geometridae, Oecophoridae, Tortricidae, Psychidae and Lymantriidae, and all are polyphagous taxa that have adopted pines as an addition to their already broad host range. For some, pines seem to be an inferior host, as explored for two species of *Chlenias* (Geometridae) in Victoria (Britton and New 2004), for which growth and development on pines was below that on two taxonomically disparate native hosts (*Eucalyptus obliqua, Acacia mearnsii*). Despite the ability of some native insects to thrive on alien pines, it remains unknown how many more specialised species, unable to make such a switch to a novel host, have been lost through losses of native forests for plantation establishment.

Native insects can thus reach outbreak pest levels on introduced tree species, but trees closely related to native trees tend to attract more native colonisers, whether generalists or specialists (Dalin and Bjorkman 2008). Their example, of Lepidoptera colonising the North American lodgepole pine (*Pinus contorta*) in Sweden and England, reflected the different pools of potential colonisers in the two countries, and only three of the total 39 colonising species were recruited in both countries. Most colonisers in England were generalist species from the background moorland habitats, whilst those in Sweden were mainly specialist species from local conifers. Almost every 'monophagous'lepidopteran (nine of the 10 species) feeding on Scots pine in Sweden colonised *P. contorta*, a trend paralleled by the sawflies (Symphyta), of which 11 of the 13 species on *P. sylvestris* colonised this novel host. More generally, specialist feeders on close relatives of the introduced trees are likely to be in some ways pre-adapted to switching or expanding their host range when opportunity occurs.

Key forestry pests, requiring interventionist or preventative management, may thereby comprise alien and native species in both managed and more natural forests. There is clear potential for the latter to be invaded by alien pests, creating novel interactions and possible threat to native species present. The 'switching' of Australian native herbivore insects to feed on *Pinus radiata* exemplifies capability to exploit host plants far different from any previously available.

It is far more common, of course, for insects to move to other tree species closely related to their normal hosts: most pests of *Eucalyptus globulus* in Western Australian plantations, for example, are insects found on native eucalypts in the region (Loch and Floyd 2001). More unusually, the African black beetle (*Heteronychus arator*, Scarabaeidae), normally a pasture pest, also causes severe damage to *E.globulus* there, when adult beetles stress or kill seedlings by girdling them near ground level. However, Monterey pine is also attacked by alien pest insects, most notably the woodwasp *Sirex noctilio*, that causes immense economic losses through killing trees (p. 66). Vulnerability to the wasp may be increased by previous activities by moths – in Tasmania, severe defoliation of pines (with complete defoliation over about 300 hectares) by an outbreak of *Chlenias* sp. over three years (1968–1971), led to the death of many trees and predisposed many surviving trees to woodwasp attack (Madden and Bashford 1977).

The converse 'flow', of native Australian insects becoming forest pests, mainly of Australian trees, in other parts of the world, and some of the species involved themselves being biological control agents imported deliberately to counter pest herbivores or to attack pest tree species also raises many relevant issues and the knowledge gained continues to augment that available from within-Australia studies. Thus, eucalypt plantations anywhere in the world can be attacked by both native and alien insects, with impacts ranging from severe to innocuous; many native insects may simply be visitors to those plants, as is the case with many Australian insects found in Australian pine plantations (p. 40). Moths in *Eucalyptus* plantations in Minas Gerais (Brazil) were sampled by light-traps over five years, yielding a collective 1356 species of which 29 were regarded as primary (12 species causing severe damage when populations were high) or secondary (17 species occurring in lower numbers but sometimes associated with primary pests) pests. The other species lacked any defined pest importance or, simply, could not be identified readily (Zanuncio et al. 1998). The two pest categories declined in relative abundance from 39.4% of catches (year 1) to 24% (year 5), perhaps related to growth of the initially young and susceptible trees. The trends of (1) relatively small proportions of the total samples being pest species and (2) importance of Geometridae as pests (the most abundant pests were *Stenalcidia grosica* and *Glena unipennaria*) parallel pine plantation findings for Australia.

Eucalypts have increased in extent in New Zealand since they were introduced about 170 years ago and, as well as for timber and woodchip production plantations, have older uses as firewood and shelter or amenity trees. Since the 1860s, however, they have been colonised by Australian insects, many of them crossing the 1800 or so Km from the Australian mainland on wind systems, and many others by assisted transport through trade and tourism, or by accidental introductions in produce. Withers (2001) noted that many of these species were reported initially from around Auckland, the country's most important trade hub. About a third of the 26 eucalypt-feeding specialist insects have become sufficiently serious pests to warrant control, with some managed through classical biological control. A total 57 Australian insect species may feed on eucalypts in New Zealand, and Withers forecast that this number will continue to increase.

The variety of interactions, novel or re-established in different evolutionary and ecological contexts, furnishes the basis for an equally broad array of impacts and outcomes, many with direct effects on native forest insects.

The resulting novel ecological interactions can become complex. Two introduced invasive Australian psyllid species (*Glycaspis brimblecombei, Eucalyptolyma maideni*) have become important pests of eucalypts in California, where their copious honeydew production facilitates predominance of invasive ant species such as

Argentine ant, *Linepithema humile*. The abundance of arthropods on ground beneath the trees was associated positively with ant and psyllid abundance (Jones and Paine 2012). One inference from this is that increased ant activity on the trees may hamper biological control attempts against the psyllids, the ants disrupting parasitoid wasps and lowering their attack rates, and at the same time exacerbating the pests' impacts. Another is, simply, that increased ant abundance can favour native arthropods on ground beneath the tree canopies.

7.3 Ecological Patterns

7.3.1 Pollination Systems

Disruption of long-coevolved functional patterns between native forest trees and forest insects is exemplified by the large radiations of Australian native bee pollinators associated with eucalypt forests. Widespread use of the imported European honeybee (*Apis mellifera*) for crop pollination and honey production, and commonly dependent on 'migratory beekeeping' with hives moved to track seasonal nectar supplies (such as in the Jarrah [*Eucalyptus marginata*] forests of Western Australia) and crop pollination needs have created concerns over such measures increasing alien bee effects in native forest, with consequent competitive impacts on native bees and increasing their vulnerability as access to key resources is thwarted. They link with concerns over alien predatory vespoid wasps colonising native forests as generalist predators, a syndrome studied most intensively in New Zealand (Beggs et al. 1998) where dramatic effects on native insects and trophic interactions have been described.

Maintenance of pollination systems is a key need in preventing or slowing rates of plant extinctions in changed environments (Pauw 2007), so that the assured presence of the most significant pollinator groups such as native bees is critical. Strong changes in bee assemblage composition were linked to forest fragmentation in Costa Rica (Brosi et al. 2008). Bee diversity and abundance did not respond greatly, but the tree-nesting meliponines were associated with larger fragments, greater proportions of forest around the sampled plots, and smaller edge:area ratios. Meliponines, the only eusocial bees in the areas surveyed, require forest for effective conservation – as do the euglossines (orchid bees). Even small forest fragments benefit these important pollinators. Taxon-specific differences in responses to land use change were clear: in pastures, euglossines were absent, and the introduced *Apis mellifera* was abundant.

Forest pollinator communities benefit from open canopies, moderate levels of shrub cover leaving some open ground, and increased diversity of herbaceous plant cover. Such areas can be created by a variety of forest management practices, including fire, physical removal of shrubs, and harvesting or thinning, to create gaps (Chap. 10). Hanula et al. (2015) found that thinning combined with shrub control provided good habitat for bees, with both richness and abundance lowest in dense pine forest and highest in clearcut areas. In comparisons of bees across several forest types in Georgia (United States), higher total basal area of overstorey trees was correlated with decreased bee diversity, and 12 of the 128 bee species captured were indicators of forest type – most for recently cleared pine forest.

More widely, the conservation of pollinators, notably bees and butterflies, in forests is important both for the plant species continuity and as reservoirs for restoration/production in surrounding altered habitats. Need for pollinator conservation in managed forests is recognised widely (Hanula et al. 2016). Both these insect groups in North America tend to be more abundant and/or rich in open forest conditions, so that creation of that generalised structure by interventions may be valuable – but with the interventions also potentially harming rare species with small scattered or localised populations.

7.3.2   Dieback

The term 'dieback' has been used widely to designate losses of eucalypts in Australia, and is applied commonly to deaths of individual trees, patches of trees or decline in condition in both pastoral and forest environments. The process follows the general sequence for rural dieback. That sequence (Heatwole and Lowman 1986) commences with initial foliage dieback on the outer growth of the trees, which progresses gradually towards the trunk, with dead branches then protruding beyond the foliage. Most branches die, but epicormic growth on and close to the trunk provides a flush of fresh foliage before this, too, dies and trees are wholly dead. Causes are varied and complex, and broadly involve 'complex interactions between biotic and abiotic factors' (SoFR 2013). Large-scale defoliation by native insects, most usually scarab beetles, Lepidoptera (perhaps most notably by the Gum-leaf skeletoniser, *Uraba lugens* [Nolidae]) or psyllid bugs, is often implicated as contributing to the process. However, the role of such insects, and of fungi (notably *Phytophthora cinnamomi* and *Armillaria luteobubalina*), is often assessed as contributory rather than being the primary cause of dieback.

The complexity of causes of such eucalypt declines (Jurskis 2005) embraces that trees stressed by some form of environmental change/s, including poorly drained soil, eutrophication, human management, and others may have increased susceptibility to a variety of insect and fungal attack, perhaps by increased attraction and nutritional value of foliage for insect herbivores. Many of the 'theories' on the causes of dieback were discussed by Heatwole and Lowman (1986), who emphasised the needs for investigative studies to ascertain the real roles of taxa such as scarab beetles and others in the 'dieback syndrome'. They also differentiated 'rural dieback' and 'forest dieback', which appears distinct in that the former is more the outcomes of stress from pressures of land clearing, whilst local damage (such as logging) in forests may attract defoliating insects to the stressed trees. That dichotomy is oversimplistic, and a wide range of dieback syndromes occur - more

Fig. 7.1 Possible eucalypt dieback scheme depicted by Jurskis and Turner (2002), and indicating the interacting processes and management influences that are linked with tree responses and fate

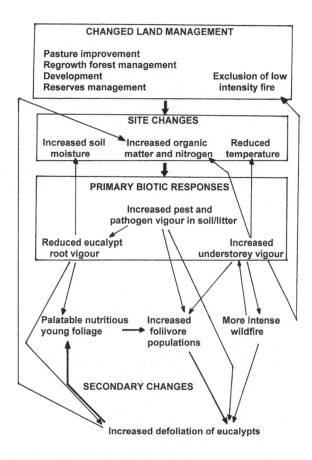

generally, dieback of eucalypts is widespread and its progress is complex – but with repeated defoliation by insects implicated frequently in its occurrence, and their attack probably facilitated by 'stress' in some form. Initially, many of the distinct categories of dieback were attributed directly to pathogens (above), but are now recognised as outcomes of varied physiological stresses from a variety of causes (Ciesla and Donaubauer 1994). Exploring causes of eucalypt dieback in both forest and pastoral environments, and affecting a range of different tree species and with symptoms/causes including intensive defoliation by insects, predators of insects (most notably feeding on psyllids by the bell miner bird, *Manorina melanophrys*) or foliivory by koalas (*Phascolarctos cinereus*) adding to the complexity. A number of schemes have been advanced in attempts to interpret the syndrome. That discussed by Jurskis and Turner (2002) (Fig. 7.1) exemplifies the many participating and inter-actng factors that may become involved.

Amongst those factors, the complex roles of fire in promoting forest health in Australia, and subsequent manipulations of burns for forest conservation management are intricate and remain controversial. 'Unnatural' fire regimes, some a deliberate consequence of forest management and constituting a disturbance to which the affected forests may not have become adapted, have led to declines of some eucalypts, and changed resource base for their consumers (Jurskis 2005). Jurskis noted that reduced use of low-intensity fires has led to declines in forest health, increased disease and pest problems, invasions by understorey plants, and build-up of fuel to lead to more extensive high-intensity fires. Excluding fire from forests can lead to tree stresses and physiological changes that render them more susceptible to pest attack – so that several groups of phytophagous insects increase markedly in local declining eucalypt forests, and accelerate tree decline – Jurskis (2005) summarised such occurrences for scarabaeid and chrysomelid beetles, psyllids, sawflies and a range of Lepidoptera, so that pest outbreaks may be integral components of dieback, and also link directly with changed soil and subsoil conditions contingent on the fire regime.

References

Beggs JR, Toft RJ, Malham JPO, Rees JS, Tilley JAD, Moller H, Alspack P (1998) The difficulty of reducing introduced wasp (*Vespula vulgaris*) populations for conservation gains. N Z J Ecol 22:55–63

Britton DR, New TR (2004) Exotic pine plantations and indigenous Lepidoptera in Australia. J Insect Conserv 8:263–274

Brosi BJ, Daily GC, Shih TM, Oviedo F, Duran G (2008) The effects of forest fragmentation on bee communities in tropical countryside. J Appl Ecol 45:773–783

Ciesla WM, Donaubauer E (1994) Decline and dieback of trees and forests. FAO Forestry Paper no 120, Rome

Dalin P, Bjorkman C (2008) Native insects colonizing introduced tree species – patterns or potential risks. In Paine TD, Lieutier F (eds) Insects and diseases of Mediterranean forest ecosystems. Springer, Cham, pp 63–77

Hanula JL, Horn S, O'Brien JJ (2015) Have changing forests conditions contributed to pollinator decline in the southeastern United States? For Ecol Manag 348:142–152

Hanula JL, Ulyshen MD, Horn S (2016) Conserving pollinators in North American forests: a review. Nat Areas J 36:427–439

Heatwole H, Lowman M (1986) Dieback. Death of an Australian landscape. Reed Books, Sydney

Jones ME, Paine TD (2012) Associations between invasive eucalyptus psyllids and arthropod litter communities under tree canopies in southern California. Entomol Exp Appl 143:280–291

Jurskis V (2005) Eucalypt decline in Australia, and a general concept of tree decline and dieback. For Ecol Manag 215:1–20

Jurskis V, Turner J (2002) Eucalypt dieback in eastern Australia: a simple model. Aust For 65:87–98

Kohler G, Pasalic E, Weisser WW, Gossner MM (2015) Beech forest management does not affect the infestation rate of the beech scale *Cryptococcus fagisuga* across three regions in Germany. Agric For Entomol 17:197–204

Loch AD, Floyd RB (2001) Insect pests of Tasmanian blue gum, *Eucalyptus globulus globulus*, in South-Western Australia: history, current perpectives and future prospects. Austral Ecol 26:458–466

Madden JL, Bashford R (1977) The life history and bahaviour of *Chlenias* sp., a geometrid defoliator of radiata pine in Tasmania. J Aust Entomol Soc 16:371–378

Neumann FG, Marks GC (1990) Status and management of insect pests and diseases in Victorian softwood plantations. Aust For 53:121–144

New TR (2015) Alien species and insect conservation, Cham. Springer

Pauw A (2007) Collapse of a pollination web in small conservation areas. Ecology 88:1759–1769

Rieske LK, Buss LJ (2011) Effects of gypsy moth suppression tactics on litter- and ground-dwelling arthropods in the central hardwood forests of the Cumberland Plateau. For Ecol Manag 149:181–195

SoFR (State of the Forests Report) (2013) Department of Agriculture, Australian Bureau of Agricultural and Resource Economics and Sciences, Canberra

Withers TM (2001) Colonization of eucalypts in New Zealand by Australian insects. Aust Ecol 26:467–476

Zanuncio TV, Zanuncio JC, Miranda MMM, de Barros Medeiros AG (1998) Effect of plantation age on diversity and population fluctuation of Lepidoptera collected in *Eucalyptus* plantations in Brazil. For Ecol Manag 108:91–98

Chapter 8
Saproxylic Insects and the Dilemmas of Dead Wood

Keywords Biological legacies · Boreal forests · Dead wood succession · Ecological traps · Logging residues · Red-listed species · Salvage logging · Tree stumps · Woody debris

8.1 Introduction: The Conservation Significance of Dead Wood

Trees die, branches fall, and forestry operations generate discarded branches, remnant stumps, logs and smaller woody materials that may be abandoned or for which harvest is delayed. Changing attitudes toward woody debris in forests have increasingly incorporated considerations of its wider values in conservation. It has thus moved from the earlier perspective of being considered simply as 'a symbol of waste' that had to be removed in forest harvest (Harmon 2001), to realisation of need to sustain supplies across the entire decomposition process – sometimes involving what Harmon (2001) called 'morticulture', planning for those resources to be available.

'Dead wood has been identified as a crucial component for forest biodiversity' (Jonsson et al. 2005), and growing appreciation of its importance has led to its active consideration in forest management. Whilst numerous insects depend broadly on dead wood in some form, many have very specialised needs, and creation of regimes to supply those needs can become complex. Many of those species are naturally rare, poor dispersers and characteristic of forest interiors and, as Jonsson et al. noted, creating suitable habitat for these is a considerable challenge. Because of the diversity and ecological variety of needy insect species, any single embracing general prescription will be inadequate, and four more general goals for management for saproxylic insects in forests (after Jonsson et al. 2005) are (1) to counter the widespread shortage of dead wood availability; (2) to plan at landscape scale, to reduce edge effects and isolation; (3) to create a variety of dead wood species, sizes and ages; and (4) to use whatever quantitative analytical tools, such as modelling, are available to help this.

© Springer International Publishing AG, part of Springer Nature 2018
T. R. New, *Forests and Insect Conservation in Australia*,
https://doi.org/10.1007/978-3-319-92222-5_8

Manipulation of dead wood supply can affect many saproxylic insect species but, as Gibb et al. (2006a, b) commented, many such manipulations have been undertaken optimistically but with little scientific evidence of benefits occurring. Benefits of both logs and snags, for example may link strongly with their associated fungal biota and not become evident for many years. Renewed focus on conservation values of coarse woody debris, in particular, has concentrated on three main topics, as (1) understanding patterns of availability of dead wood in relation to forestry effects, stand dynamics and the disturbances that are imposed or occur naturally; (2) the role of the woody debris in nutrient dynamics; and (3) its importance for the large array of wood-dependent species, most notably saproxylic taxa (Jonsson and Kruys 2001; Grove and Hanula 2006). In addition to the abundant information on the terrestrial saproxylic insects that prevail in many forest entomology considerations, immersed wood provides refugia and food for a considerable array of aquatic macroinvertebrates (Cranston and McKie 2006), spanning representatives of Coleoptera, Diptera, Plecoptera and Trichoptera. Some are restricted to particular species or decay stages so, although Chironomidae have adopted the habit across three of 10 subfamilies, the requirements of different taxa differ considerably.

Although insects associated with dead wood, predominantly truly saproxylic taxa, have received far more attention, subcortical insects, spiders and other invertebrates are also important components of the biodiversity often linked (and overlapping) with dead wood. The high incidence of dehiscent bark characteristic of many eucalypts in Australia constitutes habitat and retreats for numerous such species. As such flammable bark is a substantial component of forest fuel, many such invertebrates may become susceptible. However, a comment on bark-dwelling spiders of northern boreal forests, that bark habitats are crucial and 'must be considered central to maintenance of spider diversity' (Pinzon and Spence 2010) can be aptly applied to many other taxa and places in both managed and unmanaged forest. Flat bugs (Hemiptera, Aradidae) feed on fungi under bark, many of them mostly on dead or dying trees, and some species associated clearly with burned forest (Sweden: Johansson et al. 2010). Several aradid bugs, including two red-listed species, were primarily or predominantly found in burned stands (Hagglund et al. 2015), leading to suggestion that creating dead wood, as well as conducting prescribed burns, might benefit the bugs.

The relative importance of fallen dead trunks in forests (length 1.5 m, diameter >20 cm) and fallen dead limbs (branches 1–1.5 m long, diameter 5–10 cm, with no side branches) for saproxylic insects was compared for beech, *Fagus sylvatica*, in Sweden (Schiegg 2001). Eclector trap catches collectively yielded 426 species of saproxylic Diptera and 228 species of Coleoptera. The marked differences in richness from either trunks or limbs (Table 8.1), with predominance of diversity in limbs, was unexpected, and most pronounced amongst Coleoptera for those species that were not host plant specific (Fig. 8.1). Overall, the high proportion of both orders breeding in dead *Fagus* limbs clearly endorsed the value of tree limbs as a component of coarse woody debris.

Saproxylic insects are abundant and diverse in many forest types, but their importance in conservation has developed largely from studies on temperate forest

Table 8.1 The numbers of species and individuals collected from unit amounts (1.15 m³) of trunks and limbs of beech (*Fagus sylvatica*) in Switzerland (Schiegg 2001)
Numbers in parentheses are those obtained when all trunk eclector catches, total volume of 4.66 m³, are considered

	Trunks	Limbs
Diptera		
No. species	167 (305)	347
No. individuals	3165 (14452)	15,411
Coleoptera		
No. species	70 (153)	182
No. individuals	737 (2620)	2286

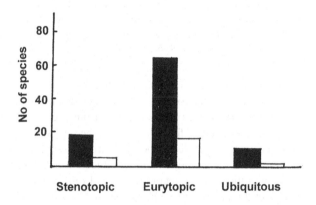

Fig. 8.1 The numbers of species of Coleoptera collected from 1.15 m³ samples of trunks (open) and limbs (black), separated by ecological categories as stenotopic, eurytopic and ubiquitous. (Schiegg 2001)

fauna. In advocating the need for saproxylic insects to also become a focus for conservation of tropical forests, Grove and Stork (1999) listed several key questions in need of research to augment awareness of the impacts of logging practices available from elsewhere (Table 8.2). All these remain highly relevant and some remain difficult to achieve. The initial points (taxonomic impediment, lack of species-level ecological knowledge, uncertainty over sampling approaches and validity) in principle appear straightforward but are formidably difficult to advance constructively. Points four-six are again hampered by the reality that the saproxylic insects of a managed tropical forest may never become known adequately, so that specific measures to assure their needs are correspondingly incomplete. All these matters need to be addressed before any real confidence in predicting long-term impacts of logging is achieved. Lack of detailed knowledge dictates that some precautionary management is wise, but may become difficult to justify without hard evidence of its benefits. Notwithstanding their widespread values in forest insect conservation issues, sampling saproxylic insects reliably is not always easy. The three main categories of sampling methods are (1) direct searching methods (such as bark stripping and beating dead wood for hand-collecting the insects found); (2) rearing techniques (using emergence traps on logs or trunks, or enclosing samples of dead

Table 8.2 The key questions posed in formulating a research agenda to conserve saproxylic insects in tropical forests (from Grove and Stork 1999)

1. Can suitable techniques be found to effectively sample and enumerate the saproxylic insects of tropical forests?
2. Do logged forests actually differ in their saproxylic insect assemblages from old-growth forests?
3. If there is an effect of logging, is this apparent in the short-term or does it take several logging cycles to develop?
4. If logging is sufficiently benign to enable survival of the tropical 'charismatic megafauna', is that sufficient to sustain tropical saproxylic insects, or is there no simple correlation?
5. Can researchers sufficiently understand what are the key dead wood habitats in tropical forests to say which habitats and associated insects are likely to be under particular threat from logging?
6. Can forestry growth and yield models be extended to model the long-term supply of overmature trees and dead wood resources under a range of silvicultural options?
7. Does any logging-induced change in saproxylic insect assemblages affect the normal breakdown of dead wood and recycling of nutrients?

wood); and (3) mass trapping. All discussed by Leather (2005), each has many inconsistencies with, for example, design details of flight-intercept traps having considerable influences on the catches (Bouget et al. 2008).

The northern boreal forests have become a major focus for conservation attention to insects frequenting dead wood as saproxylic species, amongst which beetles have become the most prominent study subjects in seeking to harmonise conservation needs with those of production forestry. Logging and wood harvesting are the predominant threats to saproxylic beetles in Europe, directly affecting some 232 species – amongst which 35 of the 57 threatened beetle species in the region are represented (Nieto and Alexander 2010). A variety of other threats were distinguished, but with the next most important (namely, agricultural intensification and expansion, urban sprawl) affecting only 25 and 26 species, respectively, but also with the major threats to 86 saproxylic beetles in the European Union area still unknown. Loss and decline of dead wood, with veteran trees a component of this, has become a major concern allied with the needs to assure continuity of old-growth forest. More broadly in Europe, the richness and high number of threatened species has endorsed the values of saproxylic beetles as a key focal group for conservation – altogether, nearly 11% of the 436 species (across 21 famiies) assessed were considered threatened in Europe, and a further 13% were 'Near Threatened' (Nieto and Alexander 2010). However, even for this well-studied group, more than a quarter of species (28%) remained Data Deficient, and the risks of their extinction could not be properly estimated. Nieto and Alexander (2010) summarised the main threats to consider as (1) loss of habitat through logging and wood harvesting; (2) decline of veteran trees (p. 205); and (3) lack of land management that promotes recruitment of new tree generations. These are augmented by a variety of more local and so more specific threats, such as fire regimes. All these factors are widespread elsewhere, but are perhaps particularly evident in the well-documented European forest landscapes with their long history of cultural changes. Comparative detail for

Australian taxa would immensely aid understanding of focused priorities for forest insect conservation.

The attention given to saproxylic Coleoptera reflects that they are relatively well known, with a substantial taxonomic and ecological framework for many families. In Finland, with about 800 saproxylic beetle species across some 60 families, they are nevertheless probably outnumbered by both Diptera and Hymenoptera, each with an estimated 500–1000 wood-associated species (Siitonen 2001). Both these orders, however, are far less understood than the beetles, and taxonomic problems that impede interpretation persist among groups such as Mycetophilidae (with an estimated 200 species of these fungus gnats associated with fungi on or in dead wood) and large groups of parasitoid wasps that attack saproxylic hosts. In Fennoscandia, again, many bark beetle species are known to each host 10 or more wasp parasitoids.

Although incomplete, Nieto and Alexander's overview remains a highly informative status statement, authoritative because the extensive workshop reviews undertaken during its preparation enabled status comments and allocations to be based on strong specialist review and consensus and, so, likely to be accepted widely. The richest areas of Europe for saproxylic insects are remnant old-growth forests rather than natural forests, and that old-growth habitat has until relatively recently persisted on 'wood pasture' now increasingly threatened by agriculture and development. Veteran trees (p. 205), especially those with some form of hollowing or heartwood decay, are amongst the most important habitats for saproxylic beetles in Europe. The Red List assembled from Nieto and Alexander's work demonstrated that nearly 14% of the saproxylic beetles they discussed had declining populations, and 57% had unknown population levels, and many of these were found only in the decaying heartwood of veteran trees.

In a few cases, and in a scenario likely to become more frequent, re-introductions or translocations of key species with limited dispersal capability to new habitats – including isolated veteran trees – may occur, as discussed for *Cerambyx cerdo* (p. 122) by Drag and Cizek (2015). In some cases, re-introductions may seem the only way to prevent extirpations by loss of isolated populations, but the fate of re-introductions of forest-associated insects of any kind is rarely reported.

Modelling of extinction risk of forest saproxylic beetles in Europe helped to suggest priority needs for conservation management (Seibold et al. 2015), namely (1) that lowland forests should be set aside for conservation in preference to the more remote montane forests which, in any case, are usually less exploited and have intrinsically greater conservation significance; and (2) the amount of large diameter dead wood, dead wood of broadleaved trees, and dead wood in sunny areas should be increased. Those proposals arose from modelling extinction risk of 1025 saproxylic beetles, collectively indicating higher risks for lowland and larger species, and for taxa that occur mostly in large diameter dead wood, in broadleaved tree species, and in open canopy areas. Such broad guidelines need investigation to determine their values in other regions. Values of dead wood extend beyond the closely associated saproxylic taxa. Presence of coarse woody debris enhanced species richness and abundance of litter-dwelling fungivorous and carnivorous beetles in Slovakia (Topp et al. 2006), and its lack might lead to local extinctions among those guilds.

Site conditions favourable for saproxylic insects are often difficult to appraise comprehensively. Site-specific temperature conditions may need to become more closely linked with management of dead wood supply, for example (Muller et al. 2015) with realisation that increasing temperatures may help to compensate for poor habitat conditions in production forests. Evidence for these assumptions, from extensive surveys of saproxylic beetles in Europe, including 791 species (excluding Staphylinidae) and using several sampling methods, demonstrated that temperature needs of the beetles should be considered more effectively as a variable linking with diversity and with supply of dead wood. Indirect effects of temperature can become complex – such as by changed composition of wood-infesting fungi along temperature gradients, and alterations to local food webs from changed relative abundance or incidence of participant taxa. However, the relative influences of increased temperature and increased dead wood are often unclear with, as above, some implications that increased dead wood supply can override warming effects. The suggestion by Muller et al. (2015) that supply of dead wood should be enhanced more in colder climates has widespread implications for saproxylic insect conservation management. A corollary is that lower amounts of dead wood may be less critical under warm than cold conditions. A further 'warning' from these authors is that old veteran trees (p. 205) may be more susceptible to climate warming than are younger 'fitter' trees, a condition regarded as a new threat to some relict sites that support notable and highly threatened species.

8.2   Coarse Woody Debris

The roles of logging residues and other coarse woody debris (CWD) such as fallen branches and fallen dead trees in forests are key conservation concerns for many invertebrates, and the fate of this resource is sometimes controversial. This, and the other major dead wood component, standing dead trees, each provide a succession of decay stages on which particular invertebrates may specialise, and depend, but which are frequently subject to 'sanitation' measures in practical forest management. The characteristics of CWD may differ according to the disturbance that produced it, but this variety has sometimes been overlooked in appraising its suitability for insects. In Canada's boreal forests the CWD in clearcut areas was largely made up of small pieces of recently downed wood, plus logs and stumps, but in burned areas was largely of standing wood in early decay stages (Podlar et al. 2002). The amount and type of CWD in either category depends on time since disturbance, but the initial differences found endorse wider needs for mosaic management of production forest for conservation.

Broadly, CWD can be defined as 'the standing and fallen dead wood in a forest' (Woldendorp and Keenan 2005), and has varied ecological roles. It participates in nutrient cycling, provides refuges for organisms and sites for seed germination and, of primary relevance here, substrates and habitat for a wide range of insects and other animals, fungi and microbes. The smaller fractions (fine woody debris, p. 175) of the forest debris include bark, twigs and small branches, fruits, foliage and other 'leaf litter' constituents. CWD is an important nesting habitat for forest ants in Canada (Higgins and Lindgren 2006), as a key component of maximising daily heat gain and as the best of the three common nesting strategies in boreal forest – namely (1) nesting under rocks, (2) constructing thatched nests from forest litter, and (3) nesting in dead wood. Nine of the 11 ant species in *Pinus contorta* stands in British Columbia nested in CWD, with one of the other two species known elsewhere to use CWD and soil nesting. Within harvested stands, ants used stumps more than fallen debris, with an average of 61% of stumps occupied. Relatively low numbers of ants were found in non-harvested stands.

The amounts and persistence of CWD – regarded widely as fuel for wildfires and likely to exacerbate their impacts – has led to formulation of forestry-based 'thresholds' of amounts that can be tolerated, and its targeting for reduction when such levels are exceeded. In Australia, Woldenkorp and Keenan (2005) noted CWD levels ranging from 19 t/hectare^{-1} in woodland to 134 t/hectare^{-1} in tall open forest, and comprising substantial proportions of above-ground biomass. They quoted approximately 18% (open forest), 16% (tall open forest), 13% (rainforest), and 4% (eucalypt plantations) and with higher levels in alien conifer plantations sometimes reflecting the residual debris from earlier forest clearing. The considerable variations reflect features such as stand structure, management history, and legacy of disturbance (Woldendorp et al. 2002). Persistence also varies greatly, but large logs and snags (standing trunks) may take well over a century to decay, and support successive communities of saproxylic and fungivorous inhabitants over that time. Many of those species do not (and cannot) thrive elsewhere. 'Dead wood' is therefore an important and specific aspect of conservation of forest insects, and seeking a balance between its removal (for management, fuel reduction, or harvest) and its retention (as critical habitat and resources) can become controversial.

Rearing of saproxylic beetles from dead wood can provide samples that (1) are highly specific to the substrate species and decay stage and (2) disclose species only rarely collected by other means (Lee et al. 2014). Relatively distinct beetle assemblages form a successional gradient as decay proceeds, with each decay stage a critical resource needed for some of the species present, and whose presence may determine suitability for later-arriving taxa. Thus, for White spruce (*Picea glauca*) in Canada, each such stage is a critical need for some beetles (Lee et al. 2014), so that the legacy of forest use should ensure continuing availability of all decay stages – collectively over 50 years or more. Logs in advanced decay stages are needed, and should be actively protected during harvest and site sanitation.

Decay stages of the wood thus constitute a succession pattern that influences which saproxylic beetles may colonise, so that the assemblage changes as each stage provides conditions suited to more, or different, species. The assemblages associated with dead snags of aspen (*Populus tremuloides*) and Black spruce (*Picea mariana*) in Quebec were compared along a gradient of four decay stages, using wood density to represent the decay gradient (Saint-Germain et al. 2007). All adults

Fig. 8.2 The successive phases in dead wood successions to show changes in the invertebrate communities exploiting dead wood; relative species richness of four communities (*A–D*) are shown in relation to period (years) from death of the tree. (After Ehnstrom 1979, as re-published in Heliovaara and Vaisanen 1984)

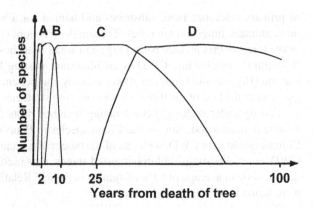

and larvae of Cerambycidae, Buprestidae and Curculionidae: Scolytinae (or Scolytidae) were collected and identified as precisely as possible. In spruce, Most Cerambycidae and Scolytinae (in all, 15 of the 17 beetle species) were found in early stages of decay, with the former including mostly 'stressed host' species, and few were found in the late decay stages. The converse pattern occurred in aspen, and stressed host species were absent. Saint-Germain et al. speculated on causes of those differences, noting possible involvement of plant defences and associated secondary compounds – but such differences urge the need for any generalisations on succession patterns to be made with caution. Habitat specificity amongst saproxylic insects links with specific responses to decay stage of the wood, with the implication that if such species are especially vulnerable to fragmentation, the distribution of that wood must be considered in conservation management. Of 175 species considered by Schiegg (2000a) in Switzerland, 30 were characteristic of high levels of dead wood connectivity – for them, conservation measures might include leaving varied pieces of dead wood with only short distances between them, so facilitating movements and, where relevant, recolonisation. A survey of eclector trap and water trap catches of Diptera also revealed the paucity of knowledge in even this well-studied fauna: 186 of the several hundred species (from the 348 to 511 species in comparative trapping programmes) were new to Switzerland, and more than 20 were new to science (Schiegg 2000b). Again, the importance of short distances between dead wood units was evident, but it is also salutary that the emphasis on studying Coleoptera has overshadowed the additional diversity of other insect groups, here Diptera, associated with dead wood.

The model of four sequential phases of insect communities successively exploiting dead wood as described for Sweden by Ehnstrom (Fig. 8.2, after Heliovaara and Vaisanen 1984) suggest analogous patterns elsewhere. The phases are (A) short term with species feeding on bark and in the cavities of those bark feeders, and their natural enemies; (B) also short, with most species subcortical or feeding in the surface layers of timber over the period in which bark loosens and falls; (C) longer term, up to several decades, with most species wood-inhabiting, and many species of this phase noted as having declined; and (D) a very long stage, over which wood-inhabiting species may give way to those living under the shelter of decaying logs as the wood breaks down.

In Sweden, de Jong and Dahlberg (2017) appraised the impacts of forest harvesting for biofuels and how this practice integrated with wide forest management for all species of conservation concern, in a study motivated by concerns that harvesting of logging residues in addition to clear-cutting might add to the overall negative impacts. Despite moves for retention of CWD through retention of logs and high stumps in Swedish forests, the amount of CWD is low in managed forests compared with old-growth forests. Their survey implied that the 25,000 Swedish forest species (all taxa) included 1880 red-listed species, and the 7000 saproxylic species included 795 of the above red-listed taxa. In principle, all wood-associated red-listed species depend on CWD, and loss of this resource is thus a major contributor to their status of conservation concern. Associated syndromes include (1) dramatic changes in habitat condition following clearcutting and, for some species, increased dependence on 'reservoir resources' such as CWD; (2) local extinctions caused by clearcutting pose increased difficulties in species re-colonising because of scarcity and increased dispersion of source habitats; (3) wind-blown trees are cleared and forest fires prevented in managed forests, so that habitats and successional stages created by such processes are also rare, especially for species with poor dispersal powers. The species of conservation concern in Sweden include numerous beetles, as the major focal group in many studies.

However, the general conclusion of de Jong and Dahlberg (2017), that the impact of harvesting logging residues is small to negligible compared with clearcutting and is thus most likely to affect mostly generalist species, may have wider relevance. Exceptions might occur, for example when large-scale residue extraction might harm more specialised species that depend on – for instance – deciduous trees that are already logged heavily in the local landscape. For most insect species, and in many forest environments, it is still unclear at what level extraction of logging residues could be undertaken without affecting significant species. A long history of interest and documentation of Scandinavian forest beetles and of forest use, with concerns over the impacts of harvesting logging residues from the 1960s on, allowed that context to be evaluated in considerable detail. De Jong and Dahlberg were able to evaluate data from 122 relevant studies in a longer perspective of clearcut forest practices covering much of the last century, so that many of the cited studies had a wide ecological and management context. Almost all forest in Scandinavia is in some way managed for timber production and, as Jonsell and Nordlander (2002) noted, to be able to preserve the threatened fauna of old-growth forests, one critical aspect is identifying those species that are harmed by forest management, and how they are affected.

Lepidoptera assemblages of North American deciduous forests were affected substantially by initial timber harvest, with (as anticipated) specialist species particularly susceptible (Parrish and Summerville 2016) as logging removed or reduced the supply of suitable host tree species. However, subsequent removal of CWD, whether complete or partial, had little additional impact (Fig. 8.3). The inferences were based on light trap catches (totaling 183 moth species) with attendant biases of

Fig. 8.3 Impacts of removal of coarse woody debris (CWD) on species richness of all Lepidoptera (black) and host plant specialist Lepidoptera (open) in three harvest regimes (complete and partial removal of CWD, and unmanaged controls) in Indiana, United states. (Parrish and Summerville 2016)

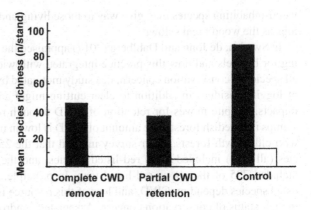

moth movements from adjacent forest stands, so that some interpretative ambiguity is inevitable. As Parrish and Summerville noted, the retention of fine woody debris (FWD, < 10 cm diameter) in their treatments may contribute suitable refuges for species that diapause in ground litter. Both living and dead woody components are essential to insect conservation in sustainable production forests – but in such forests, the structure of Lepidoptera assemblages might not depend on CWD, in contrast to those arthropods for which it is a primary resource (Hanula et al. 2006).

Similar inferences for moths in boreal forests of Ontario, where the most significant environmental correlates with moth community composition in disturbed forests included time since disturbance and the amount of woody debris and snags, also suggested that clear-felling and wildfire burning may have analogous effects (Chaundy-Smart et al. 2012). Some species were significantly more abundant in one or other regime, but the dominant species were similar in the two kinds of disturbance. Samples of moths from unlogged forests in Indiana (total of 212 species, of which 183 species were sufficiently abundant [with 20 or more individuals] for analysis) showed that most of the species were generalist feeders (Summerville 2014). The assemblages, however, were very sensitive to nearby timber harvesting. Specialist species were particularly affected by changes to food plant quality and availability. The impacts of timber harvesting in this study were not confined to the logging concessions: forest patches adjacent to these showed influences of herbaceous-feeding species and generalists after logging, and those effects extended at least 100 m into unlogged forest. Summerville suggested that the approach of using small harvest areas, as recommended widely for conserving biodiversity, might actually have the converse effect in unharvested remnants. Spatially isolated larger harvest concessions might reduce the net edge effects flowing from multiple forest openings from small coupes, and might more effectively conserve those moth species adapted to (and depending on) closed-canopy late succession forests.

The wider responses of moths to forest management in eastern North America are driven largely by changes in the plant community, with reductions in richness and changed assemblage composition correlated clearly with clear-cut harvest

Table 8.3 The key processes demonstrated to be predictors of fluctuations in forest Lepidoptera, based on North American studies (as listed by Summerville and Crist 2008)

Within and among tree crowns
Host-tree effects (leaf palatability, chemistry, position within canopy).
Relative abundance/frequency of host tree in stand/landscape.
Potential interactions between outbreak species and other Lepidoptera.
Density-dependent changes in host plant quality.
Long-term climatic variation.
Heterogeneity of forest type within regional landscape.
Oviposition behaviour, female dispersal ability, forest stand composition.
Among forest stands at local and regional scales
Floristic composition of forest stand, particularly herbaceous species richness and identity of dominant canopy tree.
Stand age, past management history, and canopy architecture.
Relative importance of invasive shrubs or herbs in forest understorey.
Floristic composition of habitat matrix surrounding forest stands, especially when stands are small.
Differences in regional biogeographic history among forested ecoregions.
Spatial geometry of forest stands within landscapes, particularly patterns of stand area and isolation.
Within managed forest ecosystems
Changes in tree species diversity within stands.
Changes in stand basal area within stand.
Changes in dominant canopy taxa within stand.
Proportion and composition of bole retained in managed landscape.
Spatial pattern of harvest in managed landscape.
Method of improvement cut used to regenerate forest stand.
Non-target effects of applications of agents used in biological control of defoliators.

(Summerville and Crist 2008). A considerable variety of ecological factors influence, and determine, the assemblage within managed forest (Table 8.3). Those moth species most susceptible to timber management and associated habitat loss tend to be those with specialised larval diets, limited adult dispersal powers, and depending on commercially valuable (harvestable) tree species. Likewise, replacement of natural forest by monocultures or plantation stands is likely to markedly reduce moth richness there. As one extreme plantation case, the richness of Geometridae and Arctiinae in Costa Rican oil palm (p. 48) plantations was severely reduced to a very impoverished representation of the intact forest understorey assemblages (Alonso-Rodriguez et al. 2017), as comparative incidence across replicated sites in each of four forest types showed (Table 8.4). Geometridae species were linked more distinctly with woody forest patches than with herbs or grasses than were arctiines, some of which are well-suited to living in more disturbed regimes. Representation of these two groups differed across the habitats, with the high arctiine abundance in oil palm reflecting incidence of few common species. However, whilst oil palm does not sustain most native moths, the values of some form of intermediate disturbance

Table 8.4 Summary of catches of two groups of moths in four environments in Costa Rica: Geometridae and Arctiinae from forest interior (FI), old-growth forest margin (FM), young secondary forest (YSM), and oil palm plantations (OPP); numbers in parentheses are % of total) (Alonso-Rodriguez et al. 2017)

	FI	FM	YSF	OPP	Total
Geometridae					
Morphospecies	113 (66.5)	93 (54.7)	90 (52.9)	31 (18.2)	170
No.individuals	570 (42.5)	321 (24.0)	314 (23.4)	135 (10.1)	1340
Arctiinae					
Morphospecies	81 (58.3)	76 (54.7)	96 (69.1)	35 (25.2)	139
No.individuals	581 (25.2)	525 (22.8)	668 (29.0)	529 (23.0)	2303

and regenerating habitats (such as young secondary forest) for moth conservation are substantial (Beck et al. 2002, Chazdon et al. 2009). The relevance of CWD to such taxa may be relatively small, but close investigation may be needed to explore its importance as a resource in any particular forest case, and its relationships to other management components.

Applications of 'new forestry practices', increasingly implemented in Scandinavia since the 1990s, intrinsically recognise the importance of CWD and stand heterogeneity. Measures such as keeping different successional stages, and retaining some live trees in clear-cut areas, are adopted purposefully to aid conservation of saproxylic insect diversity. Investigations of a range of variables in Sweden confirmed the relevance of both quantity and quality of CWD as components of the wider heterogeneity important in managing production forests also for conservation (McGeoch et al. 2007). Structural variety and diversity of wood harvesting techniques may have important conservation ramifications.

Selective logging is one of the most important alternatives to clear-felling, and can confer conservation benefits through more closely resembling natural forest processes, and creating uneven-aged stands purported to provide greater benefits to biodiversity, as 'near-to-nature forestry'. The extent of correspondence to natural processes is clearly very variable, depending on extent and form of logging undertaken. Whilst comparisons of saproxylic insects across treatments involving selective logging have often implied significant effects, with reduced abundance of beetles in logged stands compared with unlogged areas (Grove 2002), others have found little difference in richness and abundance of ground-dwelling beetles. Pitfall traps in subtropical forest in southern China (Yu et al. 2017) showed no significant effects of selective logging on the total beetle assemblage, despite some negative impacts on saproxylic taxa and some unlogged forest specialists. Selective logging, with natural regeneration, has for several decades been used in beech (*Fagus sylvatica*) forests in Europe (Gossner et al. 2013), with the expectation of negligible adverse effects on biodiversity. The practice, however, decreased the amount of dead wood and the variety of its sizes and decay stages, and effects were increased by continuing removal of old trees from sites, with these collectively influencing the saproxylic fauna. The practice was considered insufficient for sound conservation management unless supplies of old, large and well-decayed dead wood were increased substantially. However, as Ehnstrom (2001) commented for boreal forests, 'It is very difficult to specify the exact amount of dead wood needed to save both the biological diversity of organisms and all red-listed species associated with this substrate'. A combination of increased forest protection and well-considered approaches to selective harvest, salvage logging and allied tactics may go some way toward achieving this.

Variety in broad-leaved forests in Europe is exemplified by comparing the historical system of coppicing (cutting trees to ground level so that they regenerate from stumps and roots as a continuing supply of firewood or charcoal) with the somewhat more recent 'coppice with standards' approach (in which some trees are left uncut amongst coppice, essentially giving a two-storey forest of the same tree species). In France, many such sites have been abandoned, leading to presence of an over-mature coppice component combined with untrimmed trees (Lassauce et al. 2012). The beetles of managed sites (coppice with standards for ca 20 years) and over-mature sites (coppice unmanaged for at least 60 years) comprised 247 saproxylic species, including 30 red-listed species. The latter treatment stands were characterised by high volumes of fallen and standing dead wood, in part reflecting self-thinning of coppice stumps as they regenerate: higher mean richness and abundance of saproxylic beetles occurred in the over-mature stands. Recent increased commercial interest in coppicing (with short rotation times) is a consequence of increasing domestic uses of firewood and wider needs for biofuel. Safeguarding over-mature coppice with standards stands for saproxylic insects may also become more urgent, in parallel with those commercial pressures.

The extent of dead wood in tropical forests has received relatively little attention compared with recognition of its pivotal importance for invertebrate conservation in temperate forest environments (Grove 2001). Its potential as a basis for wider conservation management is correspondingly under-informed. Surveys in the Daintree area, Queensland, implied that volume and mass of dead wood was indeed relatively scarce in tropical rainforest in comparison with quantities in boreal and temperate forests elsewhere. Grove (2001) thus noted that forest management is a key driver of sustaining this resource. Logging imposed effects similar to those of some natural disturbances, such as cyclones. An important aspect of keeping dead timber on site is simply that old trees and dead timber are carriers of the 'biological legacies' (Maleque et al. 2006) from the previous stands of forest to the newer forest stands, and help recovery of structure and ecological functions after logging or other changes. Some of the values of logs for forest biodiversity (Table 8.5) show the enormous variety of resources and influences they may have (Lindenmayer et al. 2002) and evidence their important roles in forest ecology as both key habitats and utility resources (Dennis et al. 2006) that enable the normal behaviour and performance of numerous taxa.

Maintaining a supply of dead timber is thus a widespread management aim. In boreal forest, enhancing the natural supply of decaying wood in managed or plantation stands where this is scarce may be achieved by the seemingly contrary process of deliberately killing trees, a process signalled by Martikainen (2001) as a potentially valuable adjunct in managing supplies of coarse woody debris. He referred

Table 8.5 The broad values of logs for forest biodiversity, as listed by Lindenmayer et al. (2002) Adapted with permission from CSIRO Publishing	1. Provide nesting and sheltering sites for biota.
	2. Provide foraging substrates for predators (vertebrates and invertebrates).
	3. Provide hibernation sites for biota.
	4. Provide basking sites for reptiles.
	5. Provide places for key social behaviours.
	6. Act as plant germination sites.
	7. Provide substrates to promote growth of fungi.
	8. Provide mesic refugia for many organisms during drought and/or fire.
	9. Contribute to heterogeneity in the litter layer and patterns of ground cover.

also to the artificial creation of aspen stumps by explosives, following Nitterus' (1998) finding that the beetle faunas of artificial and natural stumps were similar. 'Sudden death' of trees may be far easier to achieve, but in nature the protracted death of trees can take up to many years of decline, a process which is far more difficult to mimic.

8.2.1 Saproxylic Beetles in Tasmania

Tasmania's strategy for research on CWD in production forests is a basis for discussion and possible emulation elsewhere, with recognition that local conditions are likely to dictate the primary needs of any such agenda. One important component of invertebrate surveys at the Warra Long-Term Ecological Research Site, and aimed at better understanding in developing management principles for wet eucalypt forests (Bashford et al. 2001) is to assess the long-term survival of invertebrates in single logs as these decay, using paired logs from old-growth and regrowth trees. The project links with others (Yee et al. 2001), collectively aiming to improve knowledge of conservation values of decaying wood on the forest floor. The study is founded in the reality that the complexity of saproxylic beetle assemblages and their ecological characteristics can be clarified fully only by long-term surveys tracing the incidence and richness of beetles in logs of named host tree species over considerable periods – even up to century and more as the logs decay. One such survey was introduced in Oregon, United States (Harmon 1992), with the purpose of monitoring log decomposition over up to 200 years. Understandably, such surveys are extremely rare and difficult to support.

Fig. 8.4 Proportional representation of ecological categories of saproxylic beetles according to extent of saproxylicity (black, obligately saproxylic; open, at least facultatively saproxylic); flightlessness (black, winged and presumed capable of flight; open, functionally flightless); and larval feeding category (*De*, detritus feeder; *Fu*, fungus feeder; *Pr*, predator; *Wo*, wood-feeder, *Ot*, other/ unknown). (Grove and Forster 2011a, b)

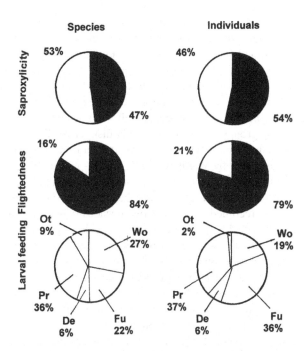

The Warra long-term experiment in lowland *Eucalyptus obliqua* forest was initiated in 1999 to help improve ecological understanding of values of logs derived from two tree cohorts: large diameter logs were fom mature tree, and smaller diameter logs were from regrowth trees which had regenerated since the most recent wildfires, in a design described by Grove and Bashford (2003). The Warra forest is about 15,000 hectares in area.

Beetles were collected by eclector traps, and some outcomes from the first decade of the project were discussed by Grove and Forster (2011a, b). The total captures over that period were 17,194 beetles, representing 453 species. The level of association of each species with dead wood enabled their allocation as 'obligately saproxylic', 'facultatively saproxylic' (also using litter or fungi not associated directly with dead wood), or 'non-saproxylic' (with no direct breeding association with wood or wood-associated fungi). Each species was also allocated to feeding guild and categorised by dispersal ability (flight, flightless). The taxonomic problems were substantial: across the 59 families represented, 229 species could not be named to species. Only 43 species were represented by more than 50 individuals, and 107 species occurred as singletons. The most species-rich families were Staphylinidae (109 species) and Curculionidae (77 species). Many species were regarded as truly saproxylic – 195 obligately so and 217 as facultatively saproxylic. Ecological compartmentalisation is summarised in Fig. 8.4.

This first decade of survey yielded much original insight on the beetles associated with *E. obliqua* logs, even though drawn from only 12 logs. One possible change over this period was an increased proportion of flightless species, and predators predominating over other larval feeding guilds. The two classes of logs supported different assemblages, and those assemblages also changed seasonally and from year to year. Logs from mature trees were particularly valuable for conservation: of the 392 species from one data set presented, more species were recorded from mature logs than from regrowth logs (335 cf. 285) and more species occurred only in mature logs (107, compared with 57 only in regrowth logs). Several species thus appeared to be associated only with one or other log category, suggesting that the log groups may be ecologically distinctive (Grove and Forster 2011b), through the mature trees having accumulated, over time, more species of fungi and beetles – as suggested by Wardlaw et al. (2009) – and perhaps reflecting a wider range of microhabitats allowing for greater species packing.

A recent update on the Warra beetle survey (Grove et al. 2017) noted about 1722 species recorded in the first two decades of study. Many were 'rare' – 350 were represented by singletons and 513 species were collected on only one occasion. About two thirds were recorded in association with logs or 'mature timber habitat' on living trees – but the complexities of interpretation are indicated by only about a third of the species being formally named.

Details of any such pattern are likely to vary across different tree species, and for which 'time-since-death' may have different rates and levels of influence. Trials such as the one noted here can gradually illuminate understanding of the successive saproxylic assemblages and their needs and, in turn, guide development of forest management optimal for their conservation.

8.3   Tree Stumps

Use of tree stumps, which often comprise a high proportion of the remaining dead wood after clear-cutting, by saproxylic invertebrates has led to considerations of 'retention levels' and reduction of stump harvesting for biofuels. Optimal levels of stump removal that will conserve saproxylic beetles are difficult to assess, but Victorsson and Jonsell (2012) suggested that clear-cutting in Swedish boreal forest harvested at currently recommended rates (retaining 15–25% of the stumps) will still lead to losses of beetle species. Stump-harvesting was established from the 1970s, initially in Sweden and later far more widely, with the primary purpose of producing biofuels being complemented by reduced site preparation costs and reduction of breeding sites for some important pest insects (such as the large pine weevil, *Hylobius abietis*). The impacts on local beetle faunas in Ireland are sometimes rather transient and relatively small: initial stump-harvesting led to noticeable declines in beetle richness and abundance, but these were not evident a year later, and changes were also site-specific (Shevlin et al. 2017). In forests in which only low amounts of CWD are allowed to persist, remnant stumps in clearcut areas may be 'the last outpost for some warmth-demanding saproxylic species' (Sweden: Andersson et al. 2015), with a combination of high and low stumps (below) needed to sustain the different beetle assemblages.

Using eclector traps over residual spruce stumps in Sweden, Work et al. (2016) compared the beetles from stumps from clearcut areas from which stumps had been harvested and others where stumps had been left, as controls. Together, the ten sites of each treatment yielded 9821 individual beetles, including 253 saproxylic species. Separation by major guilds gave 102 species of fungivores (19.2% of individuals), predators (76 species, 21.1%), and cambium/wood feeders (43 species, 55.2%). These divisions represented responses of beetles 2–3 years after harvest, and further compositional changes are likely to occur with further drying and decomposition. Many species were represented only by singletons, and 19 red-listed species were trapped. Stump harvesting had little impact on saproxylic beetle richness, abundance or assemblage composition and it seemed that, simply, the overall abundance of beetles will be proportional to the number of stumps retained. Stump volume was also important, with larger stumps commonly yielding more beetles – up to stump diameters of 30–45 cm.

In another study, the saproxylic beetle assemblages of Norway spruce (*Picea abies*) differed between 'high stumps' (around 3 m high) and horizontal logs (about 4 m long) (Andersson et al. 2015). Three of each, together with 'low stumps' were arranged on clearcut areas and all were sampled over a three year period approximately 5–10 years after clear-cutting. Beetles from low stumps were intermediate between those of the other treatments, and that investigation drew attention to need for varied substrates to be available to sustain most saproxylic beetle diversity, with logs supporting a more distinctive assemblage than high stumps. The pool of 195 beetle species trapped included 62 species (low stumps), 72 species (high stumps) and 50 species (logs). Each substrate also yielded species not found elsewhere: 27 (low stumps), 48 (high stumps), and 34 (logs). Several species were markedly more abundant on either low or high stumps and – as in some earlier studies – implied that some species cannot adequately substitute one substrate type for another. Andersson et al. (2015) recommended that stump harvesting should be undertaken only cautiously, especially until the roles of low stumps for known red-listed beetles are much better understood.

Benefits of retaining high stumps may be enhanced through combining creating these (at 3–5 m high) with slash harvesting, by which the tops and branches of cut trees are harvested after felling. In Sweden, combination of the two processes gave outcomes better for both biodiversity and profitability than from doing neither of these (Ranius et al. 2014). Separately, 'slash harvesting' was more profitable than stump harvesting, with the latter more costly because of the larger wood volumes needing more intensive processing.

High cut stumps have become a valuable conservation resource for saproxylic beetles in boreal forests, both for beetle diversity and for individual notable species. For Norway spruce (*Picea abies*) in southern Sweden, comparison of beetles from artificially created high stumps (3–5 m high) and low stumps (ca 0.5 m from ground level) by sieving bark from each gave totals of 53 (of a total 67) species in high stumps at ground level, 38 species from high stumps at approximately 1.3 m above ground ('breast height'), and 39 species from low stumps (Abrahamsson and Lindbladh 2006). The height differences between assemblages in high stumps implied some form of differentiation within the stumps, and the assemblages of low stumps also differed somewhat from these. Creation of high stumps might lead to increased populations of those beetles that 'prefer' that substrate, and lead to generally increased insect richness. The beetle fauna under the bark of high stumps of *P. abies* and *Pinus sylvestris* in Sweden included 87 broadly subcortical species, 65 of them saproxylic (Schroeder et al. 2006).

For one of the four red-listed species, *Hadreule elongatula* (Ciidae), high stumps were the major recruitment source in the wider landscape. In all, 776 individuals of this species were obtained from high stumps, and only a single specimen was retrieved from other coarse woody debris sampled over the same season. Its incidence was correlated strongly with presence of the polypore fungus *Fomitopsis pinicola* (p. 103). An earlier study (Jonsell and Weslien 2003) had also found that *H. elongatula* occurred at much higher densities (about six-fold greater) in high stumps than in logs, and the beetle is also scarce in low stumps. Assembled records implied that *Hadreule* depends on sun-exposed standing coarse woody debris. In contrast, for the other 28 beetle species appraised, less than 1% of the populations were found in high stumps in clearcut areas, so that the stumps contribute little to the species' wellbeing. Thus, because high stumps seem to confer only marginal benefits to most saproxylic beetles, but still be critical needs for a few, they should be treated as one of a varied portfolio of measures needed to conserve saproxylic insects in managed forests.

In Abrahamsson and Lindbladh's study, no red-listed beetle species were encountered. However, the importance of high cut stumps of *P. abies* was demonstrated for the red-listed *Peltis grossa* (Trogositidae), which is regarded as at risk of extinction in both Sweden and Norway. It reproduces only in standing dead wood, and prefers sun-exposed wood. The characteristic oval emergence holes of the beetle provide a useful tool for detecting and monitoring *P. grossa*, and investigations by Djupstrom et al. (2012) involved annual inspections of high cut stumps on all available clearcuts in randomly selected forest transects and in forest transects near the experimental clearcuts, all in southern Sweden. The high stumps were created in 1994 and 1995, and 425 stumps were inspected infrequently between 1994 and 2007. The first emergence holes were found in 2003, ten years after the clear cutting, and all holes were thereafter counted each autumn from 2003–2010. In 2005, all 1219 emergence holes found were on clearcuts (so that none was found in forest stands) and almost all in clearcuts that were 10 years or more old. The relatively recent practice of creating high stumps of *P. abies* has led to large increase in population size of *P. grossa*, with predictions that this will continue to increase as the practice persists. In that study, standing dead wood in closed forests was not confirmed as a suitable substrate for the beetle, and the sun exposure afforded by clearcuts appears to be critical. This example demonstrates how a forest landscape may be modified to favour an individual species of concern – and in this context may also prove a valid alternative to the expensive measure of setting aside areas of old forest, in which gap creation then depends on natural disturbances. However, the latter 'set aside approach' remains critical for numerous other and lesser-known saproxylic insects. Nevertheless, the association of other significant saproxylic insects with disturbed

Table 8.6 Numbers of saproxylic beetles from different high stump categories and ages, (*Picea abies* in Sweden)

Abbreviated from Lindbladh and Abrahamson 2008, see text

Regime			
	First thinning	Third thinning	Final felling
One summer old			
No. stumps	90	90	64
No. beetle species	31	31	34
Three summer old			
No. stumps	60	45	64
No. beetle species	32	26	18

habitats such as clearings implies that the process of stump formation might indeed have wider values. Djupstrom et al. (2012) suggested *Aradus erosus* (Hemiptera: Aradidae), *Orussus abietina* (Hymenoptera: Orussidae) and *Ipita binotata* (Coleoptera: Nitidulidae) as possible beneficiaries amongst the Swedish red-listed forest insects. Some saproxylic beetles, in contrast, are more abundant in shaded rather than exposed dead wood, as revealed for some Scolytidae from logs in northern Sweden (Johansson et al. 2006) in a study undertaken to determine the conservation values and amounts of dead wood to be left as an obligation under the widespread certification system that governs forestry practices. Twenty-two species of Scolytidae were captured in emergence traps or window traps compared across experimental logs that were artificially shaded (by shade cloth), naturally shaded, burned, an isolated log, and a snag (a high stump, cut to ca 3 m high), all replicated across old growth forest, old managed forest, and clearcut areas. That some Scolytidae are serious pests necessitates reducing risks of outbreaks by measures that include reducing the amount of fresh conifer wood that can be exposed to attacks – so that some level of limiting quantities of retained dead wood is economically necessary. Although the logs promoted some species that are potential pests, many other taxa were also present, together with a variety of natural enemies that might contribute to suppression of true pest species, whilst the logs also support and enhance populations of rare or threatened species.

High stumps do not necessarily originate from mature harvested trees. Thinning of production forests produces stumps of far smaller diameters, and such thinning – sometimes repeated at intervals – is an integral component of much forest management. In Sweden, beetles associated with first year thinning (stump mean diameter 13.6 cm) and third year thinning (mean diameter 22.3 cm) were compared with those from final tree felling stumps (mean diameter 41.2 cm) after one and three years of stump exposures (Lindbladh and Abrahamson 2008), from which bark sieving yielded a total of 73 saproxylic species, distributed as in Table 8.6. The large number of beetles in the thinnings high stumps compared with the final felling stumps was unexpected, and the assemblages showed some clear differences in composition. No red-listed species were recovered. One fungivorous rove beetle, *Placusa complanata* (Staphylinidae), was not found in the final felling stumps but occurred in both thinning treatments. Although only the first succession of bark-frequenting species was assessed in this relatively short duration study, the findings again emphasise the values of providing a variety of dead wood categories for insect conservation.

A somewhat similar study by Jonsell (2008) flowed from the increasing demand for small-diameter logging wastes, which were for long not commercially harvested. This demand was largely for biofuel and was perceived likely to increase threats to beetles and others using or preferring those substrates. As exploration of which species might prove susceptible, Jonsell (2008) compared beetles reared from four tree genera (*Populus, Betula, Quercus, Picea*), each with three diameter classes (0–4, 4–8, 8–15 cm) and two age classes (cuts at one year and 3–5 years old), in Sweden. The 160 species reared included 22 currently or in the past red-listed. Thirty-five species showed significant relationships with diameter, 21 of these with smaller diameter wood, and each tree genus had at least some beetle species specifically associated with it. The 'small diameter species' included three Buprestidae (*Chrysobothris affinis, Agrilus angustulatus, A. betuleti*), and Jonsell suggested that these species might decline if those logging residues were harvested more intensively.

Comparative rearings of a wider array of insects from high stumps and low stumps show that many species do not discriminate between these – but also that some, such as some parasitoids of bark beetles, are more frequently encountered in high stumps (Hedgren 2007). Most of the hymenopterous parasitoids recorded by Hedgren are generalists, affecting mainly secondary bark beetle species but with potential to switch to attack economically important species nearby. Continuing supply of high stumps may provide useful reservoirs of such natural enemies.

Decay rates of stumps, in part reflected by the moisture levels of the sites, have led to implications that stumps in wet positions may constitute inferior habitat in relation to those in drier areas (Ols et al. 2013). Comparing such regimes for birch (*Betula* spp.) and spruce in Sweden showed (1) beetle richness in the birch stumps were very similar, but (2) fewer species occurred in wet than in dry spruce stumps. No beetle species showed any positive association with wet stumps, and three beetles found in birch were more common in dry stumps. Some species appeared to avoid wet stumps.

Investigating stump age, also reflecting decay stage, Persson et al. (2013) compared the fauna extracted (by Tullgren funnels) from stumps and nearby soil for *Pinus sylvestris* and *Picea abies*, 5, 10 and 20 years old. Millipedes (Diplopoda) were the most abundant arthropod group, followed in sequence by Coleoptera and Diptera. None of the major taxa showed significant differences between the two tree species – so that pine might indeed substitute functionally for spruce stumps if the latter were harvested. Fig. 8.5 summarises the changes found over time for several taxa. Arthropod abundance was higher for bark of 10 year-old stumps than for any other substrate or age class.

Changes to soil structure and quality through harvesting of stumps, such as through increased exposure of mineral soils (Kataja-aho et al. 2016), can influence the decomposer communities, in conjunction with forest floor features and time

Fig. 8.5 The proportional abundance of selected groups of arthropods across various substrates in relation to stump age to indicate successional changes over time. Bark, open; periphery, diagonal hatch; wood, black; soil, dotted (Persson et al. 2013)

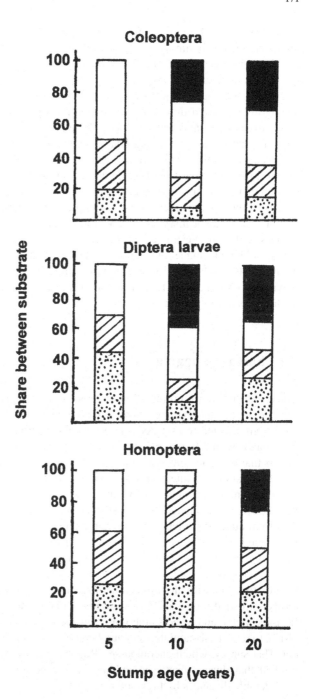

since harvesting. Different groups of arthropods show different responses, but recommendations to retain as much intact forest floor as possible at harvested sites, together with need for further long-term studies of impacts, were likely to have beneficial outcomes.

After harvesting, stumps can be stored in piles for future use. These may remain among the harvested timber and pulpwood for a year or more, over which period they attract insects and are colonised. Those piles can thereby become 'ecological traps' for saproxylic beetles and others that colonise the stored stumps but with the occupied stumps then removed for fuel (Victorsson and Jonsell 2013), and all individuals killed. Comparison of beetle densities in storage piles and 'normal' stumps showed that severe trap effects can occur for a number of beetle species in Sweden. Insects colonising the outer layers of stumps, such as bark, may be the most vulnerable, and were the primary focus of Victorsson and Jonsell's study. This led them to emphasise that their estimates of trap effects may be conservative, because allowances could not be made for interior occupants. Most obvious mortality will thereby be of insects that have colonised stumps within the storage piles rather than those previously present, with further studies needed to clarify impacts on the wood-feeding species, in particular.

8.4   Salvage Logging

Considerations of stump retention or removal are simply one aspect of the wider practice of 'salvage logging' involving more widespread removal of dead or dying trees from affected forests. Assessing the conservation impacts of salvage logging from stands with different management histories can become complex, because the factors that influence the assemblages of saproxylic beetles present are themselves not always simple to interpret. For example, in a Canadian study Cobb et al. (2010) revealed synergistic impacts of wildfire and salvage logging on those beetles – so that the effects of salvage logging after burning could not be predicted reliably by data from either disturbance alone. Large amounts of woody debris are an almost inevitable legacy of fires or wind storms over forest areas, and are the focus of salvage logging for firewood or other processing. That wood, possibly extracted from forest over several years, may also be transported over considerable distances to market, together with any insects that may have colonised them. Fallen trees from tornado-affected forests in Massachusetts (United States) were collected each year for three years, and insects emerging over the next year documented (Dodds et al. 2017), to reveal substantial insect numbers (even after three years) for five tree genera. That survey, which encompassed Buprestidae, Cerambycidae and Scolytidae, implied that diverse beetle assemblages could be relocated unintentionally over considerable distances by firewood transport.

The economic importance of such salvage following wildfires is considerable, and such operations are an important (even, necessary) counter to the massive economic losses flowing from forest wildfires in production forests in North America.

Fig. 8.6 A conceptual, model to illustrate ecological linkages between saproxylic beetles and nutrient cycling in forest ecosystems. Post-fire salvage logging (top right) may disrupt this pathway by severing the link between dead wood and saproxylic beetles in forests recovering from wildfire. (Cobb et al. 2009)

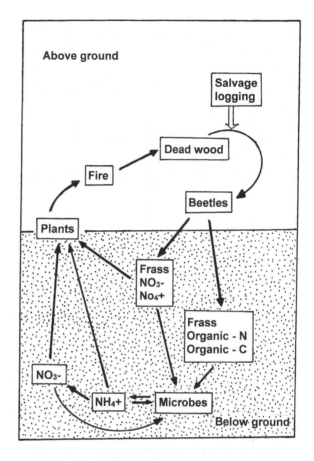

Salvage logging of boreal forests has thus increased due to economic pressures to increase overall returns from the forest crop. Post-fire salvage logging is an important context for this, and increasingly acknowledges the values of dead wood retention.

A few key beetle species have been appraised in detail to help establish the roles of salvage logging and its consequences for the insects. The widespread pyrophilic cerambycid *Monochamus scutellatus scutellaltus* (White-spotted sawyer) in North America, for example, was studied in a large area (ca 120,000 hectares) of varied forest in Alberta, Canada, burned by wildfire (Cobb et al. 2009). Removal of dead wood in post-fire salvage can here change links between the beetles and the forest nutrient dynamics. Larvae of *Monochamus* feeding on burned wood initiate the process of returning nutrients to the soil from burned conifers, implying the model shown in Fig. 8.6. Reduced densities of beetle larvae thereby influence soil nutrient dynamics and growth of colonising plant species. Impacts of wood removal extended well beyond the most obvious one of biodiversity loss, to influences on basic ecological processes.

Beetle richness (from a total catch of 245 taxa) was lower in salvage sites than in any other treatment (Fig. 8.7), and 44 species were regarded as significant indicators

Fig. 8.7 Mean catch rates of beetles associated with dead wood compared across forest treatments in Canada. Total beetles and three ecological groups are shown separately for undisturbed stands (U), clearcut harvested stands (C), stands burned by wildfire (B), and salvage-logged stands (S). (Cobb et al. 2010)

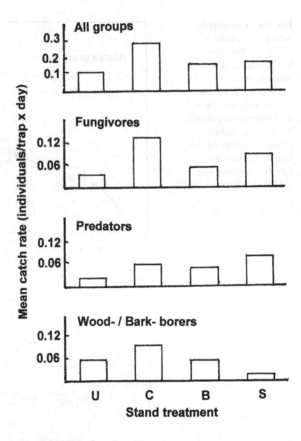

of stand treatment, with some available for each treatment appraised (Cobb et al. 2010). The consequences of multiple disturbances to forests remain very poorly understood, and Cobb et al. noted that post-fire management should be well-informed and not driven solely by economic considerations.

Major natural disturbances to forests are viewed commonly as catastrophes within the natural sequences of succession and production – as Noss and Lindenmayer (2006) put it, 'They create messes that need to be cleared up'. Salvage logging and removal of dead wood, such as logs and stumps, from disturbed production forest is a key component of that endeavour and those measures were undertaken historically with little, if any, consideration for biological outcomes and conservation of resident species. Enlarging on that perspective, they noted that although 'salvage' implies saving or recovery, this is only rarely so from an ecological perspective (Lindenmayer and Noss 2006). For salvage logging to be ecologically valid (Lindenmayer and Noss used the term 'ecologically defensible'), eight broad measures should be included, building on existing awareness and realisation that additional knowledge is urgently needed to refine the procedures (Table 8.7), together with recognition that the effects of salvage logging fall into three broad themes. These are (1) impacts on particular components of biota and species assemblages; (2) impacts on key ecosystem processes; and (3) impacts on the physical

Table 8.7 Suggested components for an ecologically defensible policy on salvage logging: the measures listed by Lindenmayer and Noss (2006)

1. Exclude salvage logging from some areas (nature reserves, water catchments, steep slopes with erosion-prone soils, some old-growth forests)
2. Ensure that unburned/partially burned patches are either exempt from burning or subject to low level harvesting with high retention levels.
3. Ensure that certain biological legacies (such as fire-damaged trees) are retained.
4. Modify salvage policies to limit amounts of biological legacies that are removed from certain areas, such as burned old-growth stands within production areas.
5. Schedule salvage logging so that effects on natural recovery of vegetation are limited.
6. Ensure future maintenance or creation of particular habitat elements for species of conservation concern within burned areas.
7. Ensure adequate riparian buffers are in place within areas where salvage logging occurs.
8. As effects of ground-based logging on soil and water can be great, limit extent of harvesting and wherever possible, employ aerial systems for removal of the trees.

structure of forests and integrating aquatic ecosystems. In the more specific context of the montane ash forests of the Central Highlands of Victoria, dominated by *Eucalyptus regnans* and *E. delegatensis* (Mountain ash, Alpine ash), recommendations for salvage harvest policies were modified somewhat, but strongly echo the sentiments of Table 8.7. The six provisions listed by Lindenmayer and Ough (2006) were (1) varied salvage logging intensity even within large areas of intensively burned forest; (2) increased retention of various forms of 'biological legacies', including large living and dead trees, large logs, and patches of unburned and partially burned understorey; (3) minimise mechanical disturbance; (4) minimise seed-bed preparation; (5) consider timing and seasons for salvage logging to occur; and (6) gather data that will better quantify the impacts of salvage logging.

8.5 Fine Woody Debris

As demonstrated above, the various forms of coarse woody debris are critical habitat needs for saproxylic insects, many of them depending wholly on these resources. Much other debris in forest and forestry, in contrast, is regarded as 'fine', comprising twigs, smaller branches and slender trees and the category sometimes delimited as that with diameter 5–9 cm, in contrast to the '> 10 cm' separation of coarse woody debris. The importance of the 'fine woody debris' (FWD) for insects has received relatively little attention. One factor endorsing why large diameter logs host saproxylic beetle assemblages that are largely distinct from those of smaller diameter logs is simply the pattern of log decomposition. In Tasmania, Yee et al. (2006) found that the brown rotten heartwood of large logs was relatively rare in smaller logs and supported beetles not found in the latter. Some of the Tasmanian beetles involved were of genera with European representatives that have declined drastically in range. Large diameter logs are an important landscape feature for conservation.

A study on the importance of FWD for cryptogams in Sweden, however, demonstrated the likely values of this category for insects, by comparing their diversity on coarse and fine logs of *Picea abies* (Kruys and Jonsson 1999) in boreal forest in which more than half the dead spruce trees were in the smaller diameter range – so that a considerable amount of potential habitat for organisms is involved. All 43 taxa of lichens, mosses, hepatics and fungi occurred on CWD, and 28 of these on FWD – so that the latter had considerable relevance in providing substrates. Diameter of logs is an important component of habitat suitability for many saproxylic beetles, but more studies are needed to clarify the roles of FWD in their ecology.

That importance was indicated in a survey of the beetles using dead twigs from a single tree of Southern red oak (*Quercus falcata*) in Lousiana (Ferro et al. 2009). Twigs averaged 36 cm long and 14 mm in diameter, and were cut and randomly allocated to one of 27 ten-twig bundles, which were subsequently placed across three treatments and sites in closed forest. At each site, three tied bundles were placed on top of the leaf litter on the ground, three propped at approximately 45° against a living tree, and three tied horizontally to the limb of a small tree or shrub about 1.5 m from the ground. After about 10 months, the bundles were retrieved, and beetles allowed to emerge over the following six months. The 414 adult beetles included 35 species (33 genera, 16 families), with 130 individuals (ground), 93 (propped bundles) and 193 (above-ground bundles). Richness was lowest for the ground bundles, and highest for those in trees – but several of the species obtained appeared to be rare (or, at least, elusive) as they were not represented in the large Lousiana State Collection. That survey indicated the presence of a rich and poorly documented beetle fauna associated with FWD, and also that the siting of the debris may affect the assemblage.

Gathering of roadside and forest floor dead wood, mostly FWD but sometime using chainsaws to dismember and remove larger pieces, for firewood is widespread, often not regulated or policed, and difficult to deter – not least because the material is often deemed untidy and its removal seen to improve the local environment and even encouraged by local authority. Occasionally, specific conservation needs arise over such removals. The specific obligate host ants of rare lycaenid butterflies in Australia are one such case. The rare 'Coconut ants' (*Papyrius* spp.) nest in dead wood, and their strong characteristic scent attracts female 'Ant-blue' butterflies (*Acrodipsas* spp.) to oviposit nearby. Young larvae are carried into the nests by worker ants and pass their whole development there. In Victoria, one ant-blue, *A. myrmecophila*, depends on a coconut ant, and a conservation plan for the butterfly at its then only validated locality in the state included measures to support the ant. These included local restrictions (with roadside signage) prohibiting collection of firewood for several Km around the butterfly population (New 2011).

Visual impressions of forest are enhanced by public perception of 'neatness', but providing information on the values of dead wood for conservation can lead to more tolerant attitudes (boreal forests: Gundersen et al. 2017). Linked with this, removal of dead wood might be more intense in areas which are primarily used for recreation. Dead wood, indeed, can have complex influences on how people perceive forests, in part by increasing the complexity of what is seen so that, while small amounts might be acceptable, large amounts of dead wood create too complex an environment, sometimes regarded as progressively 'unsafe'. In urban forest environments in Finland, visitors readily accepted downed logs as natural features, and sites with fresh logs were considered more appealing than either sites with old logs or without logs (Hauru et al. 2014). That survey generated suggestion that decaying logs should be left in urban forests as acceptable, even appealing, features.

Whether, and how, to manipulate coarse woody debris for conservation of saproxylic insects continues to attract attention in northern forests, in studies with increasing detail, and a variety of contexts and purpose. Many studies illuminate further dimensions and considerations – but many conclude that increased knowledge of local context and the key local target species is necessary in relation to characteristics of the stands, sites and substrates being investigated. Different trapping or investigation methods are also influential – in one example, window traps can reflect seasonal flight periods not detected from emergence traps used earlier in the season (Johansson et al. 2006). Influential variables, limited only by scope of any study on the beetle assemblage, can include any feature of the substrate and its condition and environment – for examples, snags/logs, bark presence/thickness, diameter, decay stage, fungal presence/absence, sun/shade exposure, humidity of wood and underlying soil, stand context, disturbance history, and so on. Collectively, these render derivation of general inferences difficult, but also confirm that 'both forest-management practices and substrate characteristics are important in determining the distribution and population densities … and both clearcut and closed forests provide important habitat for many species' (Johansson et al. 2006, on Scolytidae). That principle extends to considerations of stump/tree retention as a major source of dead wood, with the general implications of tree retention summarised as (1) primarily substrates needed by early successional beetle species; (2) alleviating the most serious consequences of clear-cutting; and (3) not maintaining the characteristics of initial mature forest (Gustafsson et al. 2010). Again, new insights into ecological outcomes continue to accumulate, but Gustafsson et al. also emphasised the practical difficulties of studying some crucial aspects, such as insect dispersal from retention areas and the duration of studies needed to appraise stand dynamics. The almost inevitable reliance on simulation and modelling studies necessarily includes simplifications, but can help to affirm that measures such as tree retention and other regulated supplies of woody debris should ideally be designed for the greatest benefit within a given forest management framework.

References

Abrahamsson M, Lindbladh M (2006) A comparison of saproxylic beetle occurrence between man-made high- and low-stumps of spruce (*Picea abies*). For Ecol Manag 226:230–237

Alonso-Rodriguez AM, Finegan B, Fiedler K (2017) Neotropical moth assemblages degrade due to oil palm expansion. Biodivers Conserv 26:2295–2326

Andersson J, Hjalten J, Dynesius M (2015) Wood-inhabiting beetles in low stumps, high stumps and logs on boreal clear-cuts: implications for dead wood management. PLoS One 10(3):e0118896. https://doi.org/10.1371/journal.pone.0118896

Bashford R, Taylor R, Driessen M, Doran N, Richardson A (2001) Research on invertebrate assemblages at the Warra LTER site. Tasforests 13:109–118

Beck J, Schulze CH, Linsenmair KE, Fiedler K (2002) From forest to farmland: diversity of geometrid moths along two habitat gradients on Borneo. J Trop Ecol 18:33–51

Bouget C, Brustel H, Brin A, Noblecourt T (2008) Sampling saproxylic beetles with window flight traps: methodological insights. Rev Ecol (Terre Vie) suppl. 10: 21–32

Chaundy-Smart RFC, Smith SM, Malcolm JR, Bellocq MI (2012) Comparison of moth communities following clear-cutting and wildfire disturbance in the southern boreal forest. For Ecol Manag 270:273–281

Chazdon RL, Peres CA, Dent D, Sheil D, Lugo AE et al (2009) The potential for species conservation in tropical secondary forests. Conserv Biol 23:1406–1417

Cobb TP, Hannam KD, Kishchuk BE, Langor DW, Quideau SA, Spence JR (2009) Wood-feeding beetles and soil nutrient cycling in burned forests: implications of post-fire salvage logging. Agric For Entomol 12:9–18

Cobb TP, Morisette JL, Jacobs JM, Koivula MJ, Spence JR, Langor DW (2010) Effects of postfire salvage logging on deadwood-associated beetles. Conserv Biol 25:94–104

Cranston PS, McKie B (2006) Aquatic wood – an insect perspective. In Grove SJ, Hanula JL (eds) Insect biodiversity and dead wood. Proceedings of a symposium for the 22nd international congress of entomology. USDA General and Technical Report SRS-93, Ashville, NC, pp 9–14

de Jong J, Dahlberg A (2017) Impact on species of conservation interest of forest harvesting for bioenergy purposes. For Ecol Manag 383:37–48

Dennis RLH, Shreeve TG, Van Dyck H (2006) Habitats and resources: the need for a resource-based definition to conserve butterflies. Biodivers Conserv 15:1943–1966

Djupstrom LB, Weslien J, ten Hoopen J, Schroeder LM (2012) Restoration of habitats for a threatened saproxylic beetle species in a boreal landscape by retaining dead wood on clear-cuts. Biol Conserv 155:44–49

Dodds KJ, Hanavan RP, DiGirolomo MF (2017) Firewood collected after a catastrophic wind event: the bark beetle (Scolytinae) and woodborer (Buprestidae, Cerambycidae) community present over a 3-year period. Agric For Entomol 19:309–320

Drag L, Cizek L (2015) Successful reintroduction of an endangered veteran tree specialist: conservation and genetics of the Great Capricorn beetle (*Cerambyx cerdo*). Conserv Genet 16:267–276

Ehnstrom B (2001) Leaving dead wood for insects in boreal forests – suggestions for the future. For Ecol Manag 16(S3):91–98

Ferro ML, Gimmel ML, Harms KE, Carlton CE (2009) The beetle community of small oak twigs in Lousiana, with a literature review of Coleoptera from fine woody debris. Coleopt Bull 63:239–263

Gibb H, Hjalten J, Ball JP, Atlegrim O, Pettersson RB et al (2006a) Effects of landscape composition and substrate availability on saproxylic beetles in boreal forests: a study using experimental logs for monitoring assemblages. Ecography 29:191–204

Gibb H, Pettersson RB, Hjalten J, Hilszczanski J, Ball JP et al (2006b) Conservation-oriented forestry and early sucessional saproxylic beetles: responses of functional groups to manipulated dead wood substrates. Biol Conserv 129:437–450

Gossner MM, Lachat T, Brunet J, Isacsson G, Bouget C et al (2013) Current near-to-nature forest management effects on functional trait composition of saproxylic beetles in beech forests. Conserv Biol 27:605–614

Grove SJ (2001) Extent and composition of dead wood in Australian lowland tropical rainforest with different management histories. For Ecol Manag 154:35–53

Grove SJ (2002) Tree basal area and dead wood as surrogate indicators of saproxylic insect faunal integrity: a case study from the Australian lowland tropics. Ecol Indic 1:171–188

Grove SJ, Bashford R (2003) Beetle assemblages from the Warra log-decay project: insights from the first year of sampling. Tasforests 14:117–129

Grove SJ, Forster L (2011a) A decade of change in the saproxylic beetle fauna of eucalypt logs in the Warra long-term log-decay experiment, Tasmania. 1. Description of the fauna and seasonality patterns. Biodivers Conserv 20:2149–2165

Grove SJ, Forster L (2011b) A decade of change in the saproxylic beetle fauna of eucalypt logs in the Warra long-term log-decay experiment, Tasmania. 2. Log-size effects, succession, and the functional significance of rare species. Biodivers Conserv 20:2167–2188

Grove SJ, Hanula JL (eds) (2006) Insect biodiversity and dead wood. Proceedings of a symposium for the 22nd international congress of entomology. USDA General and Technical Report SRS-93, Ashville, NC

Grove SJ, Stork NE (1999) The conservation of saproxylic insects in tropical forests: a research agenda. J Insect Conserv 3:67–74

Grove SJ, Wardlaw T, Forster L (2017) A megadiverse beetle fauna showing an inordinate fondness for Tasmanian forests. Tasmanian Naturalist 139:1–11

Gundersen V, Stange EE, Keltenborn BP, Vistad OI (2017) Public visual preferences for dead wood in natural boreal forests: the effects of added information. Landsc Urb Plann 158:12–24

Gustafsson L, Kouki J, Sverdrup-Thygeson A (2010) Tree retention as a conservation measure in clear-cut forests of northern Europe: a review of ecological consequences. Scand J For Res 25:295–308

Hagglund R, Hekkala A-M, Hjalten J, Tolvanen A (2015) Positive effects of ecological restoration on rare and threatened flat bugs (Heteroptera: Aradidae). J Insect Conserv 19:1089–1099

Hanula Jl, Horn S, Wade DD (2006) The role of dead wood in maintaining arthropod diversity on the forest floor. In Grove SJ, Hanula JL (eds) Insect biodiversity and dead wood. Proceedings of a symposium for the 22nd international congress of entomology. USDA general and technical report SRS-93, Ashville, NC, pp 57–66

Harmon ME (1992) Long term experiments on log decomposition at the H. J. Andrews Experimental forest. General Technical Report PNW-GTR-280. USDA Forest Service, Portland, Oregon

Harmon ME (2001) Moving towards a new paradigm for woody detritus management. Ecol Bull 49:269–278

Hauru K, Koskinen S, Kotze DJ, Lehvavirta S (2014) The effects of decaying logs on the aesthetic experience and acceptability of urban forests – implications for forest management. Landsc Urb Plann 123:114–123

Hedgren PO (2007) Early arriving saproxylic beetles (Coleoptera) and parasitoids (Hymenoptera) in low and high stumps of Norway spruce. For Ecol Manag 241:155–161

Heliovaara K, Vaisanen R (1984) Effects of modern forestry on northwestern European forest invertebrates: a synthesis. Acta Forest Fenn 189:1–32

Higgins RJ, Lindgren BS (2006) The fine scale physical attributes of coarse woody debris and effects of surrounding stand structure on its utilization by ants (Hymenoptera: Formicidae) in British Columbia, Canada. In Grove SJ, Hanula JL (eds) Insect biodiversity and dead wood. Proceedings of a symposium for the 22nd international congress of entomology. USDA general and technical report SRS-93, Ashville, NC, pp 67–74

Johansson T, Gibb H, Hilszczanski J, Pettersson RB, Hjalten J, Atlegrim O, Ball JP, Danell K (2006) Conservation-oriented manipulations of coarse woody debris affect its value as habitat for spruce-infesting bark and ambrosia beetles (Coleoptera: Scolytinae) in northern Sweden. Can J For Res 36:174–185

Johansson T, Hjalten J, Stenback F, Dynesius M (2010) Responses of eight boreal flat bug (Heteroptera: Aradidae) species to clear-cutting and forest fire. J Insect Conserv 14:3–9

Jonsell M (2008) Saproxylic beetle species in logging residues: which are they and which residues do they use? Norw J Entomol 55:109–122

Jonsell M, Nordlander G (2002) Insects in polypore fungi as indicator species: a comparison between forest sites differing in amounts and continuity of dead wood. For Ecol Manag 157:101–118

Jonsell M, Weslien J (2003) Felled or standing retained wood – it makes a difference for saproxylic beetles. For Ecol Manag 175:425–435

Jonsson BG, Kruys N (2001) Ecology of coarse woody debris in boreal forests: future research directions. Ecol Bull 49:279–281

Jonsson BG, Kruys N, Ranius T (2005) Ecology of species living on dead wood – lessons for dead wood management. Silva Fennica 39:289–309

Kataja-aho S, Hannonen P, Liukkonen T, Rosten H, Koivula MJ, Koponen S, Haimi J (2016) The arthropod community of boreal Norway spruce forest responds variably to stump harvesting. For Ecol Manag 371:75–83

Kruys N, Jonsson BG (1999) Fine woody debris is important for species richness on logs in managed boreal spruce forests of northern Sweden. Can J For Res 29:1295–1299

Lassauce A, Anselle P, Lieutier F, Bouget C (2012) Coppice-with-standards with an overmature coppice component enhance saproxylic beetle biodiversity; a case study in French deciduous forests. For Ecol Manag 266:273–285

Leather S (2005) Sampling insects in forest ecosystems. Blackwell, Oxford

Lee S-I, Spence JR, Langor DW (2014) Succession of saproxylic beetles associated with decomposition of boreal white spruce logs. Agric For Entomol 16:391–405

Lindbladh M, Abrahamsson M (2008) Beetle diversity in high-stumps from Norway spruce thinnings. Scand J For Res 23:339–347

Lindenmayer DB, Noss RF (2006) Salvage logging, ecosystem processes, and biodiversity conservation. Conserv Biol 20:949–958

Lindenmayer DB, Ough K (2006) Salvage logging in the montane ash eucalypt forests of the Central Highlands of Victoria and its potential impacts on biodiversity. Conserv Biol 20:1005–1015

Lindenmayer DB, Claridge AW, Gilmore AM, Michael D, Lindenmayer BR (2002) The ecological roles of logs in Australian forests and the potential impacts of harvesting intensification on log-using biota. Pac Conserv Biol 8:121–140

Maleque MA, Ishii HT, Maeto K (2006) The use of arthropods as indicators of ecosystem integrity in forest management. J Forestr 104:113–117

Martikainen P (2001) Conservation of threatened saproxylic beetles: significance of retained aspen *Populus tremula* on clearcut areas. Ecol Bull 49:205–218

McGeoch MA, Schroeder M, Ekbom B, Larsson S (2007) Saproxylic beetle diversity in a managed boreal forest: importance of stand characteristics and forestry conservation measures. Divers Distrib 13:418–429

Muller J, Brustel H, Brin A, Bussler H, Bouget C et al (2015) Increasing temperature may compensate for lower amounts of dead wood in driving richness of saproxylic beetles. Ecography 38:499–509

New TR (2011) Butterfly conservation in South-Eastern Australia: progress and prospects. Springer, Dordrecht

Nieto A, Alexander KNA (2010) The status and conservation of saproxylic beetles in Europe. Cuademos de Biodiversidad 33:3–10

Nitterus K (1998) Wood dwelling insects in natural and artificially produced high stumps of aspen and birch in Southwest Sweden. Examensarbe i Entomologi, Sweden

Noss RF, Lindenmayer DB (2006) The ecological effects of salvage logging after natural disturbance. Introduction. Conserv Biol 20:946–948

Ols C, Victorsson J, Jonsell M (2013) Saproxylic insect fauna in stumps on wet and dry soil: implications for stump harvest. For Ecol Manag 290:15–21

Parrish C, Summerville K (2016) Effects of logging and coarse woody debris harvest on lepidopteran communities in the eastern deciduous forest of North America. Agric For Entomol 17:317–324

Persson T, Lenoir L, Vergerfors B (2013) Which macroarthropods prefer tree stumps over soil and litter substrates? For Ecol Manag 290:30–39

Pinzon J, Spence JR (2010) Bark-dwelling spider assemblages (Araneae) in the boreal forest: dominance, diversity, composition and lifehistories. J Insect Conserv 14:439–458

Podlar JH, Pearce JL, Venier LA, McKenney DW (2002) Coarse woody debris in relation to disturbance and forest type in boreal Canada. For Ecol Manag 158:189–194

Ranius T, Caruso A, Jonsell M, Juutinen A, Thor G, Rudolphi J (2014) Dead wood creation to compensate for habitat loss from intensive forestry. Biol Conserv 169:277–284

Saint-Germain M, Drapeau P, Buddle CM (2007) Host-use patterns of saproxylic phloeophagous and xylophagous Coleoptera adults and larvae along the decay gradient in standing dead black spruce and aspen. Ecography 30:737–748

Schiegg K (2000a) Are there saproxylic beetle species characteristic of high dead wood connectivity? Ecography 23:579–587

Schiegg K (2000b) Effects of dead wood volume and connectivity on saproxylic insect species diversity. Ecoscience 7:290–298

Schiegg K (2001) Saproxylic insect diversity of beech: limbs are richer than trunks. For Ecol Manag 149:295–304

Schroeder LF, Ranius T, Ekbom B, Larsson S (2006) Recruitmant of saproxylic beetles in high stumps created for maintaining biodiversity in a boreal forest landscape. Can J For Res 36:2168–2178

Seibold S, Brandl R, Buse J, Hothoen T, Schmidl J, Thorn S, Muller J (2015) Association of extinction risk of saproxylic beetles with ecological degradation of forest in Europe. Conserv Biol 29:382–390

Shevlin KD, Hennessy R, Dillon AB, O'Dea P, Griffin CT, Williams CD (2017) Stump-harvesting for bioenergy probably has transient impacts on abundance, richness and community structure of beetle assemblages. Agric For Entomol 19:388–399

Siitonen J (2001) Forest management, coarse woody debris and saproxylic organisms: Fennocandian boreal forests as an example. Ecol Bull 4:11–41

Summerville KS (2014) Do seasonal temperatures, species traits and nearby timber harvest predict variation in moth species richness and abundance in unlogged deciduous forest? Agric For Entomol 16:80–86

Summerville KS, Crist TTO (2008) Structure and conservation of lepidopteran communities in managed forests of northeastern North America: a review. Canad Entomol 140:475–494

Topp W, Kapper H, Kulfan J, Zach P (2006) Litter-dwelling beetles in primeval forests of Central Europe: does deadwood matter? J Insect Conserv 10:229–239

Victorsson J, Jonsell M (2012) Effects of stump extraction on saproxylic beetle diversity in Swedish clear-cuts. Insect Conserv Divers 6:483–493

Victorsson J, Jonsell M (2013) Ecological traps and habitat loss, stump extraction and its effects on saproxylic beetles. For Ecol Manag 290:22–29

Wardlaw T, Grove S, Hopkins A, Yee M, Harrison K, Mohammed C (2009) The uniqueness of habitats in old eucalypts: contrasting wood-decay fungi and saproxylic beetles of young and old eucalypts. Tasforests 18:17–32

Woldendorp G, Keenan RJ (2005) Coarse woody debris in Australian forest ecosystems: a review. Austral Ecol 30:834–843

Woldendorp G, Keenan RJ, Ryan MF (2002) Coarse woody debris in Australian forest ecosystems. Bureau of Rural Resources, Canberra.

Work TT, Andersson J, Ranius T, Hjalten J (2016) Defining stump harvesting retention targets required to maintain saproxylic beetle biodiversity. For Ecol Manag 371:90–102

Yee M, Yuan ZQ, Mohammed C (2001) Not just waste wood: decaying logs as key habitats in Tasmania's wet sclerophyll production forests: the ecology of large and small logs compared. Tasforests 13:119–128

Yee M, Grove SJ, Richardson AMM, Mohammed CL (2006) Brown rot in inner heartwood: why large logs support characteristic saproxylic beetle assemblages of conservation concern. In Grove SJ, Hanula JL (eds) Insect biodiversity and dead wood. Proceedings of a symposium for the 22nd international congress of entomology. USDA General and Technical Report SRS-93, Ashville, NC, pp 42–56

Yu X-D, Liu C-L, Lu L, Bearer SL, Luo T-H, Zhou H-Z (2017) Does selective logging change ground-dwelling beetle assemblages in a subtropical broad-leaved forest of China? Insect Sci 24:303–313

Chapter 9
Forest Management for Insects: Issues and Approaches

Keywords Climate change · Control burning · Ecological traps · Fire regimes · Forest corridors · Forest reserves for insects · Plantations · Retention trees · Riparian forest · Sacred groves · Tree hollows · Urban forestry · Veteran trees

9.1 Introduction

The complexities of forest insects and their needs present complicated scenarios for their effective conservation, whether efforts are directed to single species or wider communities. Forest management can hamper or support practical insect conservation in many ways, and some of the major themes and issues are outlined in this chapter, with examples of how they may contribute to wider perspectives.

9.2 Fire and Management

Natural fires are a regular feature in many forested areas, but management plans to protect timber crops (including plantations), and reduce the wider risks of wildfire to people and property commonly include some form of 'control' or 'fuel reduction' burning, and the suppression of chances and severity of natural fire in the interests of imposing more regulated or targeted burning. Collectively, the ecological roles of forest fires vary greatly and impacts are viewed along a gradient from highly beneficial to catastrophic disasters that pose severe threats, and enormous losses of lives and property. In Australia, and despite wide appreciation of values of prescribed fire for conservation purposes, its use for conservation is generally secondary to safety and economic concerns.

Natural wildfire disturbances have helped to shape the ecology of Australian forests, and can kill a high proportion of litter and subcortical invertebrates, in particular. Mosaics of microhabitats are an almost universal consequence of the uneven impacts of low-intensity fires across forest landscapes, with irregularities in aspect, topography, ground cover and microclimates influencing this. Deliberate burning

© Springer International Publishing AG, part of Springer Nature 2018   183
T. R. New, *Forests and Insect Conservation in Australia*,
https://doi.org/10.1007/978-3-319-92222-5_9

Table 9.1 Then current (C) clear-felling practices and suggested possible improvements (M) to more closely resemble natural the wildfire regime in wet eucalypt forests of Tasmania (as listed and discussed and referenced by Baker et al. 2004)

Disturbance regime	Current/Modification descriptor
Area disturbed	C: 10–100 ha, average 50 ha.
	M: greater variability in coupe sizes.
Distribution	C: probably more fragmented.
	M: use maps of previous fire distribution and modelling in long-term coupe scheduling
Frequency	C: nominally1/100 years.
	M: longer (e.g. 100-300 years) rotation lengths for some coupes.
Predictability	C: predictable, together with time of year and fuel conditions. M: more variation in rotation length, regeneration burns not feasible for much of year.
Rotation period	C: 90 years. M: lengthen in some areas.
Intensity of fire	C: generally uniformly high.
	M: cool regeneration burns for some coupes, consider some redistribution of logging debris and protecting clumps of vegetation with firebreaks to increase variability within coupes.
Severity	C:P high, all trees felled, understorey destroyed, fine fuel burned. M: retain clumps of overstorey and understorey and scattered single trees, consider alternatives to produce multi-aged stands.

has been incorporated into silvicultural practices that purportedly mimic the natural processes and emulate the patterns of ensuing natural regeneration. In wet eucalypt forests of Tasmania, the tripartite process sequence of clear-felling, burning residue on site, and sowing by broadcasting eucalypt seeds has been adopted widely, with projected rotation times between tree harvests of 80–100 years. Whether the practice is a valid substitute for regeneration following natural wildfire was assessed by Baker et al. (2004), by comparing the litter-frequenting beetles in logged and burned regeneration forest and forest regenerated naturally after wildfire, using paired adjacent sites, all at 33 years since the disturbance. Beetle assemblages (total of 179 morphospecies in 30 families) could not be separated consistently, and species richness of the 'rare' taxa (the 151 morphospecies found in only very low numbers) was also similar: clear-cut logging had apparently not harmed the rarest components of beetle diversity after this interval. Baker et al. concluded that, at least at the coupe level, long-term differences in the beetle faunas were not obvious, and the effects of logging appeared to mimic the effects of fire. However, this inference might not extend to a wider landscape level, and several possible modifications in silviculture might improve the level of correspondence (Table 9.1).

In other contexts, the primary purpose of deliberate burning is to reduce fuel loads. Low intensity fires are the major category sought in deliberate or 'prescribed' Australian forest management burns, and are usually also very patchy, leading to mosaics in which many unburned or lightly burned microhabitats (such as near logs or rocks) persist and which may serve as refuges for animals. Thus, refuges associated with logs can contribute to conservation of ant diversity (Andrew et al. 2000).

In both that study and a related earlier one (York 2000), ant richness in leaf litter was similar across burned and unburned patches in both open and log-sheltered habitats, and most ant species occurred in all four of those treatments. Some of the rarer ant species were more closely linked with litter of single treatments, and availability of refuges might contribute to this – their low numbers rendered any such inference tentative.

The combinations of prescribed fires to reduce fuel and retention of dead wood supplies for insect conservation can pose further dilemmas for management. Suppression of fires in the boreal forests of Fennoscandia, for example, has induced declines in some pyrophilous insects associated with dead wood. Notable amongst these are several species of red-listed Flat bugs (Heteroptera: Aradidae), which are increasingly threatened by lack of fire. Experimental trials in which impacts of fire suppression and reduced supply of dead wood were compared (Heikkala et al. 2017) showed that burning increased richness and abundance of such species – with *Aradus angularis* and *A. laeviusculus* becoming abundant after fire. The bugs can locate recently burned forest rapidly – but the beneficial effect is rather transient, so that conservation measures should include recently burned forest with sufficient retention trees across the landscape.

Although aradids have attracted specific attention, the same principle is likely to apply to other insects, with management involving fire suppression and post-fire salvage logging leading to decline of areas burned annually and the amount of dead and injured trees, and declines of pyrophilous saproxylic insects. The rationale of prescribed burning in these boreal forests is in part to promote biological legacies similar to those of natural forests. Emphasis is very different from the priority use of prescribed burning in Australia, in which fuel reduction and 'insurance' against severe wild fires is more prominent. From Heikkala et al.'s (2017) study, the major points for aradid conservation include (1) good colonisation ability enables the bugs to exploit burned wood substrate rapidly, but (2) for only a short period, so that (3) a mosaic of suitable substrates at the landscape scale is needed, with supply of dead wood facilitated by retention trees. If sufficient burned wood is available, the pyrophilous bugs can be conserved within managed forests.

Proliferation of wood-boring beetles attracted to burned trees may affect values of those trees for harvest (Costello et al. 2011, on Ponderosa pine), with impacts of beetles reflecting intensity of fire and time since burning (Fig. 9.1), with higher beetle densities in more severely burned sites.

However, many forest biota have a long history of coadaptation to 'natural' fire regimes and the mosaic habitats that commonly result. A possible consequence of wildfire suppression is the loss of so-called natural pyrogenic landscapes that (1) contribute to mosaic heterogeneity across affected forest areas, and (2) support particular suites of fire-adapted and early successional species that cannot thrive in more mature forest. In Canada, the 'fire residuals' are key habitats for some glacial relict carabid species restricted to the locally suitable forest area. Preservation of fire residuals (as well as of harvest residuals) in managed forests may be an essential component of conserving those species (Gandhi et al. 2011).

Fig. 9.1 The mean incidence of wood borers (as totals of egg niches, larvae and borer perforations in the xylem) in Ponderosa pine trees (in South Dakota) with green (black), scorched (dotted) and fire-consumed (open) needles within plots of different times-since-fire, indicating effects of burning intensity. (Costello et al. 2011)

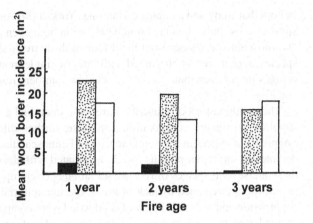

In many cases, management or other fires create dead wood – but existing supplies may also be depleted by hot fires, and both fallen and standing trees, as well as ground litter of all kinds, affected. Studies in boreal forest have emphasised the values of continuing supplies of dead wood for insects, with fires affecting that supply considerably. Scale and intensity of burning can vary enormously, and management by mosaic ('patch') burning is recommended widely, rather than the burning of single large forest areas. The latter may result from wildfires, but vagaries of topography and microclimate may still impose some form of mosaic intensity and impact. Burned forests thereby provide (1) large amounts of dead wood and (2) open stand conditions, both situations exploited by numerous saproxylic insects. Several studies in Canada have reported this bias – more than 40 beetle species on unburned Black spruce (*Picea mariana*) in Quebec were not recovered from unburned forests (Saint-Germain et al. 2004), for example. Direct comparisons of impacts of fire and other disturbances on saproxylic assemblages are often incomplete, but clear-cutting seems not to emulate fire closely in the responses induced. Langor et al. (2008) emphasised the variety of responses of the beetles to forest changes, but also that species-level responses to disturbances and mitigation measures need far more study to improve understanding of those responses. In general, and as discussed for Swedish forests (Hedwall and Mikusinski 2015), the importance of old-growth forest features of old, large trees and large amounts of dead wood – features that markedly enhance conservation values for insects – are widely lacking in intensively managed forests. The roles of regulated fires in that supply are commonly overshadowed by the dangers perceived.

Decisions over how, what, where, and when to burn each pose problems for optimal definition and guidance within different contexts. For many insects, the season of burning may also be critical in relation to susceptibility of different life stages. Deliberate use of fire for habitat management of individual threatened species may need to be planned carefully in relation to the species' phenology, and be viable for only a short period each year (New 2014). More generally, the above variables are components of a fire 'regime', with facility to induce a mosaic of outcomes that assure heterogeneity across the landscape and increase conservation benefits, within a wider pattern of burned/unburned areas. Factors that dictate thresholds for burning vegetation often give priority to plant responses to fire, with 'time-since-fire' and fuel loads noted as important variables for considering faunal responses. Floristic composition is driven by many factors other than fire (Spencer and Baxter 2006) and these are often unclear for any given site. However, as noted for New South Wales (Croft et al. 2016), faunal considerations may not coincide with plant responses, and upper thresholds derived from the latter are often too low to assure habitat needs for animals. Tree hollows and log hollows, as important nesting and roosting sites for vertebrates, and amounts of fallen timber were all more abundant in long-unburned forests, and the more pristine areas are regarded as a significant conservation asset for protection. Although emphasising needs of vertebrates, simply that McElhinny et al. (2008) could distinguish 34 structural features of eucalypt forests and woodlands that provide faunal habitat, and that many of these are affected directly by fire, underlines the complexity likely in describing invertebrate requirements.

Four broad habitat components helped to characterise needs of invertebrates in forests within the three elevational strata (overstorey trees, understorey shrubs and grasses, woody debris and bare earth) and in each of which species may be extremely localised in relation to factors such as individual plant species, the condition or growth phase of these, or the character of dead wood present. Those four components (listed by McElhinny et al. 2008) are (1) overstorey foliage and flowers; (2) bark; (3) shrubs and ground vegetation; and (4) litter and woody debris, in each of which changes can reduce invertebrate diversity and abundance. However, very few of the attributes identified by McElhinny et al. have been used formally to quantify faunal habitat.

One consequence of 'too frequent fire' is simply that the key structural attributes of mature forest are eliminated by continual forest rejuvenation. The currently recommended upper threshold intervals for burning the New South Wales forests (that is, 30–50 years) should be at least 50 years if faunal habitat needs are to be included meaningfully (Croft et al. 2016).

Fuel reduction burns in Australian forests are undertaken frequently to decrease ground debris and reduce the intensity and potential hazards from later wildfires. They are typically of low intensity, and produce mosaic ground-level conditions in which large logs and similar objects and unburned litter patches may remain. Unburned litter, and unburned or lightly burned wood, may then function as refuges for fauna, and the extent to which this scenario is effective has been evaluated in a few studies in which the insects or other fauna of experimentally burned and unburned (control) sites have been compared.

The extent to which ants used unburned litter associated with logs after burning was assessed in New South Wales by comparing ant richness and abundance in litter near and distant from logs (Andrew et al. 2000) in both burned and unburned areas. A high proportion (94%) of total ants was made up of the 15 (of 42) species occurring in all four treatments, and few substantial differences in representation of functional groups occurred (Fig. 9.2). Many of the common ant species were thus found

Fig. 9.2 Percentaqe composition of ant species among seven functional groups, from samples from leaf litter in March 1996 in four different habitats (*B/O* burned/open, *B/L* burned/log, *U/O* unburned/open, *U/L* unburned/log) at Bulls Ground State Forest, New South Wales (groups are specialist predator, diagonal hatch; generalist, open; opportunist, dense dots; cryptic, black; climate specialist, cross hatch; subdominant, dashed hatch; dominant, sparse dots. (Andrew et al. 2000)

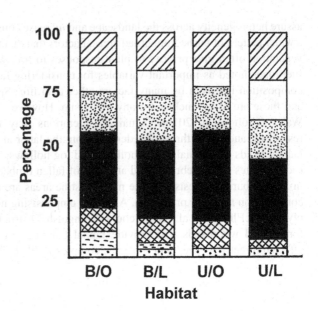

to be widely distributed and scarcely affected by burning or presence of logs. However, amongst the low abundance taxa, each of the four habitats yielded species not found elsewhere – but their scarcity precluded any understanding of why this was so, or whether the impression is simply due to chance. Leaf litter left sheltered near woody debris after control burning may indeed contribute to sustaining ant diversity, and even their parallels to unburned areas endorse the value of retaining logs and sheltered litter as refuges in frequently burned areas.

Fire is used deliberately in other forest management measures. The traditional fate of 'windrows', the line piles of coarse woody debris left after clearing land and preparation for afforestation or alternative use, is to burn them after drying out, in which eventuality the numerous saproxylic insects that have colonised that wood will be destroyed. They thus have clear potential to become ecological traps.

However, interpreting the impacts of forest fires on the ecology of even conspicuous insect taxa can be complex. Fruit-feeding Nymphalidae (p. 126) have been used as indicators for a range of forest disturbances (such as selective logging) that alter the physical structure of the forest, and a study in Mato Grosso, Brazil, demonstrated substantial differences in their assemblages between burned and unburned forest plots, but not between plots burned annually and every three years (de Andrade et al. 2017). The local nymphalid butterfly assemblage of 56 species had similar richness in each treatment with major compositional differences reflecting presence of open ground 'savanna specialist' species in both burned regimes and their absence from unburned control sites (Fig. 9.3), together with reduced richness of forest specialists in the burned plots. That difference correlated with canopy cover, and implied that microclimate changes resulting from fires affected the assemblage composition and loss of forest species – forest specialists are perhaps more generally susceptible to fire impacts.

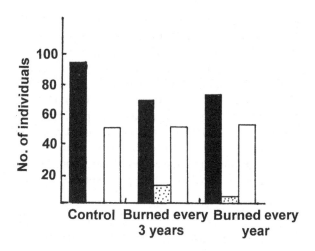

Fig. 9.3 Abundance of nymphalid butterflies of three categories (forest specialists, black; savanna specialists, dots; generalists, open) and fire in an Amazon forest. Numbers of individuals are shown for a control site and two burning regimes. (de Andrade et al. 2017)

9.3 Ecological Traps

Any forest products left stored in or near the forests – stumps, logs, bark and others – also offer a mixture of benefits and accelerated losses to saproxylic insects. Stumps of beech (*Fagus sylvatica*) in Europe are attractive for oviposition by the Rosalia longicorn beetle (*Rosalia alpina*, p. 124). The timber, however, is normally stored in the forest for a year or less, a period much shorter than the period of *Rosalia*'s larval development (three years) (Adamski et al. 2016). Parallels are likely to be widespread, simply because many beetles seek out dead wood and depend on it for reproduction: stored wood provides a concentrated and attractive resource for many species.

R. *alpina* oviposits on standing timber, and transport of that timber from the forest directly removes those potential offspring from the breeding locality, preparatory to the timber being processed. Short timber storage times are usual in surveys in Poland, so that large numbers of *Rosalia* may not accumulate – but the main seasonal storage period coincided with the beetle's main flight period. Nevertheless, and despite fulfilling the twin major requirements of an ecological trap (active attraction to reproduce, impossibility of completing development) Adamski et al. (2016) noted some doubt over the extent to which *Rosalia* populations were affected. Even if the adults do not attempt to breed on the stumps, their very presence there reduces opportunity to breed in more natural habitats. Their major conclusion was simply that it is not advisable to store stumps, even for short periods, in or close to locations where this notable beetle had been recorded.

For another notable saprophage, *Morimus asper* (p. 124), the balance between circumstances of logs being a valuable monitoring tool or an ecological trap was evident in northern Italy (Chiari et al. 2013), where freshly cut log piles of oaks, *Quercus* spp., and *Carpinus betulus* were inspected for the beetle over a short (six day) monitoring period. Beetles were marked and released on the pile where they were initially found. Over this short period, 118 beetles were found, and 15 of the 54 recaptures were on a different log pile. The log piles were considered a useful rapid census technique for *Morimus*, but in order to avoid their becoming ecological traps, the piles should ideally be left alone until they have decayed completely.

Piles of oak wood stored in the forests of southern Sweden yielded 39 beetle species, six of them red-listed (Hedin et al. 2008). The collective 3528 individuals were found in wood samples taken from the top, middle and base levels of the piles. Most beetles were found in the top layer, presumed to colonise that region, and leading to recommendation that many of the beetles exploiting the log piles can be saved if the top 10% of piles are retained rather than removed. Depending on the season, rapid removal of the remainder of the piles would also reduce the ecological trap effect.

9.4   Forest Reserves and Landscape Structure

The small proportion of global forest area designated for conservation is considered highly insufficient to protect the biodiversity that depends on forest environments. Schmitt et al. (2009) noted that only 7.7% of that area is designated under the World Conservation Union's major classes (I-IV) of protected areas (Table 9.2). With inclusion of categories V and VI, that portion increased to 13.5%. That review by Schmitt et al. (2009) sought to assess how the world's major forest types were represented in protected areas, especially in relation to the stated then current 2010 target that 'at least 10 % of each of the world's forest types should be conserved effectively' (CBD 2009), with the principles of giving priority to representativeness, intactness and security (freedom from threat). Serious gaps persisted in achieving this level of protection, and even reaching that level will not adequately conserve the biodiversity those forests contain. Forests within many 'biodiversity hotspots', for example, have already been reduced significantly, so that a 10% target actually represents a much smaller proportion of the original forest. The CBD target was described as a 'political compromise', which 'does not mask the need for each region to undertake significant increase and planning of reserve areas'. Widespread advocacy to increase the extent of reserved forests continues in many parts of the world, with a variety of conservation targets expressed.

Formal criteria for defining 'Forest' and a 'Forest Protected Area' are integral to understanding and conformity of such ventures, but the first four of the above categories are the most significant in a sequence from strict protection without unauthorised disturbance to areas in which biodiversity protection is combined/intermeshed with human traditional or cultural uses and may be highly uncertain. In the latter (mainly categories V, VI), trade-offs and some site degradation or losses are perhaps inevitable. One important need, for example, is for effective buffering of sites to increase their biological integrity and protection of the 'core' (Shafer 2015). Local governance and cooperation is wise, and relevant values have been debated strongly, with comment regarding those categories as being only 'soft

Table 9.2 IUCN protected area management categories, which should be based around the primary management objective(s), which should apply to at least three-quarters of the protected area
Reproduced, with permission, from Dudley (2013)

Ia Strict nature reserve. Strictly protected for biodiversity and also possibly geological/geomorphological features, where human visitation, use and impacts are controlled and limited to ensure protection of the conservation values.
Ib Wilderness area. Usually large unmodified or slightly modified areas, retaining their natural character and influence, without permanent or significant human habitation, protected and managed to preserve their natural condition.
II National park. Large natural or near-natural areas protecting large-scale ecological processes with characteristic species and ecosystems, which also have environmentally and culturally compatible spiritual, scientific, educational, recreational and visitor opportunities.
III Natural monument or feature. Areas set aside to protect a specific natural monument, which can be a landform, sea mount, marine cavern, geological feature such as a cave, or a living feature such as an ancient grove.
IV Habitat/species management area. Areas to protect particular species or habitats, where management reflects this priority. Many will need regular, active interventions to meet the needs of particular species or habitats, but this is not a requirement of the category.
V Protected landscape or seascape. Where the interaction of people and nature over time has produced a distinct character with significant ecological, biological, cultural and scenic value: and where safeguarding the integrity of this interaction is vital in protecting and sustaining the area and its associated nature conservation and other values.
VI Protected areas with sustainable use of natural resources. Areas which conserve ecosystems, together with associated cultural values and traditional natural resource management systems. Generally large, mainly in a natural condition, with a proportion under sustainable natural resource management and where low-level non-industrial natural resource use compatible with nature conservation is seen as one of the main aims.

protected areas'. Terborgh (2004) noted a 'pronounced trend' toward their creation, at the expense of further higher quality protected areas in categories I-IV.

One major hindrance in calling for increased forest reserves is simply that adequate stocks of all successional stages needed by some insect groups or guilds, such as saproxylic taxa, may require very large (several hundred hectares: Speight 1989) size in order to be effective. Speight (1989) also pointed out that most protected forests in Europe were already too small for the full range of natural forest conditions to re-establish, so that management to enhance those conditions becomes mandatory. Managed forest production landscapes in exploited areas can include Protected Forest Areas (PFAs), as discussed for Sweden by Hedwall and Mikusinski (2015) in a model which is useful for careful study for parallel moves elsewhere. PFAs may be small remnants of unmanaged forest with varying extents of influence from past changes in land use, but that also represent high conservation values. Their long-term development within a surrounding matrix relies on forest management in adjacent areas and understanding the impacts of local disturbances. Larger PFAs are effective in biodiversity conservation (Geldmann et al. 2013) – but understanding the features that enable this, and planning for them to be available, is a central need. Thus, the roles of mixed-species plantations, of old-growth and mature trees, the differing successional stages and supply of dead wood, and imposed and natural disturbance regimes, are all key features in forest biodiversity conservation.

The wide effectiveness of formal Protected Areas occurs through countering impacts on habitat cover and species populations and, in their overview of these from published studies, Geldmann et al. drew three general conclusions: (1) that PFAs have indeed preserved forest habitat; (2) that evidence that they have effectively maintained species populations, although with some positive outcomes reported, is inconclusive (but possible bias may be present, as only two of the 42 studies addressing this theme in their review noted insects, neither study in forests, and with the remainder dealing primarily with mammals [32 studies] or other vertebrates); and (3) that causal connections between conservation outcomes and management inputs to PFAs are rarely evaluated. The widespread dismissive assessment of Protected Areas as 'paper parks', although confronting, may be far less evident for insects than for trends among the more influential vertebrates, simply reflecting the capability of numerous insect taxa to thrive in small areas and with restricted resource supply. However, as Geldmann et al. (2013) commented 'testing how and why they [protected areas] are effective is of critical importance to conservation science.'

Creation of reserves in areas without silviculture, however desirable, is increasingly difficult to achieve and in most regions improved management for biodiversity in forest is the predominant – even only – option. Future establishments of large forest reserves, in particular, are becoming extremely difficult to pursue, despite wide recognition that the survival of much diversity, from primates to invertebrates, depends on such measures. However, setting aside large reserves is no longer even an option in some regions, because virtually the whole landscape has already been modified by people. That reality renders smaller areas, reserves or remnant forest fragments in anthropogenic landscapes increasingly important – both for themselves and as providing connectivity across those landscapes. Those areas may remain fortuitously, or because they are for some reason too difficult or remote to alter, or because they are actively protected from change because of some acknowledged local values or significance.

This is the case for sacred groves, fundamentally protected patches of natural forest vegetation in agricultural or urbanised landscapes, and which occur in many parts of the world. Ray et al. (2014) noted their preservation in the name of deities in Africa, Asia, America, Australasia and Europe, unified by being the purported residence of spiritual beings and in which ordinary activities – including tree felling and gathering or removal of wood – are prohibited. With the notable exception of West Africa, below, invertebrates have received little attention in studies of their biodiversity. The only insects noted in the 75 studies in India reviewed by Ray et al. were butterflies (and these only rarely noted), whilst surveys of plants by far exceeded all faunal studies combined.

The sacred forest groves of West Africa, for an example with close parallels elsewhere and in areas of considerable biological diversity, have considerable conservation value and have been protected over long periods by local indigenous culture.

Their roles as reservoirs for butterflies have long been appreciated (Larsen 2005), and explored in more detail recently (Bossart and Antwi 2016). Level of local care for sacred areas now varies greatly. For some, it has diminished, with one of Bossart and Antwi's study groves in Ghana (Kajease) now largely engulfed by urbanisation and degraded considerably by unregulated human intrusions. It is now essentially a rather low quality 'urban island' with surrounding urban development unlikely to allow much further colonisation by forest butterflies and others. Most sacred groves in Ghana apparently date from the early twentieth century, linked with the wider establishment of forest reserves around that time. In contrast to Kajease, Bossart and Antwi (2016) noted that some groves are still well protected and respected by their local village community, and survive because of these values shared between residents and authority, and based in cultural values that are traditionally well-embedded. Even small groves contribute to protection of much of the insect fauna found in larger forested areas – and Bossart and Antwi believed that they will be 'a critical element of any effective conservation management strategy'. Their survey, based on fruit-feeding butterflies, compared diversity in five groves within two larger forest reserves to demonstrate that even small sacred groves may mirror the conservation values of much larger forest patches.

Riparian strips (p. 214) have parallel values as contiguous corridors within forest and across changed areas between forest patches.

9.4.1 Fragmentation

The better-documented effects of forest fragmentation, of increasing vulnerability to resident species and facilitating invasions by alien species (p. 142), are important impetus to increase connectivity and preserve larger forest patches wherever possible, in the expectation of conserving some level of ecological integrity. However, in addition, the population dynamics of forest insects may be affected, because the resulting spatial heterogeneity may change the relative dispersal rates of the focal species and their natural enemies, so in some cases enabling 'escape' from the impacts of predators or parasitoids and permitting greater survival. These scenarios are invariably difficult to evaluate, but a study of outbreaks of a major North American pest species, the Forest tent caterpillar (*Malacosoma disstria*, Lasiocampidae), in Ontario helped understanding. The moth undergoes cyclic outbreaks, with severe defoliation of its boreal forest environment, in which the principal host tree is Aspen (*Populus tremuloides*) (Roland 1993). *Malacosoma* populations are driven largely by parasitoid and pathogen impacts, and Roland's survey suggested that increased forest fragmentation affects the interactions amongst these, and that the increased clearing of boreal forest may be exacerbating outbreaks. Duration of outbreaks was linked with increased structural heterogeneity of forest and non-forested land, expressed as 'Km of forest edge per Km^2', with outbreaks rising from 1–2 years in continuous forest to 4–6 years with increased forest edges.

Although isolated forest fragments have been likened to 'habitat islands', true islands surrounded by a water barrier are functionally likely to be more isolated than equivalent areas contagious with others on a continuous land mass. Few island forest faunas have been appraised, not least because of lack of good historical information and documentation of the pre-disturbance fauna. However, on the small (ca 4.2 Km^2) Malaysian island of Mengalum, some 56 Km from Kota Kinabalu, Sabah, Sodhi et al. (2010) showed that the transformation from being almost entirely forested in 1928 to entire loss of closed-canopy forest by the time of their survey in 2007 was indeed correlated with losses of some previously recorded Odonata and butterflies. They believed that five (of nine) and six (of 15) butterfly species recorded in 1928 had become extinct there. The losses included the two forest-associated dragonflies and two of four forest butterflies. Nevertheless, a number of 'new' species (14 Odonata, five butterflies) were found in 2007, and these included forest species (one dragonfly, three butterflies). Many of these are likely to be recent colonists – but the completeness of the 1928 survey inevitably remains uncertain. The study implied that tropical insects may be susceptible to continuing deforestation.

Isolated islands incur conservation problems through alien plant invasions or spread. Needs may arise to eradicate alien trees, as explored for the introduced forestry tree *Bischofia javanica* (Euphorbiaceae) on the Ogasawara Islands, Japan (Sugiura et al. 2009). Since the early twentieth century, this invasive tree has partially replaced native forests there, and is of concern because the islands support a highly endemic and diverse suite of wood-feeding beetles. Artificial gaps in the alien forest, created by ring-barking *Bischofia* trees, yielded more beetles of several families than did parallel Malaise trap catches in closed forest. More endemic Mordellidae were captured in the gaps, and most Cerambycidae and Scolytidae were more frequent in gaps than in closed canopy sites. Eradication of *Bischofia* was considered desirable in order to conserve native insects on the islands but, and despite the short-term nature of Sugiura et al.'s study, the artificial gaps emulating natural disturbances that increase heterogeneity may also have a role in conserving native endemic insects there.

'Gap width' between fragments may be critical, as creating access thresholds related to a potential coloniser's dispersal prowess. In Costa Rica the large damselfly *Megaloprepus caerulatus*, which breeds in tree-holes in mature forest, did not colonise isolated forest fragments. Trials on its dispersal range demonstrated that adults could cross gaps of 25 m, but were far less successful when confronted with gaps of 50–100 m (Khazan 2014). Individual marked damselflies tethered by a fine (0.07 mm) nylon thread were tested for success in dispersing across distances of 25, 50, 75 and 100 m from pasture to forest (Fig. 9.4). Limitation of dispersal in regions where desirable habitats such as patches of forest in a pasture landscape are isolated may be a critical factor affecting movement and colonisation. For *M. caerulatus*, the limited perceptual range affects the functional connectivity of the landscape, and Khazan noted that this aspect of fragmentation impacts has very rarely been studied. Inadvertent isolation of populations or metapopulation segregates from this cause may have increasingly intensive impacts as it progresses.

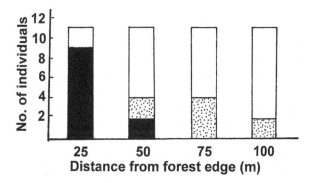

Fig. 9.4 The success of crossing pasture gaps in forest by a damselfly, *Megaloprepus caeruleus*, in Costa Rica at different distances from the forest edge (successfully flew to forest, black; perched for >3 minutes on living fence, dotted; perched for >3 minutes in pasture, open); (Khazan 2014)

Colonisation of forest fragments reflects both colonisation ability and the availability of critical resources, either or both of which may be limiting. In Venezuela, colonisation by the frugivorous nymphalid butterfly *Hamadryas februa* was not limiting, but lack of appropriate larval food plants in some forest patches restricted butterfly abundance across the metapopulation (Shahabuddin et al. 2000). The general trend, as predicted by Spence et al. (1996) for carabid beetles, that short-term impacts of fragmentation on forest assemblages will flow from increased diversity from invasion of the fragments by colonising habitat generalists and open habitat specialists from clearcut areas has much wider application. In Korea, continuous forest had higher richness and abundance of Carabidae than forest patches (Jung et al. 2018), and both 'small' (1.1–9.6 hectare) and 'medium' (12.8–51.2 hectare) patches conserved the beetle biodiversity less effectively than continuous forest. Medium remnants, however, were more suitable for forest specialist species than were the small remnants, as a trend consistent across the deciduous and pine forests sampled. Forest patch size had significant effect on the assemblages, and emphasise the importance of undisturbed 'core areas' for protection of the more specialised forest species.

9.4.1.1   Monarch Butterflies in Mexico

The 'Monarch Butterfly Biosphere Reserve' in Mexico is focused on one of the most important overwintering sites for the migratory Monarch butterfly (*Danaus plexippus*, Nymphalidae). Adult butterflies migrate from much of eastern North America in late summer to hibernate in massed gatherings clustered on trees of Oyamel fir (*Abies religiosa*) in a small montane area of Mexico. The area has for long been threatened by deforestation, with the almost inevitable consequences of mass butterfly loss and potential extinction of one of North America's most iconic and spectacular insects; those overwintering aggregations have for long been cited, together with those of California, as a 'threatened phenomenon' within the World Conservation Union. The Mexican overwintering sites occurred in 11 different mountains in the region, where suitable microclimates enabled butterflies to survive and conserve energy over winter, before they flew northward. Initially, five of those areas were protected by Presidential Decree and designation of the Biosphere Reserve in 1986, but the resulting conflicts between loggers and conservationists led to considerable tensions, and some re-design of the original reserve (Bojorquez-Tapia et al. 2003). Using expert opinion and knowledge as bases for spatial modelling, the prime overwintering habitats were mapped more precisely, and enabled Bojorquez-Tapia et al. to define a core area that in principle would minimise inclusion of the commercially desirable forest stands under greatest local pressures for logging.

In California, plantations of the Australian *Eucalyptus globulus* have helped to compensate for the loss of native trees used as roosting sites by *D. plexippus* (Lane 1993), but constraints on deforestation in Mexico have proved difficult to implement. This is despite the high formal status of the area as a Biosphere Reserve, and recognition of the urgent need for forest conservation there. Series of decrees sought to strenuously protect 'core roosting sites', and their buffer zones. The history of impacts on the area includes social conflicts, with economic activities now restricted to the buffer zones and the core used for scientific research activities only. Changing boundaries and extents accompanied by a range of support programmes in turn support a burgeoning ecotourism programme that accommodates the many visitors who arrive each year to view the butterflies (Manzo-Delgado et al. 2014). Local income from this activity is viewed as a potential alternative for economic development.

9.4.1.2 The Wog Wog Experiment

Fundamental understanding of impacts of forest fragmentation on insect communities is still rather sparse, and their integration with forestry impacts even more so. One of the most informative studies on effects of eucalypt forest fragmentation in Australia is that known widely as the 'Wog Wog habitat fragmentation experiment' (Margules 1992), in which *Pinus radiata* plantations were established within the area and beetles sampled to assess richness and community changes across time and treatments. The experimental design (Fig. 9.5) comprised six replicates, each of three plots – small (0.25 hectare), medium (0.875 hectare), and large (3.062 hectare). Four replicates of each plot size became 'fragments' after the embedding eucalypt forest was removed (1987) and planted to *P. radiata* in 1988. The other two replicates of each size remained in uncleared forest and were treated as unfragmented controls. The rationale of the Wog Wog experiment (discussed by Margules 1992) set out to test two hypotheses related to habitat fragmentation: (1) that habitat fragmentation reduces biological diversity, and (2) that the reduction in diversity depends on the fragment size – hence the graded size series of plots.

Sampling of beetles over seven years included two pre-fragmentation years, using pitfall traps for a week each season, for a total of 28 sampling occasions. By 1995, catches included 655 beetle species, many of them unnamed, and more than a third of them represented by only one or two individuals. However, 325 species were followed over the seven years, with changes suggesting effects of fragmentation

Fig. 9.5 The Wog Wog experimental design to study fragmentation impacts on forest beetles in New South Wales (see text); survey sites of three size classes are shown within eucalyptus forest (shaded) and as remnants within pine plantations (open, in which isolated study sites are also indicated by separate spots). (Margules 1992)

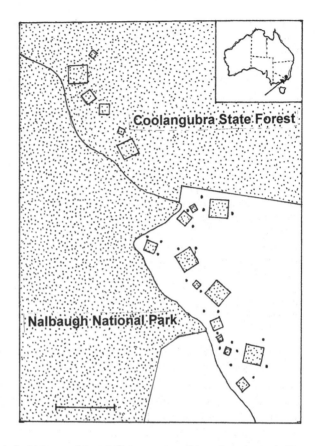

(Davies et al. 2001). Both 'within-patch' and 'between-patch' scales showed distinct species-level and community-level responses (Table 9.3). Within patches, two edge effects were discernible, with different penetration distances – richness at fragment edges increased from a 'shallow effect' (20 m or less), and relative abundance and composition responded to a 'deeper' edge effect (of around 100 m).

After about a quarter of a century, microclimate differences in the three fragment sizes remain (Farmilo et al. 2013), and the smallest fragments were considered too small to buffer any interior area against edge effects. The plantation matrix was demonstrated to have substantial impacts on ecological processes – Farmilo et al. reported changes in canopy cover, litter cover, soil moisture and temperature within fragments compared with the control plots within continuous forest. Altered conditions within small fragments, however, have implications for management to conserve native diversity and suggest that very small fragments may not achieve this purpose.

Four main findings are relevant considerations for forest management in Australia and, perhaps, more widely. First, the two different edge effects with different outcomes demonstrated greatest changes at the edges of small fragments, and only negligible changes in the interior of large fragments. Second, fragmentation reduced

Table 9.3 The impacts of forest habitat fragmentation on beetle assemblages found at Wog Wog, New South Wales, after seven years of survey (see text)
Abbreviated from Davies et al. (2001)

Community structure responses:
Species richness – significant effect of fragmentation: no effect of fragment size, but edge effects to 20 m or less associated with increased richness
Species composition – changed significantly in fragments compared with continuous forest.
Relative abundance – changes due to habitat fragmentation were more pronounced than changes in species composition, with biggest effects in slope habitats.
Turnover – fragmentation reduced species turnover significantly in slope habitats but not in drainage line habitats.
Individual species responses:
Rarity – natural abundance affected responses of species to fragmentation: abundant species increased in occurrence in fragments compared with continuous forest, rare species were not affected.
Isolation – species that were not isolated increased in occurrence in fragments, most at small edges and least at large interiors. Species isolated on fragments were only slightly affected by fragmentation.
Trophic groups – some significant effects of trophic group on response to fragmentation: detritivores and fungivores occurred at more sites post-fragmentation, but occurrences of herbivores were variable; predators occurred at fewer sites post-fragmentation.

species turnover, with this and and other changes in community structure declining over the extended survey period. Third, 'matrix sites'resembled fragments more than continuous forest sites, and contained fewer beetle species than either of these. Last, individual species' changes were associated with rarity, isolation and feeding habits – and the patterns paralleled the edge impacts noted above. Edge effects were considered the most important outcomes of habitat fragmentation, and were the major influences on community change within fragments.

More specific focus on Carabidae at Wog Wog (45 species trapped: Davies and Margules 1998) emphasised trends among the eight most frequently collected species – together comprising about 92% of pitfall-trapped ground beetles – and their responses to habitat fragmentation. Overall species richness did not decline with fragmentation, but the eight species exhibited three diffent responses. Six species showed very clear responses, three decreasing in remnants compared with continuous forest, and three showing the converse response, with increased abundance in fragments. The other two species showed no such responses. Of the 'responders', three responded further, and individualistically, to fragment size: one was most abundant in small remnants, one most abundant in large remnants, and the third was equally abundant in small and large remnants but less abundant in the intermediate remnant size.

Near Tumut, New South Wales, Buccleuch State Forest contains many (192) eucalypt forest patches within a large (> 50,000 ha) plantation area of *Pinus radiata*, with the pines abutting large areas of native forest. Genetic study of two species of flightless tenebrionid beetles (*Adelium calosomoides, Apasis puncticeps*) associated with rotting logs involved direct collections from dead logs in eucalypt patches

(21–36 years after the patches were created) and the native forest (Schmuki et al. 2006). Both species revealed reduced mobility and gene flow in fragmented forest – trends present, surprisingly, after only about 10–15 generations. *A. calosomoides* was the better disperser, but nevertheless had mobility impeded through patch and pine areas relative to within continuous forest.

9.5 Corridors and Connectivity

Linear strips of forest have potential to function as corridors that enable or facilitate dispersal of insects between larger forest patches, and so reduce isolation of populations in fragmented environments. However, that appealing and simplistic ideal is often difficult to validate, and many authors have commented on what Varkonyi et al. (2003) referred to as the 'surprisingly little evidence supporting that corridors are effective dispersal routes'. An important consideration, and bias, for studies of insect movements in forest corridors is simply that true forest-dwelling insects have received rather little attention in relation to forest-avoiding species that disperse through cleared corridors within the forest matrix. Studies on forest insects in forest corridors are scarce.

The twin functional roles of corridors need to be differentiated clearly in how they may benefit species, as (1) with continuing habitat destruction, reduction of remaining habitat patch isolation by landscape-level planning beforehand may be possible by incorporating corridors, and (2) when remaining habitat patches are small and isolated from each other, corridors may effectively reduce chances of stochastic losses and inbreeding depression. Both outcomes, of course, depend on corridors being used effectively by the species of concern!

Most studies on the potential values of corridors for insects have involved carabid beetles or butterflies, the latter mostly in open habitats through forests rather than in forest itself. The incidence and relative abundance of moths in boreal spruce forest corridors in Finland was evaluated by bait traps across replicates of four treatments, namely (1) forest interiors (> 110 m from the nearest edge); (2) forest edge (17–25 m from edge of mature spruce forest); (3) corridors (average 50 m wide, surrounded by recent clearcuts and secondary growth); and (4) clearcut areas (at least 50 m from the nearest mature forest) (Monkkonen and Mutanen 2003). Many of the 57 moth species caught were represented by singletons or very low numbers, and nine species, each represented by more than 100 individuals, accounted for 92% of the 6485 specimens. All these were Noctuidae or Geometridae, and *Xestia speciosa* was by far the most abundant species (with 3260 individuals). Most species occurred relatively commonly in corridors, and none showed lower abundance in corridors or at edges than in forest itself. Two species (*Arichanna melanaria*, *Chloroclysta citrata*) were more abundant in corridors than in forest interiors. This might represent dispersal movements along the corridors, but that behaviour cannot be confirmed without further assessments. Their presence, however, implies that the moths indeed 'used' these riparian forest corridors.

The limited information on corridor use by, and conservation values for, forest insects is partially a consequence of very limited knowledge on how those insects disperse (Varkonyi et al. 2003). Their mark-release-recapture study of two closely related noctuid moths (*Xestia speciosa*, *X. fennica*) in a fragmented old-growth spruce forest (including corridors) in Finland revealed some of the complexities of interspecific behavioural and ecological differences. *X. speciosa* is more generalised in its habitat use and occurs in natural and managed spruce forest, whilst *X. fennica* is more clearly restricted to old-growth forest. Colour-marked moths were released by night and recaptures sought by sugar bait traps and traps with simple sugar attractant – the latter less rewarding, and collectively with some strongly differential captures between males and females. Both species, however, used the corridors and, in general, avoided entering the open matrix. For both species, the forest corridors were both dispersal routes and breeding areas. As for other organisms, it was probable that the moths, as forest interior species, may be disadvantaged by narrow corridors in which edge effects constitute barriers.

Intensity of management of 'open' corridors within forests is often very high, with roads and powerline easements widespread examples of this. Powerline corridors are of particular interest, because they are not used intensively by traffic, have usually not been paved or otherwise surfaced, and are simply linear strips kept open by regular thinning and clearing of shrubs and young trees to create conditions in which open vegetation and wild flowers may thrive. In Sweden for example, those corridors are described as undergoing management that 'creates conditions for a species-rich plant community similar to that in extensively managed grasslands' (Berg et al. 2016). They are thus important habitats for butterflies and others, independently of their wider context.

9.6   Retention Forestry

Retention forestry has been adopted progressively in many places as a measure toward conserving biodiversity, specifically with the purpose of promoting heterogeneity in forest composition and structure. Standing living or dead trees (below) can be left as patches or smaller dispersed groupings. Advantages claimed for the former (Lee et al. 2017, in relation to White spruce [*Picea glauca*] forests in Canada) include (1) retention of overstorey trees and multiple canopy levels; (2) maintaining undisturbed leaf litter and soil, so helping to sustain the large guilds of invertebrates using those resources; (3) ameliorating local microclimate conditions, so that tree mortality to wind-throw may be lessened; and (4) conserving structurally and biologically complex old-growth features, together with original flora and fauna. For dispersed retention patterns, Lee et al. (2017) noted (1) enhanced connectivity of underground biota across blocks harvested; (2) promotion of seedling growth; (3) benefits to species with any 'edge preference' (p. 9); (4) increased evenness of coarse woody debris distribution; and (5) increased aesthetic perceptions of harvested landscapes. Natural protection of aggregated retention may be improved by the roles of dispersed retention as windbreaks, and the two approaches may be used together for block management. The above listing may lead to definitions of parallel advantages for other forest types and regions, but these are only rarely set out in such form. Lee et al.'s study focused on saproxylic beetles in *P. glauca* forest management, but the principles listed are likely to have much wider application.

The conservation benefits of tree retention over more traditional clear-felling devolve largely on the concept of the retained trees being 'life-boats' for species over the periods needed for wider forest regeneration, sustaining microhabitats for those species to remain continuously present and enabling them to constitute reservoir populations over that period. They also enhance connectivity and so may also enhance species' dispersal. The topic of life-boating has attracted considerable attention. Rosenvald and Lohmus (2008) assessed 214 relevant studies from Europe and North America but in some cases assessments were difficult or impossible to make: saproxylic insects and their predators, for example, could not be appraised realistically because green tree retention was often mixed with simultaneous retention of dead wood, so that the exact contribution of the trees was unclear. Only 20 studies, 16 of them from boreal forests, involved arthropods. The vulnerability of scarce threatened species is especially difficult to assess because of their very low incidence at the scale of individual sites and within the individual history of those sites. The concept is attractive, perhaps a core basic paradigm, but necessarily relies largely on generalities for predicting its effects.

Retention forestry also incorporates considerations of landscape design and forest dispersion and for forest establishment. In one sense, it overlaps with 'set aside' forestry, in which forest blocks are reserved and protected from exploitation. Large-scale ecological networks are used in South Africa to offset the impacts of commercial afforestation on biodiversity. These, after Samways (2007), have considerable conservation importance in changed landscapes, both in directly providing 'habitat' for native species and in facilitating their dispersal among more persistent natural areas. They effectively mitigate some effects of habitat fragmentation (Jongman and Pungetti 2004), but there is rather little evidence for their worth for insects, although a study on grasshoppers supported their value (Bazelet and Samways 2011). That study took place across a network of 15 semi-natural grassland sites within an alien *Eucalyptus* timber matrix in Kwa-Zulu Natal, South Africa, in which 18 species of grasshoppers (of 38 collected) were sufficiently common for appraisal. Sampling was by direct collection and 'flushing' observations. The responses varied across higher taxa, but species' responses to the three categories of environmental variables were consistent over the two years of study. 'Area of site' and whether the site had been burned three years before grasshopper sampling were the most influential variables, and the major response (in only one sampling year) was to the proportion of short grasses at a site. The potential of ecological networks for mitigating losses of local insects will vary amongst taxa, but it seemed that the grasshopper assemblage in this Natal study was more sensitive to such internal patch variations than to patch isolation.

Populations of threatened beetles associated with aspen (*Populus tremuloides*) in Finland can be conserved by a 'retention harvest system' (Martikainen 2001,

Martikainen et al. 2006). Many beetles associated with aspen can tolerate clearfelling, and retention of both live and dead trees on largely cleared areas may be beneficial. Following Franklin et al. (1997), Martikainen (2001) considered three main functions of retained trees in that context as (1) safeguarding species and processes over the period from logging/clearing until forest cover has again developed; (2) ensuring the structure and resources of any re-established stand on the area; and (3) enhancing connectivity in the managed landscape. The first of these supposes that the retained trees act as refugia from which a re-established forest may gain source populations of beetles and others from nearby. Martikainen suggested, for example, that large aspens could function as refugia for some threatened beetles (such as *Cossus cylindricus*) for up to a century, over which time new hollow trees may have developed. For some species, the retained trees may become preferred habitat after surrounding disturbance, rather than simply remnant refugia. However, as Martikainen cautioned, in some regions threatened aspen specialist beetles may already have vanished – so that retention trees can only help to sustain an unknown proportion of the original assemblages. The second function above simply ensures the presence of large aspen trees which would otherwise take many decades to develop, as mosaic components in managed forest stands; and the third function is somewhat similar in providing suitable habitats across the beetles' wider dispersal range. An important role of retention trees is simply to assure future supplies of dead wood (Freedman et al. 1996). Green tree retention thus attempts to imitate the conditions that occur in forest after moderate disturbance (Vanha-Majamaa and Jalonen 2001), with the two main approaches being to retain scattered individual trees or to retain aggregated patches of trees, with the latter suggested to be more effective in maintaining an array of features and resources. Patches of retained trees can sustain 'key habitats' – such as patches of moist forest likely to be relatively resistant to fires – as refuges for insects. Prescriptions for numbers or density of trees to be retained vary greatly

Other studies in Fennoscandia have also supported the values of retention aspen trees for beetles, not least for their use by the numerous saproxylic species (p. 151). Thus, larger patches of aspen, with both living and dead trees, were correlated positively with beetle richness in Sweden (Sahlin and Schroeder 2010) and deliberate retention of aspens in Norway was endorsed by Sverdrup-Thygeson and Ims (2002).

Retention of both living and dead trees within logged or felled coupes ('harvest residuals') can be viewed also as providing analogues of 'fire residuals' in forests with long history of natural wildfire disturbance, to which the resident biota have adjusted in some way to become resilient. Thus, Gandhi et al. (2004) defined fire residuals as 'forest patches left unburned during wildfire', and hypothesised that these patches retain the structural elements of the forest that are sufficient to conserve populations of the local fauna, even acting – if large enough – as reserves or reservoir habitats for such species. Comparative surveys, using pitfall traps, of epigaeic beetles (Carabidae, 32 species; Staphylinidae, 46 species) assessed values of putative residuals of retention trees, as one of the series of treatments (harvest residuals [mainly of Lodgepole pine, *Pinus contorta*], true fire residuals, clearcut areas, uncut forest) in Alberta, Canada, and gave the information on distributions

Fig. 9.6 The captures of (**a**) Carabidae and (**b**) Staphylinidae in mature forest (open), burned forests (dotted) and fire residuals (black) at study sites of two ages in Alberta, Canada. (Gandhi et al. 2004)

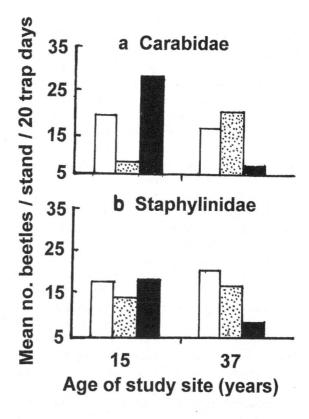

summarised in Fig. 9.6. Assemblages in the harvest residuals were similar to those in uncut forest, but one-year residuals did not retain higher diversity than in surrounding uncut or cut forest. Harvest residuals contained more beetle species than did the fire residuals, reflecting immigration of open-habitat species. However, earlier suggestions that larger residuals have higher beetle richness and abundance than smaller residuals were not supported for species diversity, total beetle catches, or catches of the more abundant forest beetle species. Whilst harvest residuals in that study retained some of the elements of uncut forest, it remained unclear whether harvest residuals fully equated to fire residuals in conserving biodiversity, or if the inferences are of more general relevance. That study was in high elevation (1525–1710 m) subalpine forest, with trapping over the three months (June–August) when it was snow-free.

The roles of retention trees, whether single or in groves, include providing some of the substrate needed by early-successional saproxylic insects, of which beetles have become the predominant representatives in management plans, and to thereby alleviate some of the consequences of clear-cutting on those taxa. However, they cannot wholly emulate the benefits of intact mature forests (Gustafsson et al. 2010), but are valuable participants in sustaining early successional progress and ecosystem functions. The 'life-boat' role of retention trees is an attractive practical concept, but may simply manifest through the assured continuation of coarse woody debris. Comparison of the carabid beetles and spiders (ground-active taxa, assessed by pitfall trap catches) in smaller and larger groups of retention trees was founded in the assumption that larger groups of trees should retain the original fauna more effectively than smaller groups of trees, because (1) the 'normal' light intensity and moisture regimes would be better retained, and (2) edge effects are not as severe in larger groups (Matveinen–Huju et al. 2006). As in some parallel situations, small groups of trees might be entirely edge-affected. No unambiguous support was found for a primarily 'life-boat role' of retained trees from the original forest species, and for some trials it seemed that the simple presence of dead wood may be more important than microclimate regimes.

Delaying harvest of trees, particularly of over-mature trees, has thus been proposedly as a measure to maintain diversity of saproxylic insects associated with old-growth forest, but can lead to managerial conflict in that, whilst older trees have greater ecological value, larger trees have higher economic value (Bouget et al. 2014). However, the practice may benefit markedly by active augmentation to this 'passive self-retention' approach, to promote development of what Bouget et al. referred to as 'old-growth features', rather than simply delaying harvesting. Those features (including stands of dead trees, tree cavities and large fallen logs) can be generated artificially, and may collectively be more beneficial to saproxylic insects than relying on their more natural development through extended rotations and reservation. Those active management complements deserve wide consideration, perhaps especially when known threatened insects depending on those features become increasingly vulnerable.

Retention forestry has expanded as a conservation tool in many contexts, but assessing its effectiveness is complex. The unifying belief and principle of retaining biological legacies across forest generations in different forest types and at different scales is, perhaps, a major driver of its use. Reviewing numerous examples of the practice, Mori and Kitagawa (2014) noted the perception that seeking only a low level of retention was undesirable but 'at least support the notion that any level of retention is better than leaving nothing', and advocated the perspective of retention practices focusing more on what species are left in the post-harvest forest stands. Retention of dead wood for saproxylic insects (p. 156) is a widely accepted need, but wider retention forestry can foster conservation of species richness which is largely equivalent to that of primary forest stands. Mori and Kitagawa drew attention to the different management foci of retention forestry (focusing on what is, or should be, retained) and selective logging (with focus on what is logged or removed), respectively.

The unifying and essential principle of retention forestry is that, at the time of tree harvest, selected important structures are retained intentionally on the site, with conservation of biodiversity one of the explicit purposes of this. The major objectives of the practice are varied (Gustafsson et al. 2012, Table 9.4) but link fundamentally with assuring continuity of key forest resources and supporting their biological legacy through providing the structures and organisms essential for forest communities. Some form of retention forestry can be applied to all silvicultural

Table 9.4 Retention forestry: the varied objectives of use to maintain structures and organisms from the pre-harvest forest ecosystem, as listed by Gustafsson et al. (2012)

1. Maintain and enhance the supply of ecosystem services and provisioning of biodiversity.
2. Increase public acceptance of forest harvesting and the options for future forest use.
3. Enrich the structure and composition of the post-harvest forest.
4. Achieve temporal and spatial continuity of key habitat elements and processes, including those needed by both early and late-successional specialist species.
5. Maintain connectivity in the managed forest landscape.
6. Minimise the off-site impacts of harvesting, such as on aquatic systems.
7. Improve the aesthetics of harvested forests.

systems and forest types, with the form and extent of forest retention open to considerable variation. These can differ across sites and forest categories, and reflect local practices and conditions: they may also be subject to thresholds defined under some form of local or national regulation or forest certification need. The long-term studies needed to validate the various practices continue – there are relatively few of these, but Gustafsson et al. noted the example of the state forests of Tasmania, and also that information on 'best practice' from such studies will help to evaluate and refine practicalities for the future. Retention is an important tool in forest restoration.

The ideal of retention forestry – and, indeed, of wider management of production forests – is in principle simple: that timber harvesting and related extraction operations should not reduce the sustainability and future wellbeing of biodiversity and ecological functions of the forest.

9.7 Scattered and Veteran Trees

'Retention', as above, links strongly with the importance of older trees for insects in the landscape. One of the most important habitats for saproxylic beetles in Europe, old mature or post-mature trees, have suffered extensively through lack of appreciation of their ecological importance (Nieto and Alexander 2010a). Many have been removed as 'untidy' or posing potential danger to public safety through falling, or dropping branches (p. 229). The term 'defective trees' used in some harvest logging contexts (Lindenmayer et al. 2002) can also be applied to old trees. Nieto and Alexander (2010b) pointed out that old trees and dead branches are often 'automatically removed', without proper assessment of any actual threat they might pose, from areas frequented by people. Whilst understandable in an era of growing litigation in case of accident, this syndrome is ecologically unfortunate.

The remaining scattered large trees are important features of many formerly forested landscapes, and may survive broader forest clearing by design or simple chance. Scattered large eucalypts were habitually left on land cleared for farm stock in Australia, for example, as providing shade for the animals, and some individual trees have cultural values to early Australians. Whatever the reason for their persistence, such old trees can provide local reservoirs for insect diversity, and are recognised in helping to counter losses of insect species to intensive agroecosystems. Some are recognised also as potentially harbouring the natural enemies of insect pests harming nearby crops, and providing alternative food for these over non-cropping seasons, so have tangible 'applied values' to local farmers. As remnants of the original forest present, those trees have conservation values of their own as well as through general features such as augmenting local heterogeneity and facilitating connectivity (Athayde et al. 2015).

Declines of large trees in forests throughout the world cause considerable concern (Lindenmayer et al. 2014, Lindenmayer 2017). Whether in production forests or as remnants on open savanna or farmland, they have varied and important ecological roles, either as single trees or as small remnant groups, but are also susceptible to a range of threats, so that conservation may need almost individual attention to reflect their local environment and context. The characteristics of, and conservation needs for, large old trees vary greatly depending on taxon and ecological context (Lindenmayer and Laurance 2017). Lindenmayer (2017) noted three general principles for their conservation, each likely to apply widely, as (1) protecting existing large old trees; (2) reducing the risks of mortality among large old trees; and (3) ensuring the presence of sufficient recruits among younger trees to replace current older trees as they progressively disappear. Protection of existing trees was considered the most important single management tool, and is the option most frequently considered for conservation of saproxylic invertebrates. For the Mountain ash (*Eucalyptus regnans*) forests in central Victoria, large old trees are increasingly vulnerable and their loss has considerable negative ecological impacts (Lindenmayer et al. 2015): protection measures recommended include imposing 100 m buffers around large old trees, as well as around remnant old-growth forest patches and as riparian strips along streams.

Both within forests and on cleared ground, large old 'solitary trees' (Sebek et al. 2016) or 'veteran trees' are significant landscape features for insect conservation, but their individual position in the landscape may influence their functional roles. Comparisons between the insect faunas of solitary trees and forest regimes – specifically solitary trees in open woodlands and habitats of nearby closed canopy forest – in the Czech Republic showed that solitary trees harbour the greatest species richness of Coleoptera and aculeate Hymenoptera (bees, wasps, and ants), with canopy edge regions also of high value for bees and wasps. Forest interior assemblages were largely nested subsets of the solitary tree fauna. Beetle species richness declined from the solitary trees to the forest interior, and compilation of the overall 'conservation value' for each of the insect taxon groups showed highest values for the solitary trees. For beetles, bees and wasps, the conservation values were computed as the sum of abundances weighted by the individual species' status in the national red list. Weightings used were: regionally extinct (5), critically endangered (4), endangered (3), vulnerable (2) and near threatened (1). For ants, abundance is an unreliable parameter because of nest-based foraging patterns, and conservation values were expressed simply by presence of species (Fig. 9.7). Sampling was by

Fig. 9.7 The 'conservation values' modelled for (**a**) beetles, (**b**) bee and wasps, and (**c**) ants for different forest strata (C, canopy; U, understorey) and habitats in the Czech Republic. Conservation values were computed for each trap as a sum of abundances (for ants, presence/absence data were used) weighted by the species' status in national red lists. (Sebek et al. 2016)

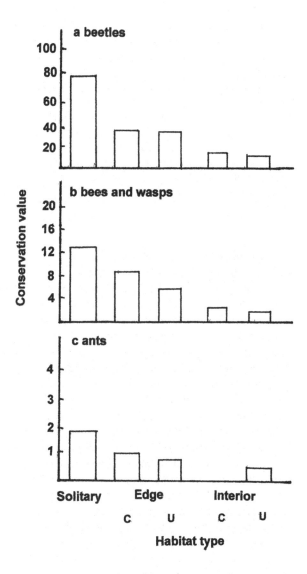

flight interception traps only, but this survey demonstrated both the richness and the irreplaceability of the assemblages on solitary trees, with their exposure aiding establishment of insect species that could not thrive in closed forests. Sebek et al. (2016) claimed that the biodiversity of temperate woodlands is enhanced greatly by presence of solitary trees.

On a similar theme, Horak (2017) noted that 'veteran trees' in woodlands and more open landscapes are important for saproxylic organisms – and also that 'large' trees are not necessarily 'veterans', which are described by, at least, components of size (diameter), age, and the microhabitats present. Thus, age may be associated

Table 9.5 Recommendations for assessing hazards posed by trees in urban green areas, to help promote conservation of saproxylic insects in harmony with public safety (Carpaneto et al. 2010)

1. Reduce cuts of secondary tree branches to conserve a canopy cover and optimal microclimate conditions within tree cavities.
2. Leave the largest branches and trunks on the ground after cutting, in order to increase development sites for saproxylic taxa.
3. Use steel ropes and other shores to hold collapsing trees.
4. Apply 'natural fracture techniques' such as pruning methods to mimic fractured ends that naturally occur on trunks and branches.
5. Remove inadequate reinforcement materials used in the past to strengthen old trees or to occlude hollows.
6. Remove trash and other objects left in hollows by people, in order to recover habitat conditions.

with development of tree hollows, as a clear need for nesting sites for a range of birds and arboreal marsupials in Australia and which also harbour a substantial variety of insects and other invertebrates. However, whilst age and microhabitat features may only be estimated, tree diameter can be measured directly and accurately, so is often the 'preferred variable' in surveys.

Perhaps especially in parkland and urban forests, veteran trees can have considerable values in harbouring small populations of saproxylic insects, not least because those trees are indeed allowed to age naturally rather than succumb to commercial harvesting at some earlier stage. However, as Carpaneto et al. (2010) discussed, those same urban trees can become a public danger due to falling branches and progressive weakness from boring and tunneling activities. Nevertheless 'an urban park entrapped inside a big city can be a reservoir for saproxylic beetles' and for species such as *Osmoderma eremita* in Europe may be more hospitable than the diminished host tree supply in the wider countryside. There is little doubt of parallel significance, as yet undocumented, of veteran urban native trees in Australia – and little doubt also of the hazard potential sometimes present from old eucalypts that are notorious for dropping limbs! Carpaneto et al.'s (2010) application of 'Falling Risk Category' to individual trees, and derived from visual assessments and penetrometer measurements to categorise trees from 'A' (healthy tree) to 'D' (severely damaged) led to a series of management recommendations to promote conservation of saproxylic insects in urban areas in conjunction with measures to protect public safety (Table 9.5).

Linking with the extensive work in Scandinavia on values of dead wood for beetles, Jonsell (2012) explored the values of 'old trees' in Parks in Sweden – sites where management usually includes sanitation measures such as removal of dead wood to maintain an aesthetically appealing 'tidy' garden appearance. In addition, the local supplies of future dead wood may be very low, and Jonsell reiterated the themes that the fauna of ancient trees is threatened in Europe because those trees have become increasingly scarce. In Sweden, many of the remaining ancient trees are in parkland, notably former estates of the nobility. More widely in Europe, such

Table 9.6 The numbers of saproxylic beetles obtained from a study of their occupancy of substrates on lime trees (*Tilia* spp.) in parkland sites in Sweden (Jonsell 2012)

	All Saproxylic	Hollows	Wood/bark	Sap runs
All species				
No. individuals	14,460	5352	8862	246
No. species	323	56	259	8
Red-listed species				
No. individuals	1429	331	1098	0
No. species	50	17	33	0

parks were established around notable houses from the late seventeenth century onwards. Jonsell used lime trees (*Tilia cordata*), a species which was planted widely to form park avenues around 250–300 years ago, from which to sample beetles by window traps placed near tree hollows. Trees in three site categories were compared: 'open' (grazed wooded meadows), 'regrown' (as open, but with regrown trees from 40–70 years old), and 'park'. Four trees, preferably coarse and hollow, were selected on each of 27 sites. Almost 15,000 individuals, of 323 species of saproxylic beetles, were collected (Table 9.6). Numbers of hollow-dwellers and red-listed species did not differ greatly between parks and more natural sites (Fig. 9.8), implying that the old park trees, despite intensive local management, still harbour significant fauna, although with significantly fewer taxa associated with bark and wood. No sites were categorised as 'forest', reflecting that in the region lime trees had almost always been grown within agricultural landscapes. Old park trees were a valuable reservoir for saproxylic beetles, but in many areas are still targeted for sanitation removal. Jonsell (2012) noted the possible conservation focus of retaining old trees (and/or fallen or cut dead wood) in 'tree graveyards' in less conspicuous parts of parks, in conjunction with assuring a long-term continuous supply of old trees.

A characteristic feature of many older trees, 'tree hollows' are most noted in Australia as nesting or roosting sites for birds and arboreal marsupials and, in that role, are firmly on the agendas of many conservation plans for forest vertebrates. Less heralded, and far less explored, for their values in saproxylic insect (and other invertebrates) conservation, that parallel role also appears likely to be significant. Elsewhere, saproxylic beetles (in particular) have been associated with tree holes. Cavities in several broad-leaved trees (*Quercus*, *Fraxinus*) in Spain supported assemblages of saproxylic beetles and hoverflies (Syrphidae), with larger hollows associated with higher insect richness and abundance (Quinto et al. 2012). In those Mediterranean woodlands, tree hollows are regarded as one of the most important microhabitats for those insects. Surveys and protection of insects living in tree hollows may necessitate exclusion of vertebrates seeking such hollows for nests or roosts – a black acrylic mesh was used by Schauer et al. (2017) to safeguard insects in beech forest in Germany, for example. Such protection is necessary for some of the rarer beetles that depend on decaying wood mould within tree hollows. The

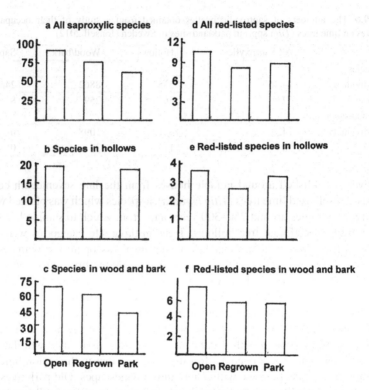

Fig. 9.8 The numbers of saproxylic beetles found with *Tilia* in three stand types as
'Open' (8 sites), 'Regrown' (11 sites) and 'Park' (8 sites), with numbers of red-listed species and
incidence in tree hollows and wood/bark samples shown separately. (Jonsell 2012)

Violet click beetle (*Limoniscus violaceus*, Elateridae) is associated strongly with
advanced decay stages in France (Gouix et al. 2015), and occurs in the basal hollows of a range of broadleaved trees once a suitable bed of wood mould has been
formed. Netting of hollows over tree trunks may be needed also to protect saproxylic beetles from predation by birds (*Cerambyx cerdo*: Iwona et al. 2017).

Hollow oaks have received considerable attention in planning conservation of
associated beetles in Europe, and such data have contributed to the wider debates
over optimal conservation targets – whether proactive or reactive conservation may
be more effective. Gough et al. (2014), for example, sampled 307 beetle species
from oaks in Norway in assessing common ground between these conservation
approaches. Proactive conservation gives priority to irreplaceability – exemplified
by the hundreds of beetle species associated with hollow oaks and that are not
threatened at present but are rare by being locally endemic or having very limited
distributions and small populations. In contrast, reactive conservation gives priority
to red-listed species already needing urgent measures to prevent their extinction; it
thus gives priority to vulnerability.

The surroundings of hollow, or other veteran, trees can have wide influences on associated fauna. Again for oaks in Norway, comparison of the rare and threatened Coleoptera from trees in open parkland and in forest showed some clear differences, despite general similarity in their diversity of red-listed species (Sverdrup-Thygeson et al. 2010). Oaks in parks had greater proportions of the species associated with hollows, whilst forest oaks had higher proportions of those species that depend more generally on dead wood. In general, parkland oaks were larger and had less canopy cover, but the smaller forest oaks still supported similar beetle species richness. As well as clarifying which features of the trees and adjacent areas were important influences (tree circumference, cavity decay stage, proportion of oaks in the surroundings, and amount of coarse woody debris all affected richness of red-listed beetles), Sverdrup-Thygeson et al. also demonstrated that the two tree categories can not be wholly substituted in seeking to conserve all the region's associated beetles. They also noted, however, that high amounts of coarse woody debris may compensate in part for the negative impacts on species richness associated with smaller trees and less sun exposure of forest oaks.

A wider perspective on management of tree hollows in eucalypt forests (Gibbons and Lindenmayer 1996) emphasised the diversity of roles, and that meeting the needs of all cavity-nesting animal species across logged sites would need considerable changes in forestry practices, moving to partial cutting systems rather than short rotation clear felling. They noted that approximately 180 species of Australian vertebrates use eucalypt tree hollows, together with a wide array of invertebrates. Suitable hollows for some vertebrates take a century or more to develop, with invertebrates a contributor to this process: Gibbons and Lindenmayer pointed out that Australia has no primary excavating vertebrates (such as woodpeckers) whose activities can accelerate hollow development. Retention of hollow-bearing trees is prescribed widely by management agencies, as widespread acknowledgement of their conservation significance – albeit acknowledged mainly for vertebrates. As well as conserving selected hollow-bearing trees, long-term management needs include assuring the supply of hollows through continuing to protect the trees and recruiting younger trees and allowing them to mature and develop hollows in the future.

As a practical parallel to supply of nest boxes for birds and mammals, and with earlier advice (Jansson et al. 2009) that such structures may contribute to conservation of saproxylic beetles, Hilszczanski et al. (2014) explored the use of wooden boxes mimicking tree cavities for the endangered beetle, *Osmoderma barnabita*, placing larvae of the beetle as innoculants into boxes. This beetle is a characteristic inhabitant of oak tree cavities, and has been found also in hollows that contain bird nests. The boxes used contained a mixture of oak sawdust and oak leaves, and were installed by tying to trees about 4–5 m from the ground. Colonisation by local beetle populations was relatively low, but colonists included other beetles, Cetoniidae. Further study is needed to clarify values of nest boxes for *O. barnabita*, but Hilszczanski et al. noted the possibility that they could function as stepping stones in fragmented landscapes. Hollows in oak trees can persist for a century and more, over which they accumulate 'wood mould' of fine, often fungus-infected, wood fragments, insect frass and remains of any vertebrate or ant nests at any time constructed within the hollow. Cavities are thus regarded as an environmentally stable persistent microhabitat that is largely free from external disturbances. 'Wood mould' is a key nutritional resource for *Osmoderma eremita* in hollows (Jonsson et al. 2004) but, more widely, those hollows have been considered keystone structures for saproxylic fauna, and are known to harbour many endangered species in Europe (Mico et al. 2015). The intriguing finding that oak trunk hollows with *O. eremita* present contained higher richness of saproxylic beetles than unoccupied hollows might in part reflect increased nutritional values of the frass of *O. eremita* larvae (Jonsson et al. 2004).

Altogether, Jansson et al. (2009) retrieved 105 saproxylic beetle species from the 47 boxes they deployed over three years. These included 70% of the species found in more natural hollows in oak trees in Sweden. Additions (such as a dead hen) to the box contents increased the richness of beetles retrieved.

In addition to allowing trees to age naturally, more dramatic management may be needed to promote development and continued availability of tree hollows needed by some rare saproxylic beetles. For example, Ranius and Jansson (2000) noted possible use of explosives or chain saws to shatter branches or branch stumps of younger trees and initiate formation of tree hollows from those damaged sites. Their study in Sweden showed that richness of saproxylic beetles in old oak (*Quercus robur*) trees was greatest in large free-standing trees, with most species also 'preferring' trees in intermediate or late decay condition in having living but hollow trunks. Any measures to delay death of such trees could also decrease risks of extinction of some endangered beetle species.

The extensive studies of beetles and veteran trees in Scandinavia has led to such trees being considered an 'excellent model system' for comparing responses of specialist and generalist beetles, in particular in relation to influences of habitat connectivity, as mirrored in the different spatial contexts of parks and forests (Sverdrup-Thygeson et al. 2017), and involving oak trees in Norway. Extensive knowledge of the beetle fauna enabled clear distinction of the specialist species that depend on old veteran oaks and the remaining species with varying levels of association with those trees. Species richness of specialists increased with tree size, and negative effect on species richness were evident with regrowth of surrounding trees. That trend was independent of park or forest surroundings, and differences in response to tree (habitat) connectivity related partially to specialist beetles being larger and feeding on larger and more decayed wood in cavities – beetle traits that may relate to increased susceptibility to forest fragmentation.

9.8   Urban Forests

Tree planting in urbanised areas, including roadsides and divides, gardens and parklands, is widespread, in conjunction with increasing calls to preserve remnant natural forest patches in and around highly urbanised areas. 'Urban forests' are an increasingly important conservation theme both as direct habitat and in promoting connectivity, and also for their amenity values to people. However, urban secondary forests generally receive little attention for their roles in sustaining forest diversity and in insect conservation, but can still support high richness of native species. The primary motivation to establish many urban forests is simply aesthetic, but each may become an important habitat for insects and other wildlife.

Values in insect conservation have been investigated by comparative inventory studies, and also by 'gradient studies' that examine changes in assemblages along series of sites representing the transition from rural to urban environments, such as from outer city areas to the most intensively urbanised central district. A format for such studies formulated by Niemela et al. (2000) for carabid beetles provoked much later work – with this case emphasising need for standardised pitfall trapping as a replicable basis for comparisons at sites along the gradient, in which the most urban sites may be relatively isolated fragments subject to strong edge effects.

Level of urbanisation, rather than distance from the forest edge, had major influence on species richness of ground beetles in Brussels (Belgium) (Gaublomme et al. 2008), and forest patch size was influential only in more rural sites – where larger patches yielded more species than smaller patches. Abundance of both forest generalist and forest specialist species increased with increasing distance into the forest from the forest edge. This survey confirmed that the more disturbed (more heavily urbanised) sites were significantly poorer in forest specialist carabids – so that restoration of high quality habitat in urban environments is more important for their conservation than is promoting habitat connectivity. In addition, those specialists may be unable to disperse by means of stepping stones or corridors, because smaller forest fragments subject to severe edge effects may be able to support only more generalist taxa.

Richness of insects in urban forests was demonstrated by beetle catches from flight interception traps on dead or dying trees in the city of Ostrava, Czech Republic (Horak 2011). Of the 203 beetle species collected, 119 were obligately saproxylic; 26 red-listed species were retrieved, and richness was overall higher on beech (*Fagus sylvatica*) trees than on mixed stands of other trees. Beech trees also had higher beetle abundance.

Pest or pathogen outbreaks or invasions of small urban forest areas can pose conservation issues. The Oak processionary moth, *Thaumatopeia processionea* (Thaumatopeidae), occurs widely on mainland Europe, where it causes severe defoliation of oaks (*Quercus* spp.) – it is also a severe threat to public health through the larval urticating hairs. Discovery of the moth in London in 2006, apparently imported with a consignment of oak trees grown in Italy and planted in London, prompted a major eradication campaign (Tomlinson et al. 2015) to clear moths from the urban trees – many of which were in private gardens and collectively posed some difficulties of access. Several years later, the campaign was declared impossible, and emphasis switched to containment to minimise spread and impacts of the moth, but Tomlinson et al. noted that the moth continues to threaten trees and people alike. The former, leading to loss of trees that support a high diversity of native insects, may have wider conservation implications. *T. processionea* is progressively moving northward in Europe, and its establishment in the United Kingdom is likely to reflect response to warming climate. Likewise, it was reported first from the Netherlands in 1991 and has also expanded northward since then. Preference for oak trees in urban environments rather than in forests may reflect these being warmer, a trend noted also for the beetle *Agrilus biguttatus* (Buprestidae) with its larval excavations on the warmer south-facing sides of oaks in the Netherlands (Moraal and Jayers op Akkerhuis 2011). Climate responses are an important, but often neglected, aspect of insect attack on trees, with temperature an influence on many saproxylic insects.

The gestations and design of urban forests in Australia have followed rather different patterns between cities, and the contrasts are exemplified by accounts of development in Canberra (Banks and Brack 2003) and Brisbane and adjacent cities (Roy et al. 2017). Both the contexts were planned rather than simply allowed to develop casually, but in rather different ways.

The original site selected for Canberra, planned as Australia's new inland capital city, was a treeless plain. Major tree plantings taking place from the 1920s led to urban forest on public land and comprising some 400,000 trees by the early twenty-first century. From the earliest periods of Canberra's development, landscaping involving trees was a central part of the planning – much of it under the inspired guidance of Charles Weston, and leading progressively to Canberra's designation as 'a city in a forest'. That forest includes numerous tree species in a variety of well-connected configurations and changing dominance from alien to native species over time. Presence of substantial tree diversity into the future seems assured.

For the Brisbane surrounds, the five constituent cities across south-east Queensland have rather different patterns of urbanisation, but are founded in deforestation and urbanisation leading recently to 'urban greening movements' (Roy et al. 2017), in which each city selects species for its own street tree plantings. Selections have varied somewhat, but with environmental and aesthetic considerations linking with site factors, management and maintenance issues and costs, and with factors such as socio-economic and community benefits scarcely acknowledged. Promotion of native species, shade, 'greenness', and other factors were highlighted amongst the responses by Shire officers to a questionnaire.

9.9   Riparian Vegetation

Forested riparian buffers have been produced or retained largely to protect the aquatic organisms and ecosystems from changes to nearby land, such as run-off and soil erosion, as well as from increased exposure, but have been advocated also to protect terrestrial biota. Their abundant putative conservation roles have led to the widespread establishment and protection of native riparian vegetation, coupled in some contexts with removal of intrusive alien vegetation. The complexities of estimating effects of salvage logging in riparian areas on the adjoining aquatic ecosystems and influencing watersheds depend on both the local landscape and its disturbance history (Reeves et al. 2006), with recognition that riparian habitats are also a mosaic resulting from various disturbances that generate opportunities for different biological communities. The direct effects of fire, for example, may be minor on benthic macroinvertebrates not exposed to flames and heat, but the ensuing increased run-off, sedimentation and channel changes may lead to faunal changes for which recovery periods are long (Minshall 2003), or recovery may not occur. A combination of loss of riparian vegetation and deposition of ash, charcoal and dead wood may produce substantial – but sometimes short-term – decreases in macroinvertebrate diversity and biomass. Heterogeneity of those impacts is normally high, so that levels of impact differ with factors such as fire extent and intensity, stream size and gradient, precipitation and run-off, vegetation cover, geology and topography. In general, Minshall (2003) concluded that fire itself is not detrimental to aquatic macroinvertebrates associated with unburned forest, but may be far more severe for already degraded watersheds.

Prescriptions for the size, width and composition of buffers of undisturbed forest – so that a widely recommended width of around 30 m on each side of running water in forests is often adopted to protect the water bodies from logging activities – are among many guidelines. Much wider buffers are recommended to sustain wider site conditions and are recommended for some terrestrial vertebrates to thrive, but their beneficial effects are often difficult to quantify. A meta-analysis of published studies (Marczak et al. 2010) showed considerable variations in faunal responses across major taxonomic groups, and implied that wider buffers did not necessarily induce greater similarity between buffers and comparative reference sites. The studies they reviewed varied greatly in scope and taxonomic focus, with arthropods a priority in only 10 of 397 estimates. However, only arthropods and birds showed overall higher abundance in buffers than in unmanaged riparian forest. Taxa regarded as 'edge specialists' showed significantly higher densities in the buffers. They purportedly included taxa (with specific mention of spiders and Odonata) that forage on flying insects that may become concentrated along forest edges.

Presence or removal of riparian trees can have very varied impacts, but light and the supply of organic matter and nutrients to aquatic ecosystems can be dramatically altered (Kiffney et al. 2003).

Recognising that a primary value of riparian buffers is to mitigate effects of clear-cutting of adjacent forest on those ecosystems, Kiffney et al. compared headwater streams with three riparian buffer treatments (buffers of 30 m, 10 m and clear-cut to the stream edge) and uncut controls in British Columbia, Canada, to demonstrate that uncut buffers of 30 m or more on both sides of the streams were needed to limit changes associated with clear-cut logging. Benthic invertebrates were assessed by monthly examination of tile traps, as substrates suitable for colonisation by a wide range of taxa, so that seasonal differences could also be examined. Abundance of Chironomidae was generally greater in the 10 m and 30 m buffer treatments than in controls, with more variable results in the controls – and differences might be associated with high sediment content of the periphyton net on the tiles in clear-cut areas, as another likely cause of differences.

Riparian strips have been valued also as possible sources of naturally occurring forest insects that can 'spill over' into adjacent modified areas (as discussed for oil palm plantations in Sabah: Gray et al. 2014) and so enhance otherwise eroded ecosystem functions. Gray et al. surveyed two key functional insect groups, ground-dwelling ants and dung beetles, from 14 sites adjacent to rivers. Seven of those sites were in continuous oil palm plantations, and the others were in areas of oil palm with a riparian reserve strip. Samples were taken along transects, and at 0, 50 and 100 m from the indicated high water line, using baited pitfall traps (dung beetles) and a combination of bait cards and direct collections for ants. Altogether, 36 dung beetle species and 58 ant morphospecies were collected. Assemblages of both taxa in riparian reserves differed considerably from those in oil palm adjacent to the reserves, and from those in oil palm plantations lacking any such reserve. Assemblages in the last regime did not differ significantly from those in oil palms abutting a reserve. Dung beetle diversity, richness and abundance declined with distance from a reserve.

Collectively, there was very little spillover of dung beetles, and no spillover of scavenging ants into oil palm plantations from adjacent riparian reserves, and little evidence of any ecological process involving these insect groups being affected. Dung beetle spillover occurred over only short distances, and it appeared that contiguous habitat corridors were important for maintaining connectivity for many invertebrate species.

9.10   Implications of Climate Change

Anticipating changes in species distribution and ecological needs with climate change involves prediction and modelling drawing largely on responses to temperature and differential impacts on the focal insect species and its key resource species. Such projections have been made in greatest detail for some key forest insect pests (such as some bark beetles: Smith et al. 2013), as potential aids to monitoring and management under a variety of possible changes and climate scenarios. Latitudinal or elevational series that purportedly evaluate assemblage changes as surrogates for temperature changes (Hodkinson 2005) have also been pursued (Chinellato et al. 2014). Effects on single species and on wider parameters such as assemblage richness may help to forecast needs for changed forest management. The likely outcome for southern European spruce forests – that those forests may experience increasing damage from aggressive pest species and the abundance of other species may decline because of host plant limitations, especially on the southern edge of current forest distribution – may have wide relevance (Chinellato et al. 2014). Possible beneficial management could be to avoid new spruce plantations at lower elevations and outside the normal climatic range as it is understood at present.

Elevational gradients in Australian forests, and changes to the composition of assemblages of sensitive insect groups may prove a useful indicator of community responses to climate change. Light-trap catches of moths (collectively including

3035 species of 'larger' [>1 cm forewing length] 'macromoths' and Pyraloidea) in tropical and subtropical rainforest in Queensland demonstrated some of the differences. Ashton et al. (2016) examined catches from two elevational transects of each forest category, with plots replicated at 200 m elevational intervals, at each of which canopy and understorey trapping was undertaken. In both these major ecosystems, moth assemblages were clearly stratified, suggesting that elevation may be an important structuring influence. Both forests also showed a mid-elevation peak in species richness. For some shared moth species, ranges shifted upward, implying their value as sensitive indicators of climate and as components of 'predictor sets' (Kitching and Ashton 2014).

Disruption of ecological interactions by differing impacts of climate change on the participating species extends to the relationships between host trees and their insect herbivores, potentially changing the dynamics of pest outbreaks and the management responses needed. Warmer temperatures can advance phenology, causing earlier hatching or breaking of dormancy, or dispersal, in spring so that flush growth on trees may not then coincide fully with insect needs. In the interaction between the forest moth *Malacosoma disstria* and its two host species (Aspen, *Populus tremuloides*, and Birch, *Betula papyrifera*) in Minnesota forests, the long-evolved synchrony was shifted unevenly in regimes with experimentally increased temperatures of 1.7 and 3.4 °C (Schwartzberg et al. 2014), with bud-break of the trees and *Malacosoma* egg hatch and pupation uneven. The asynchronies produced increased from first to second year of the trials, with trees showing greater sensitivity to temperature than the moth. Such differential 'tracking' of climate changes may influence dynamics and potential for outbreaks, not least through changes in foliage chemistry (Jamieson et al. 2015).

Phenological differences have been investigated also for another important lepidopteran defoliater of North American forests, the Spruce budworm (*Choristoneura fumiferana*, Tortricidae). If phenology is advanced by warmer climate, the normally poor quality host Black spruce (*Picea mariana*) could support much better budworm survival and undergo increased defoliation, so increasing the wider severity of pest attack (Fuentealba et al. 2017).

Climatically-related asynchrony has potential both to interrupt normally well-synchronised interactions between species and also to create opportunities for novel interactions to arise. Parallel effects have been suggested to arise from direct forest disturbances – such as between the partners in mutualisms. Although some mutualisms (such as the by-product mutualisms between ants and the epiphytic ferns in which they dwell in Borneo) seem resilient to changes such as selective logging and alien speies establishment, it has also been suggested that such resilience may be rather exceptional (Fayle et al. 2015), not least because of the participation of alien generalist ants in disturbed habitats. Predicting impacts, spread and changing distributions of such aliens with changing climates is inevitably uncertain, but many such predictions may lead to modified management to compensate for any undesirable effects thought likely to occur. Likewise, assessments of some bark beetles have suggested how synchrony with their host trees might progress. Whilst vulnerability may be revealed by climate models based on predictions to be derived from present distributions and knowledge of climate tolerances, this is not always so – for the Mexican bark beetle *Dendroctonus rhizophagus*, exploration of four different climate scenarios for the period 2040–2069 implied that the core area of both the beetle and its *Pinus* host habitat are likely to persist under any of these options (Smith et al. 2013).

References

Adamski P, Bohdan A, Michalcewicz J, Ciach M, Witkowski Z (2016) Timber stacks: potential ecological traps for an endangered saproxylic beetle, the Rosalia longicorn *Rosalia alpina*. J Insect Conserv 20:1099–1105

Andrew N, Rodgerson L, York A (2000) Frequent fuel-reduction burning: the role of logs and associated leaf litter in the conservation of ant biodiversity. Austral Ecol 25:99–107

Ashton LA, Odell EH, Burwell CJ, Maunsell SC, Nakamura A, Mcdonald WJF, Kitching RL (2016) Altitudinal patterns of moth diversity in tropical and subtropical Australian rainforests. Aust Ecol 41:197–208

Athayde EA, Cancian LF, Verdade LM, Morellato LPC (2015) Functional and phylogenetic diversity of scattered trees in an agricultural landscape: implications for conservation. Agric Ecosyst Environ 199:272–281

Baker SC, Richardson AMM, Seeman OD, Barmuta LA (2004) Does clearfell, burn and sow silviculture mimic the effect of wildfire? A field study and review using litter beetles. For Ecol Manag 199:433–444

Banks JCG, Brack CL (2003) Canberra's urban forest: evolution and planning for future landscapes. Urb For Urb Green 1:151–160

Bazelet CS, Samways MJ (2011) Grasshopper assemblage response to conservation ecological networks in a timber plantation matrix. Agric Ecosyst Environ 144:124–129

Berg SA, Bergman K-O, Wissman J, Zmihorski M, Ockinger E (2016) Power-line corridors as source habitats for butterflies in forest landscapes. Biol Conserv 201:320–326

Bojorquez-Tapia LA, Brower LP, Castilleja G, Sanchez-Colon S, Hernandez M et al (2003) Mapping expert knowledge: redesigning the Monarch Butterfly Biosphere Reserve. Conserv Biol 17:367–379

Bossart JL, Antwi JB (2016) Limited erosion of genetic and species diversity from small forest patches: sacred forest groves in an Afrotropical biodiversity hotspot have high conservation value for butterflies. Biol Conserv 198: 122–134

Bouget C, Larrieu L, Brin A (2014) Key features for saproxylic beetle diversity derived from rapid habitat assessment in temperate forests. Ecol Indic 36:656–664

Carpaneto GM, Mazziotta A, Coletti G, Luiselli L, Audisio P (2010) Conflict between insect conservation and public safety: the case of a saproxylic beetle (*Osmoderma eremita*) in urban parks. J Insect Conserv 14:555–565

CBD (Convention on Biological Diversity) (2009) Year in review 2009. Secretariat of the Convention on Biological Diversity, Montreal

Chiari A, Bardiani M, Zauki A, Hardersen S, Mason F, Spada L, Campanaro A (2013) Monitoring of the saproxylic beetle *Morimus asper* (Sulzer, 1776) (Coleoptera: Cerambycidae) with freshly cut log piles. J Insect Conserv 17:1255–1265

Chinellato F, Faccoli M, Marini L, Battisti A (2014) Distribution of Norway spruce bark and wood-boring beetles along Alpine elevational gradients. Agric For Entomol 16:111–118

Costello SL, Negron JF, Jacobi WR (2011) Wood-boring insect abundance in fire-injured ponderosa pine. Agric For Entomol 13:373–381

Croft P, Hunter JT, Reid N (2016) Forgotten fauna: habitat attributes of long-unburnt forests and woodlands dictate a rethink of fire management theory and practice. For Ecol Manag 366:166–174

Davies KF, Margules CR (1998) Effects of habitat fragmentation on carabid beetles: experimental evidence. J Anim Ecol 67:460–471

Davies KF, Melbourne BA, Margules CR (2001) Effects of within- and between-patch processes on community dynamics in a fragmentation experiment. Ecology 82:1830–1846

de Andrade RB, Balch JK, Carreira JYO, Brando PM, Freitas AVL (2017) The impacts of recurrent fires on diversity of fruit-feeding butterflies in a South-Eastern Amazon forest. J Trop Ecol 33:22–32

Dudley N (2013) Guidelines for applying protected area management categories including IUCN WCPA best practice guidance on recognised protected areas and assigning management categories and governance types. IUCN, Gland, Switzerland. (Available from https://portrals.iucn.org/library/node/30018)

Farmilo BJ, Nimmo DG, Morgan JW (2013) Pine plantations modify local conditions in forest fragments in southeastern Australia: insights from a fragmentation experiment. For Ecol Manag 305:264–272

Fayle TM, Edwards DP, Foster WA, Yusah KM, Turner EC (2015) An ant-plant by-product mutualism is robust to selective logging of rain forest and conversion to oil palm plantation. Oecologia 178:441–450

Franklin JF, Berg DR, Thornburgh DA, Tappeiner JC (1997) Alternative sylvicultural systems to timber harvesting: variable retention harvest systems. In Kohm K, Franklin JF (eds) Creating a forestry for the 21st century. Island Press, Washington, DC, pp 111–139

Freedman B, Zelazny V, Beaudette D, Fleming T, Johnson G et al (1996) Biodiversity implications of changes in the quality of dead organic matter in managed forests. Environ Revs 4:238–265

Fuentealba A, Pureswaran D, Bauce E, Depland E (2017) How does synchrony with host plant affect the performance of an outbreakng insect defoliator? Oecologia 184:847–857

Gandhi KJM, Spence JR, Langor DW, Morgantini LE (2004) Fire residuals as habitat reserves for epigaeic beetles (Coleoptera: Carabidae and Staphylinidae). Biol Conserv 102:131–141

Gaublomme E, Hendrickx F, Dhuyvetter H, Desender K (2008) The effects of forest patch size and matrix type on changes in carabid beetle assemblages in an urbanized landscape. Biol Conserv 141:2585–2596

Geldmann J, Barnes M, Coad L, Craigie ID, Hockings M, Burgess ND (2013) Effectiveness of terrestrial protected areas in reducing habitat loss and population declines. Biol Conserv 161:230–238

Gibbons P, Lindenmayer DB (1996) Issues associated with the retention of hollow-bearing trees within eucalypt forests managed for wood production. For Ecol Manag 83:245–279

Gough LA, Birkemoe T, Sverdrup-Thygeson A (2014) Reactive forest management can also be proactive for wood-living beetles in hollow oak trees. Biol Conserv 180:75–83

Gouix N, Sebek P, Valladares L, Brustel H, Brin A (2015) Habitat requirements of the violet click beetle (*Limoniscus violaceus*), an endangered umbrella species of basal tree hollows. Insect Conserv Divers 8:418–427

Gray CL, Slade EM, Mann DJ, Lewis OT (2014) Do riparian reserves support dung beetle biodiversity and ecosystem services in oil palm-dominated tropical landscapes? Ecol Evol 4:1049–1060

Gustafsson L, Kouki J, Sverdrup-Thygeson A (2010) Tree retention as a conservation measure in clear-cut forests of northern Europe: a review of ecological consequences. Scand J For Res 25:295–308

Gustafsson L, Baker SC, Bauhaus J, Beese WJ, Brodie A et al (2012) Retention forestry to maintain multifunctional forests: a world perspective. BioScience 62:633–645

Hedin J, Isacsson G, Jonsell M, Komonen A (2008) Forest fuel piles as ecological traps for saproxylic beetles in oak. Scand J For Res 23:348–357

Hedwall P-O, Mikusinski G (2015) Structural changes in protected forests in Sweden: implications for conservation functionality. Can J For Res 45:1215–1224

Heikkala O, Martikainen P, Kouki J (2017) Prescribed burning is an effective and quick method to conserve rare pyrophilous forest-dwelling flat bugs. Insect Conserv Divers 10:32–41

Hilszczanski J, Jaworski T, Plewa R, Jansson N (2014) Surrogate tree cavities: boxes with artificial substrate can serve as temporary habitat for *Osmoderma barnabita* (Motsch.) (Coleoptera, Cetoniinae). J Insect Conserv 18:855–861

Hodkinson ID (2005) Terrestrial insects along elevational gradients: species and community responses to altitude. Biol Rev 80:489–513

Horak J (2011) Response of saproxylic beetles to tree species composition in a secondary urban forest area. Urb For Urb Green 10:213–222

Horak J (2017) Insect ecology and veteran trees. J Insect Conserv 21:1–5

Iwona M, Marek P, Katarzyna W, Edwards B, Julia S (2017) Use of a genetically informed population viability analysis to evaluate management options for Polish populations of endangered beetle *Cerambyx cerdo* L. (1758) (Coleoptera, Cerambycidae). J Insect Conserv 22:69. https://doi.org/10.1007/s10841-017-0039-3

Jamieson MA, Schwartzberg EG, Raffa KF, Reich PB, Lindroth RL (2015) Experimental climate warming alters aspen and birch phytochemistry and performance traits for an outbreak insect herbivore. Glob Change Biol 21:2698–2710

Jansson N, Ranius T, Larsson A, Milberg P (2009) Boxes mimicking tree hollows can help conservation of saproxylic beetles. Biodivers Conserv 18:3891–3908

Jongman RHG, Pungetti G (2004) Ecological networks and greenways; concept, design, implementation. Cambridge University Press, Cambridge

Jonsell M (2012) Old park trees as habitat for saproxylic beetle species. Biodivers Conserv 21:619–642

Jonsson N, Mendez M, Ranius T (2004) Nutrient richness of wood mould in tree hollows with the scarabaeid beetle *Osmoderma eremita*. Anim Biodiv Conserv 27:79–82

Jung J-K, Lee SK, Lee S-I, Lee J-H (2018) Trait-specific response of ground beetles (Coleoptera: Carabidae) to forest fragmentation in the temperate region in Korea. Biodivers Conserv 27:53–68

Khazan ES (2014) Tests for biological corridor efficacy for conservation of a Neotropical giant damselfly. Biol Conserv 177:117–125

Kiffney PM, Richardson JS, Bull JP (2003) Responses of periphyton and insects to experimental manipulation of riparian buffer width along forest streams. J Appl Ecol 40:1060–1076

Kitching RL, Ashton LA (2014) Predictor sets and biodiversity assessments: the evolution and application of an idea. Pac Conserv Biol 19:418–426

Lane J (1993) Overwintering monarch butterlies in California: past and present. In: Malcolm SB, Zalucki MP (eds) Biology and conservation of the monarch butterfly. Natural History Museum of Los Angeles County, Los Angeles, pp 335–344

Langor DW, Hammond HEJ, Spence JR, Jacobs J, Cobb TP (2008) Saproxylic insect assemblages in Canadian forests: diversity, ecology, and conservation. Canad Entomol 140:453–474

Larsen TB (2005) The butterflies of West Africa. Apollo Books, Denmark

Lee S-I, Spence JR, Langor DW (2017) Combinations of aggregated and dispersed retention improve conservation of saproxylic beetles in boreal white spruce stands. For Ecol Manag 385:116–126

Lindenmayer DB (2017) Conserving large old trees as small natural features. Biol Conserv 211:51–59

Lindenmayer DB, Laurance WF (2017) The ecology, distribution, conservation and management of large old trees. Biol Rev 92:1434–1458

Lindenmayer DB, Claridge AW, Gilmore AM, Michael D, Lindenmayer BR (2002) The ecological roles of logs in Australian forests and the potential impacts of harvesting intensification on log-using biota. Pac Conserv Biol 8:121–140

Lindenmayer DB, Banks SC, Laurence WF, Franklin JF, Likens GE (2014) Broad decline of populations of large old trees. Conserv Lett 7:72–73

Lindenmayer DB, Blair D, McBurney L, Banks SC (2015) The need for a comprehensive reassessment of the Regional Forest Agreements in Australia. Pac Conserv Biol 21:266–270

Manzo-Delgado L, Lopez-Garcia J, Alcantara-Ayala I (2014) Role of forest conservation in lessening land degradation in a temperate region: the Monarch Butterfly Biosphere Reserve, Mexico. J Environ Manag 138: 55–66

Marczak LB, Sakamaki T, Turvey SL, Deguise I, Wood SLR, Richardson JS (2010) Are forested buffers an effective conservation strategy for riparian fauna? An assessment using meta-analysis. Ecol Appl 20:126–134

Margules CR (1992) The Wog Wog habitat fragmentation experiment. Environ Conserv 19:316–325

Martikainen P (2001) Conservation of threatened saproxylic beetles: significance of retained aspen *Populus tremula* on clearcut areas. Ecol Bull 49:205–218

Martikainen P, Kouki J, Heikalla O (2006) Effects of green tree retention and subsequent prescribed burning on the crown damage caused by the pine shoot beetles (*Tomicus* spp.) in pine-dominated timber harvest areas. Ecography 29:659–670

Matveinen-Huju K, Niemela J, Rita H, O'Hara RB (2006) Retention-tree groups in clear-cuts: do they constitute 'life-boats' for spiders and carabids? For Ecol Manag 230:119–135

McElhinny C, Gibbons P, Brack C, Bauhus J (2008) Fauna-habitat relationships: a basis for identifying key stand structural attributes in temperate Australian eucalypt forests and woodlands. Pac Conserv Biol 12:89–110

Mico E, Garcia-Lopez A, Sanchez A, Juarez M, Galante E (2015) What can physical, biotic and chemical features of a tree hollow tell us about their associated diversity? J Insect Conserv 19:141–153

Minshall GW (2003) Responses of stream benthic macroinvertebrates to fire. For Ecol Manag 178:155–161

Monkkonen M, Mutanen M (2003) Occurrence of moths in boreal forest corridors. Conserv Biol 17:468–475

Moraal LG, Jayers op Akkerhuis GAJM (2011) Changing patterns in insect pests on trees in the Netherlands since 1946 in relation to human induced habitat changes and climate factors – an analysis of historical data. For Ecol Manag 261:50–61

Mori AS, Kitagawa R (2014) Retention forestry as a major paradigm for safeguarding forest biodiversity in productive landscapes: a global meta-analysis. Biol Conserv 175:65–73

New TR (2014) Insects, fire and conservation. Springer, Cham

Niemela J, Kotze J, Ashworth A, Brandmayr P, Desender K et al (2000) The search for common anthropogenic impacts on biodiversity, a global network J Insect Conserv 4:3–9

Nieto A, Alexander KNA (2010a) European red list of Saproxylic beetles. Publications office of the European union, Luxembourg

Nieto A, Alexander KNA (2010b) The status and conservation of saproxylic beetles in Europe. Cuademos de Biodiversidad (2010):3–10

Quinto J, Marcos-Garcia MA, Diaz-Castelazo C, Rico-Gray V, Bustel H et al. (2012) Breaking down complex saproxylic communities: understanding sub-networks structure and implications to network robustness. PLoS ONE 7: e45062

Ranius T, Jansson N (2000) The influence of forest regrowth, original canopy cover and tree size on saproxylic beetles associated with old oaks. Biol Conserv 95:85–94

Ray R, Chandran MDS, Ramachandra TV (2014) Biodiversity and ecological assessments of Indian sacred groves. J For Res 25:21–28

Reeves GH, Bisson PA, Rieman BE, Benda LE (2006) Postfire logging in riparian areas. Conserv Biol 20:994–1004

Roland J (1993) Large-scale forest fragmentation increases the duration of tent caterpillar outbreak. Oecologia 93:25–30

Rosenvald R, Lohmus A (2008) For what, when, and where is green-tree retention better than clear-cutting? A review of the biodiversity aspects. For Ecol Manag 255:1–15

Roy S, Davison A, Ostberg J (2017) Pragmatic factors outweigh ecosystem service goals in street tree selection and planting in south-East Queensland cities. Urb For Urb Green 21:166–174

Sahlin E, Schroeder LM (2010) Importance of habitat patch size for occupancy and density of aspen-associated saproxylic beetles. Biodivers Conserv 19:1325–1339

Saint-Germain M, Drapeau P, Hebert C (2004) Comparison of Coleoptera assemblages from a recently burned and unburned black spruce forest of northeastern North America. Biol Conserv 118:583–592

Samways MJ (2007) Implementing ecological networks for conserving insects and other biodiversity. pp. 127–143 in Stewart AJA, New TR, Lewis OT (eds) Insect conservation biology. CAB International, Wallingford

Schauer B, Steinbauer MJ, Vailshery LS, Muller J, Feldhaar H, Obermaier E (2017) Influences of tree hollow characteristics on saproxylic beetle diversity in a managed forest. Biodivers Conserv 27:1–17

Schmitt CB, Burgess ND, Coad L, Belokurov A, Besancon C et al (2009) Global analysis of the protection status of the world's forests. Biol Conserv 142:2122–2130

Schmuki C, Vorburger C, Runciman D, Maceachern S, Sunnucks P (2006) When log-dwellers meet loggers: impacts of forest fragmentation on two endemic log-dwelling beetles in southeastern Australia. Molec Ecol 15:1481–1492

Schwartzberg EG, Jamieson MA, Raffa KF, Reich PB, Montgomery RA, Lindroth RL (2014) Simulated climate warming alters phenological synchrony between an outbreak insect herbivore and host trees. Oecologia 175:1041–1049

Sebek P, Vodka A, Bogusch P, Pech P, Tropek R, Weiss M, Zimova K, Cizek L (2016) Open-grown trees as key habitats for arthropods in temperate woodlands: the diversity, composition, and conservation value of associated communities. For Ecol Manag 380:172–181

Shafer CL (2015) Cautionary thoughts on IUCN protected area management categories V-VI. Glob Ecol Conserv 3:331–348

Shahabuddin G, Herzner GA, Aponter C, Gomez M del C (2000) Persistence of a frugivorous butterfly species in Venezuelan forest fragments: the role of movement and habitat quality. Biodivers Conserv 9:1623–1641

Smith SE, Mendoza MG, Zuniga G, Halbrook K, Hayes JL, Byrne DN (2013) Predicting the distribution of a novel bark beetle and its pine hosts under future climate conditions. Agric For Entomol 15:212–226

Sodhi NS, Wilcove DS, Subaraj R, Yong DL, Lee TM, Bernard H, Lim SLH (2010) Insect extinctions on a small denuded Bornean island. Biodivers Conserv 9:485–490

Speight MCD (1989) Saproxylic insects and their conservation. Nature and Environment Series no. 42, Council of Europe, Strasbourg

Spence JR, Langor D, Niemela J, Carcamo H, Currie C (1996) Northern forestry and carabids: the case for concern about old-growth species. Ann Zool Fenn 33:173–184

Spencer R-J, Baxter GS (2006) Effects of fire on the structure and composition of open eucalypt forests. Aust Ecol 31:638–646

Sugiura S, Yamaura Y, Tsuru T, Goto H, Hasegawa M, Makihara H, Makino S (2009) Beetle responses to artificial gaps in an oceanic island forest: implications for invasive tree management to conserve endemic species diversity. Biodivers Conserv 18:2101–2118

Sverdrup-Thygeson A, Ims RA (2002) The effect of forest clearcutting in Norway on the community of saproxylic beetles on aspen. Biol Conserv 106:347–357

Sverdrup-Thygeson A, Skarpaas O, Odegaard F (2010) Hollow oaks and beetle conservation: the significance of the surroundings. Biodivers Conserv 19:837–852

Sverdrup-Thygeson A, Skarpas O, Blumentrath S, Birkemoe T, Evju M (2017) Habitat connectivity affects specialist species richness more than generalists in veteran trees. For Ecol Manag 403:96–102

Terborgh J (2004) Reflections of a scientist on the world parks congress. Conserv Biol 18:619–620

Tomlinson I, Potter C, Bayliss H (2015) Managing tree pests and diseases in urban settings: the case of Oak processionary moth in London, 2006-2012. Urb For Urb Green 14:286–292

Vanha-Majamaa I, Jalonen J (2001) Green tree retention in Fennoscandian forestry. Scand J For Res 16(S3):79–90

Varkonyi G, Kuussaari M, Lappalainen H (2003) Use of forest corridors by boreal *Xestia* moths. Oecologia 137:466–474

York A (2000) Long-term effects of frequent low-intensity burning on ant communities in coastal blackbutt forests of southeastern Australia. Aust Ecol 25:83–98

Chapter 10
Forest Management for Insect Conservation in Australia

Keywords Forest conservation strategy · Forest disturbance regimes · Forest landscape design · Forest refuges · Forest policy · Forest reserves · Forest restoration · Plantations · Priority areas · Windthrow impacts

10.1 Introduction: Perspective

The widespread acknowledgements of need for forest preservation and reservation have generated numerous stated 'targets' for this to occur, with recognition that these targets are often ambitious and challenging, and that 'our ability to recover biodiversity and ecosystem functions fully is currently limited' (Jacobs et al. 2015). Progress is rendered more uncertain from increased climatic variability and the additional stresses this imposes on restoration outcomes based on current practices. Design of new resilient forest ecosystems that can adapt to such changes and changing conditions may incorporate additional species (with further incorporation of alien species sometimes controversial), and patterns such as creation of forest fragments in positions where ecosystem services and biodiversity connectivity may be best enhanced or facilitated.

Perhaps especially in the northern hemisphere, and reflecting the many centuries of anthropogenic changes to native forests, defining the forest state targeted for restoration can become difficult because of the paucity of remaining forest that is considered truly 'natural'. That dilemma is somewhat less in Australia, where patches of relatively undisturbed forests of most main forst categories persist and provide a strong indicative template of what the natural condition may be, as a target and help to direct the steps and management needed. Nevertheless, the phases of formal forest management distinguished for the United States (discussed by Coulson and Stephen 2008) provide a transition from 'more anthropogenic' to 'more ecologically based' approaches (following Yaffee 1999) which is also reflected in Australia. Those stages flow from the earliest (1) dominant-use management, with the single purpose of maximising production for economic gain; expanding to (2) multiple-use management to provide a wider variety of goods/services for human benefit but incorporating sustainability (for persistence of supply); this becoming,

© Springer International Publishing AG, part of Springer Nature 2018   225
T. R. New, *Forests and Insect Conservation in Australia*,
https://doi.org/10.1007/978-3-319-92222-5_10

gradually (3) environmentally-sensitive multiple-use management as conservation needs become more apparent; leading to (4) a more holistic ecosystem management approach in which ecological integrity and environmental protection even take priority over human needs alone; and finally (5) landscape management with the perspective increased beyond the specific forest areas managed earlier to incorporate aspects of interconnections amongst all landscape elements.

A further aspect of ambiguity in defining target conditions centres on the concept and definition of 'old-growth forest', often used as a standard for emulation or comparison, as in many of the studies cited in this book. The term, whilst conveying a general impression of maturity, climax or near-climax successional state, large old trees and accompanying characteristics infers ecological complexity, with any single definition difficult. Spies (2004) commented that 'A consensus on the wording of an ecological definition of old-growth will never be reached and may not be desirable, given the diversity of forests' and 'multiple general definitions, however, are needed'. Specific definitions have been advanced for several forest types, and draw on aspects of stand structure, composition and age, but separating old-growth from other forest categories can become complex – implied relativity is usually clear, but more precise categorisation may be wise in many comparative studies. Old-growth forest, however it is formally defined, can contain up to 50 times the amount of dead wood as present in younger managed forests, as well as much higher diversity of dead wood sizes, ages and decay stages (Gibb et al. 2006).

Referring to the development of intensive industrial forestry in Finland, Hanski (2008) wrote that 'early successional natural forests have disappeared so completely that even small stands cannot be located for research purposes'. The rapid decline of biodiversity, including insects, clearly called for substantial conservation measures to prevent the extinction debt (p. 112) being paid, and in principle necessitated a combination of preservation of large forest areas and what Hanski (2008) referred to as 'precision conservation' of selected small areas: woodland key habitats (WKHs), selected as especially important for threatened species. In Finland, many WKHs are enclaves of distinctive habitats such as open bogs or rocky outcrops within surrounding forests. The conservation values of those areas, which are by definition usually very small, are often doubtful, and Hanski regarded them as only one of a range of measures needed for practical conservation in boreal forest. As a 'stand-alone' measure, WKHs may be 'largely cosmetic'.

Forest restoration is a major component of modern restoration ecology, and Nunez-Mir et al. (2015) ilustrated its diversity by finding almost 30,000 abstracts of papers on the topic published (over 1980–2014) in 15 leading forestry journals. However, the suite of key themes for future forest restoration, and drawn from a wide array of forest managers and scientists, poses many considerations for insect and wider conservation (Table 10.1). As others (such as Stanturf 2015) have also shown, considerable uncertainties in methods and outcomes result from the local environment (notably climate) and social changes, encapsulated in changed ecological conditions. A long list of practical opportunities relevant to forest landscape restoration and increased resilience incorporates aims of afforestation that (1) achieve mitigation; (2) promote biodiversity and ecosystem services; (3) benefit

Table 10.1 The key themes emerging from an International Congress on Forest Restoration, and that were deemed to warrant consideration by scientists and managers involved in forest restoration (based on list by Jacobs et al. 2015)

Forest loss and degradation continues at high rates, but humankind is experiencing an historical momentum that favours forest restoration at global, regional and national scales.
Definition and utility of a 'reference ecosystem' must be revisited, especially for its role in distinguishing restoration and ecological rehabilitation from activities that do not give priority to historical continuity.
Global change (climate, land use changes and impacts) imparts high uncertainty to future ecological conditions of forest ecosystems to be restored.
Adaptive management implies need for a flexible management framework.
Aim to restore, rehabilitate or design forest ecosystems that are resistant and resilient to changing circumstances.
Vast amount of land requiring restoration implies need for spatial prioritisation of efforts, with selection factor including ecological risks.
Need to reconsider suite of species incorporated into restoration efforts, giving priority to native species wherever possible but not ignoring positive roles of non-native species.
Nursery propagation methods to progressively consider promotion of seedling resistance to stress, rather than solely fast growth.
Need to generalise among plant functional groups in their adaptations.
Emphasise low impact site preparation techniques on restoration sites.
Identify better-adapted species and optimise provenance selections, to counter rapid genetic responses resulting from global changes.
Multiple benefits to society from protected and restored forests require forest restoration to consider multiple objectives and tradeoffs needed to achieve those benefits.
Effective technology transfer and community-based approach to forest restoration needed.

local communities; and (4) increase the adaptive capabilities of the forest and the local community. Stanturf (2015) distinguished three active approaches to increasing forest adaptive capacity as (1) incremental (short-term, seeking to maintain forest ecosystems at their current locations, by managing for persistence); (2) anticipatory (more substantial adjustments but not transforming the system); and (3) transformational (attempting to anticipate climate changes and to respond in more intensive ways through anticipating management or key resource needs). The three approaches are compared in Table 10.2.

Management procedures must therefore consider impacts of a wide range of forestry practices and susceptibility to disturbance. Many are summarised in Fig. 10.1, after Jactel et al. (2009), who pointed out that forest management is a theme of relative benefits and risks – so, of compromise. The four main processes identified by Jactel et al. drive this balance between management and susceptibility, as (1) effect on local microclimates; (2) provision of fuel and resources; (3) biological control by natural enemies; and (4) changes to the development and physiology of individual trees. Their relationships are shown in Fig. 10.1.

Those procedures contribute to aspects of forestry in which commercial forests are managed with retention of selected living trees rather than clearcutting, creation of standing dead wood where needed, and not removing dead wood wholly

Table 10.2 Features and comparisons of the three adaptation strategies to increase resilience in forest restoration (After Stanturf 2015)

Feature	Adaptation strategy		
	Incremental	Anticipatory	Transformational
Vulnerability	Reduce to current stressors	Reduce to current and future stressors	Reduce to current and future stressors
Restoration paradigm	Ecological restoration	Functional restoration	Intervention ecology
Species	Native	Native/exotic	Native/exotic or designer species
Genetics	Local sources	Clones or provenances with adaptive traits	Transgenic for keystone species
Invasive	Prevent or remove	Accept those that are functional analogues of extirpated natives	Accept as novel
Novel Ecosystems	Prevent or avoid	Accept and manage emergent assemblages	Manage novel and emergent ecosystems

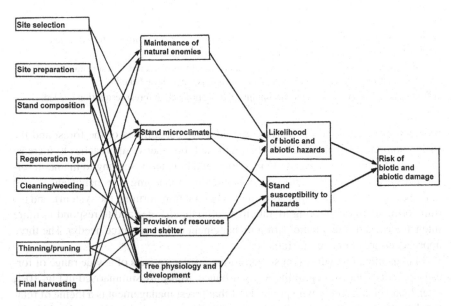

Fig. 10.1 Depiction of the key relationships between silvicultural operations and processes that drive likelihood of hazards and susceptibility of forest stands that determine risk of damage. (Jactel et al. 2009)

elsewhere. The main aim is then to maintain managed forests over sufficient periods to preserve and restore features that can sustain the expected diversity of flora and fauna. Target areas for this integrative approach can include plantation areas and mature forest areas separating protected areas (Larsson et al. 2006). Protection of resources such as dead wood may need to be regulated – but the context considered carefully. Under the New South Wales Threatened Species Act 1995, 'Removal of dead wood and dead trees' is listed as a Key Threatening Process, with removal specifically including 'burning on site'. Such demands could come into conflict with, for example, needs to reduce falling branch hazards in public areas (p. 205) and some reasoned flexibility is clearly needed in such well-motivated steps.

Australia's 'National Forest Policy Statement 1992' committed most Australian governments to undertake management of native forests 'in an ecogiclly sustainable manner, so as to conserve the suites of values that forests can provide for future generations', endorsed by specific needs for protection of wilderness values and old-growth forest. The wide range of values include habitat for nature conservation, and stated ecological attributes encompassed many roles of forests linked with human and ecological wellbeing.

The common approach to insect conservation, widespread in Australia, of focusing on individual threatened species, can easily overlook the wider variety present, and any measures to expand conservation management of forests to encompass both proactive and reactive approaches to conservation merit urgent consideration. For saproxylic beetles in Norway (Gough et al. 2014) the single measure of manipulating regrowth achieved this – in this case, clearing regrowth that shaded the trunk or canopy of a hollow oak. Clearing regrowth was predicted to increase the number of vulnerable species by 75–100% and specialist species richness by 65%. Clearing regrowth, especially trees, from around hollow oaks improved conditions for both the most vulnerable beetle species and many unthreatened specialist species. Similar considerations may become relevant in Australia.

10.2   Forest Protection

On a national level, a primary need is to slow rate of deforestation, so that remaining undisturbed and relatively little-disturbed natural forest will continue to sustain native biodiversity. Evans (2016) referred to the desirability of a 'forest transition', representing the cessation and eventual reversal of forest loss – but also pointed out that Australia's deforestation rate was the sixth highest in the world at the turn of the twenty-first century! Well-considered regulations and policy are needed to achieve any such measures.

The major conflicts and policy developments in Australia since around the 1970s have centred on the balance between conservation and economic reward, almost inevitably leading to compromises and tensions between conservationists and developers. The process of Regional Forest Agreements (RFAs) arose directly from those difficult debates, and was designed initially to satisfy both these interest groups. The history of RFAs was reviewed extensively in an Inquiry by the Resource Assessment Commission (1992) and led in turn to related national strategies, notably a series of criteria for forest conservation (Commonwealth of Australia 1995, following an earlier National Forest Policy Statement: Commonwealth of Australia 1992). The RFAs were designed as '20-year plans for the conservation and

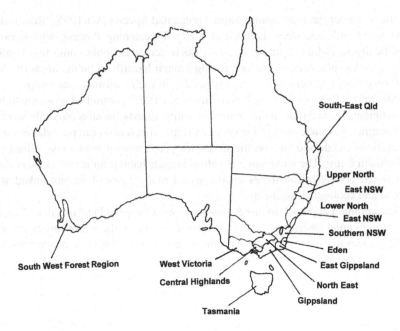

South-East Qld

Upper North
East NSW

Lower North
East NSW

Southern NSW

Eden

East Gippsland

North East
Gippsland

South West Forest Region

West Victoria

Central Highlands

Tasmania

Fig. 10.2 Distribution of the Regional Forest Agreements in Australia. (Based on map produced by Australian Government Department of Agriculture and Water Resources: Forestry, from www.agrifuture.gov.au/forestry/policies/rfa)

sustainable management of Australia's native forests', and were signed over the period of 1997–2001. Their subsequent performance and shortcomings were discussed by Lindenmayer et al. (2015), from whose account the above definition is cited. Areas covered by RFAs (Fig. 10.2) are the main production forest areas of Western Australia, New South Wales, Victoria and Tasmania, with the major purpose of assuring resource supply and economic returns to the industry whilst also maintaining favourable environmental outcomes. Lindenmayer et al. pointed out that some notable vertebrate flagship species (Leadbeater's possum, *Gymnobelideus leadbeateri*; Swift parrot, *Lathamus discolor*) have actually become more vulnerable, achieving 'Critically Endangered' status, during the life of those agreements, and that considerable re-thinking of renewed RFAs is necessary to incorporate new scientific information, newly prominent forest values (such as carbon storage), and adjusting levels of sustainable yield in conjunction with consideration of forest losses from wildfires and other disturbances. Without those considerations, Lindenmayer et al. (2015) warned that serious and harmful effects may continue.

Much of the earlier debate over forest exploitation versus conservation was summarised by Routley and Routley (1975), whose report was a 'wake-up call' for many concerned ecologists. Invertebrates figured little in those wide-ranging and influential documents, but were clearly encompassed in their intent. The Commonwealth criteria, for example, in part wished to ensure that the future of known elements of biodiversity was secured, together with 'maximising the chances of maintaining the possibly more numerous unknown elements' (Kirkpatrick 1998). All native elements of forest biodiversity should be protected within the reserve system of each region, with those reserves including 'at least 15 % of the pre-European invasion area of each forest community that it was feasible to map at a scale of 1:100,000'. Those measures thus include a considerable variety of forest types and local microhabitats.

However, later modifications weakened the primary focus on nature conservation, in part by diluting the obligations for formal targets for protection – that could, in some cases, be varied or revoked for socioeconomic priority and render some important reserve areas again vulnerable. Protests and demonstrations by concerned people continue over proposals to exploit ecologically important forest areas, and seem unlikely to abate – collectively, they reflect failure in earlier strenuous attempts to achieve greater harmony and stability. Widespread calls continue from environmentalists to 'lock up' more primary forest in National Parks and prevent its industrial exploitation. They are countered by forestry industry contentions that doing so may actually increase risk to those forests through fuel buildup, decreased trail maintenance so that emergency vehicle access is reduced, and destruction from wildfires will become more severe without enlightened management (including fuel reduction) of those forests. A policy of strict non-intervention in protected forests may thus have serious disadvantages, but polarised opinions over forest reservation will assuredly continue.

Evans (2016) commented that 'recent policy changes have raised concerns that Australia may again become a global hotspot for deforestation'. The history of deforestation (Carron 1985, Bradshaw 2012) implies that Australia still stands to lose a large proportion of its endemic biodiversity in the absence of clear policy and determination to protect remaining primary forest tracts and regenerate degraded forests to form matrix areas between primary forest patches. Bradshaw believed that major policy shifts are needed to effect those necessary changes, with the most damaged forest ecosystems (he exemplified the south west of Western Australia, central and western Victoria, the Mount Lofty Ranges of South Australia, and the southern and Wet Tropics regions of Queensland – but equally important examples could be cited from Tasmania and New South Wales) needing most urgent attention. The figures quoted by Bradshaw (2012) for accumulative deforestation levels of individual states demonstrate the likely losses and changes that have already occurred. For the Mount Lofty Ranges alone, Bradshaw quoted 'at least 132 species of animals (including 50 % of the mammal fauna) and plants have become locally extinct, and at least 648 non-indigenous species (mostly plants) have been introduced'. Such figures emphasise the likely impacts of alien species as they contribute to changes amongst native biota in forest environments, with fragmentation and increased edge effects facilitating their ingress.

However, recent policy amendments that allow greater 'self-regulation' and flexibility for landholders to retain/restore vegetation (with increasing offset policies where needed: Evans 2016) need serious evaluation and monitoring to determine if they will indeed fulfill their claimed benefits for conservation.

Advocacy and need for protected forest areas is a core element of forest policy but, as Andam et al. (2008) noted more generally for tropical forests, evaluating the 'success' of such areas is difficult. As elsewhere, some protected forests in Australia are simply those not amenable to human use by tradition, remoteness or topography, and the overall reduction in deforestation by their 'protected' designation is confounded by processes such as spillover – exemplified by more intensive forestry activities on the lands that are still accessible. In Queensland, regulation (through the 1999 Vegetation Management Act) was designed especially to provide high protection for threatened forest categories that had already lost the greatest proportions (>70%) of their area (Rhodes et al. 2017), and for which protection is most urgent – however, Rhodes et al. found little evidence that this had been successful. Although the Act has led to overall reductions in land clearing, greater level of protection for threatened than for unthreatened forest was not evident. In part, this reflected that many of the forests of highest conservation concern are found in areas where pressures for deforestation were highest, such as near urban centres and in other areas with high land values. Many current Australian forest reserves have a history of logging or other exploitation (Norton 1996, 1997), but are now important components of enhancing forest area and connectivity, in conjunction with restoration.

Australia is one of the 12 initial signatories to the 'Montreal Process' following a meeting in Canada in 1993, where a series of criteria for sustainable management of temperate and boreal forests was proposed. It included criteria for conservation of biodiversity, leading Taylor and Doran (2001) to suggest thet no monitoring of forest sustainability could be adequate without including some invertebrate taxa, including a selection of 'representative species' for each key microhabitat and which could be monitored easily to detect changes from any adverse impacts. As key microhabitats, Taylor and Doran suggested soil and litter, foliage and canopy, bark and branch, dead standing trees, and coarse woody debris – each of which is itself diverse and needing more specific prescriptions to incorporate that variety. Thus, coarse woody debris changes as decay proceeds over up to many decades (p. 156), so that beetles such as Cerambycidae and Lucanidae may be effective indicators of later decay stages: Taylor and Doran suggested the cerambycid genus *Toxeutes* as a suitable candidate for eucalypt logs in Australia. In their view, the key question to be addressed from the protocol is whether late successional species from the complete range of forest microhabitats retain viable populations within wood production landscapes. This challenge remains.

The legacy of forestry practice has led to increased interest in protection of forest fragments, with supposedly remnant isolated populations of native biota. Well-protected forest fragments in Manawatu (New Zealand) contain remnants of significant carabid beetle assemblages; compared with the nine species caught in a nearby large forest tract, the largest remnant yielded only two species (Lovei and Cartellieri 2000). Fragment assemblages were designated as 'collapsed', and this condition was attributed in part to low dispersal capabilities of the beetles, as seven of the nine species in the source assemblage were flightless, as wingless or brachypterous. Predation risk might also be an influence, with presence of predatory mammals in the larger fragment surveyed – predator control (as needed in much insect conservation practice in New Zealand) was recommended, together with possible translocations of the scarce beetles to other suitable areas.

10.3 Forest Regeneration and Landscape Design

Forest regeneration or restoration involves (1) commencing successional development in situations where no such natural recovery is then occurring, and (2) guiding and accelerating natural succession in places where it is occurring. Most commonly, this is undertaken by establishment of 'desirable' plants, whether understorey or trees, and removing undesirable elements such as invasive alien plants and other weeds. Several commentators have noted that as natural successional processes recover (or are reinstated through management), species richness increases as forest structure develops. Revegetation techniques that enable degraded forest sites to become more forest-like in structure and which are sufficiently close to intact forests as reservoirs for native species that are potential colonists appear to be 'of overriding importance in catalysing the rapid acquisition of volant rainforest beetle assemblages in the initial stages of restoration' (Grimbacher and Catterall 2007), in rainforest on the Atherton Tablelands of tropical Queensland. As reforested sites aged, their beetle assemblages increasingly resembled those of forests, and diverged from those of the previously cleared pasture on which restoration took place, and those of reforested sites adjacent to intact forest were more like rainforest assemblages than those on sites >0.9 Km away. However, despite this progressive convergence toward resembling true rainforest assemblages, significant differences persisted. Forest structure appeared to be a more important influence than age alone. Secondary forests, as they mature, may more generally come to support greater proportions of primary forest insect species, as Taki et al. (2013) implied for flower-visiting bees and cerambycid beetles in Japan.

Restoration of rain forest on pastoral land in Queensland has increased considerably, and invertebrate colonisation (and so establishing the processes characteristic of rainforest) can be accelerated by measures such as close planting of trees and mulching (Nakamura et al. 2003). The latter fosters the wellbeing of numerous litter-dwelling arthropods through providing direct resources and maintaining a relatively moist environment. In that study, depth of litter was associated with successful colonisation and increased diversity of many arthropod groups. Closer tree spacing leads to more rapid generation of canopy or tall foliage cover, and contributes to a larger local litter supply.

A healthy forest has been defined as 'one that satisfies ecological and/or economic management objectives or is structurally sustainable in relation to species of concern' (Teale and Castello 2011), and in which the balance between living and dead trees may be critical. Outbreaks of pests such as the mountain pine beetle (*Dendroctonus ponderosae*) can cause very extensive tree mortality and ecological impacts in addition to formidable economic losses in lodgepole pine (*Pinus contorta*)-dominated landscapes in North America. In outbreak conditions, the high beetle populations overcome the tree defences – in parts of Canada (where the beetle is considered invasive), *Dendroctonus* has killed 'millions of hectares of pine forest' (Cale et al. 2016), for example. Within its more limited natural range, beetle attacks and fire both contribute to more regulated forest structure.

Both imposed and natural ecological disturbance regimes vary greatly in intensity, with more extensive disturbance clearly more 'drastic' and likely to pose greater problems in undertaking forest regeneration or restoration. Regeneration is often contemplated, and sometimes assumed tacitly to proceed on a natural trajectory toward the pristine former state, from abandoned agricultural land left to its own devices. As one categorisation of impacts, in Mexico, parameters of such disturbance assessed by Zermeno-Hernandez et al. (2016) included field size, duration of agricultural use, and severity of use – this last including not only removal of tree cover but also treatments such as grazing, machinery use, agrochemicals and sizes/frequency of fires. Three major groups of land use were recognised by the overall features of disturbance, as (1) agroforestry schemes (with small size, low-intermediate duration, and low disturbance severity); (2) monocultures (typically with small size, long duration, and medium-high disturbance severity); and (3) extensive farming (large size, short-intermediate duration, high disturbance severity). In conjunction with landscape composition (proportions of the landscape used for agriculture, old-growth forest, or intermediate secondary forest), land use intensity may help to indicate the likelihood of forest regeneration, reflecting such features as seed availability and dispersal, seed predation intensity, susceptibility of propagules to fire, cattle grazing and trampling, mechanical impacts, and many others. High applications of pesticides or fertilisers to crops may also contribute to slowing or preventing forest regeneration.

Two main options for reducing the annual deforestation levels of tropical forest discussed by Stork et al. (2003) are (1) to reduce the level of timber extraction, or (2) to concentrate the timber extraction into smaller areas by measures such as replanting, so leaving large areas undisturbed whilst also recognising that intensive production areas may otherwise pose severe problems for conservation – which may be alleviated in part by measures such as border plantings and management. Young replanted forest stands provide habitat for many insects but, as shown for some forest butterflies in Cameroon (Stork et al. 2003) are not a good substitute for native forest. The level to which they fulfilled a substitute role depended on the extent and method of clearance, but that study used only small (one hectare) survey plots so that conclusions based on adults of such mobile insects may be rather tentative. Stork et al. speculated that replanted plots will over time come to resemble the unchanged forest in quality of butterfly habitat. That study focused on one plantation species, *Terminalia ivorensis* (Combretaceae), an endemic hardwood. Plantations of this species also affected ant richness in Cameroon, where Watt et al. (2002) compared canopy and leaf litter ants in relation to different methods of plantation establishment – either for plots that had been cleared completely, partially cleared by machinery (using a bulldozer to remove undergrowth and about half the large trees), and partial manual clearing, all relative to uncleared 'control' forest plots. Ants were diverse in both strata (canopy 97 species, litter 111 species), and

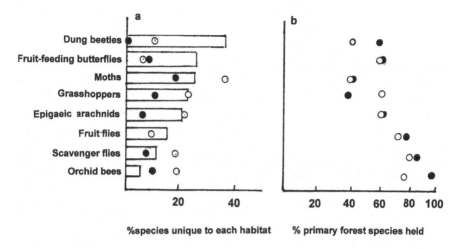

Fig. 10.3 Representation of selected insect groups in primary, secondary and plantation forests in Amazonia: (**a**) percentage of species unique to primary forest(bars), secondary forest (black spots) and plantations (open spots); (**b**) percentage of primary forest species that were also recorded in secondary forest and plantations. (Abbreviated from Barlow et al. 2007 with permission: Copyright 2007 National Academy of Sciences, U.S.A.)

strong differences were found across treatments. Species richness was implied strongly to be greater in plantations established after partial manual clearing than from complete clearance. Leaf litter ant diversity was not decreased by this transformation.

It has been claimed repeatedly that losses of biodiversity from deforestation may be offset by expansion of plantations and secondary forests (Hartley 2002, p. 35). A critical examination of those claims for tropical forest (Barlow et al. 2007) showed that primary forests are indeed irreplaceable, notwithstanding some clear benefits from plantations and regeneration of native vegetation. The latter systems, both expanding rapidly in the tropics in particular, can help to retain more forest species than alternative more intensive land uses such as agriculture. Barlow et al. compared a range of different taxa in a forestry project in Brazilian Amazonia, in which large blocks of 14–19 year-old secondary forest and 4–6 year-old eucalypt plantations were contained within a largely undisturbed primary forest matrix. Responses were highly variable – amongst insects (Fig. 10.3) fruit-feeding butterflies were richest in primary forest and declined through secondary forest to plantations; dung beetles had very similar richness in the two modified regimes, both with far fewer species than in primary forest; moths were least rich in primary forest; and several groups of Diptera showed little difference across the three categories. Inferences from taxon richness also depended on uniqueness and were highly variable depending on both the taxon and the parameters selected for appraisal. Addressing apparently simple questions related to forest conservation values in comparison with those of degraded habitats should ideally incorporate study of a range of different taxonomic groups – but Barlow et al.'s study emphasised the unique importance of undisturbed primary forests and of ensuring comprehensive reserve networks to save these across the forest landscape.

Naturally regenerating secondary forests in northern temperate China support high diversity of geometrid moths, leading Zou et al. (2016) to urge their protection, especially in areas where mature forest has disappeared, and also to plant local trees to promote connectivity. Those secondary forests support geometrid assemblages that resemble those of mature forests in richness and phylogenetic variety, but with species overlap of only about 30% with the moths of a mature temperate forest reserve. However, the latter forest type has been almost completely lost. A second study, incorporating Carabidae (Zou et al. 2015), also emphasised scarcity of the region's mature forests and noted that 'high diversity' in itself may be insufficient to indicate a 'good' forest ecosystem, because species composition must also be considered. Zou et al. noted, for example, that the few remaining forests in the region are potentially source areas for a high proportion of predatory carabids amongst the diverse forest specialist species, and that ecological complexity heightened their conservation significance.

The functional capabilities of insects can also change in regenerating forests. The ants of naturally regenerating *Eucalyptus regnans* forest in central Victoria showed increased foraging levels even on sites regenerating for only a few months following a fire (Neumann 1992). This trend occurred in both predominant ant species (*Anonychomyrma biconvexa* [then known as *Iridomyrmex foetans*] and *Prolasius pallidus*, both members of species complexes). Both are seed-feeders, and *P. pallidus* was earlier known to be a principle seed-harvester in *E. regnans* forest that had been burned by intensive wildfire and subsequently logged. For aided regeneration, aerial broadcasting of approximately 200,000 *E. regnans* seeds/hectare was anticipated to satiate the ant predators so that successful regeneration would occur despite the large ant numbers foraging for the seeds. At low seed demsities, *P. pallidus* might hamper successful regeneration. Neumann suggested the parallels between 'more natural' impacts (high intensity wildfire killing forest, salvage logging, natural regeneration induced by fire) and 'more imposed' commercial harvesting/ regeneration (clear-felling, log extraction, slash burning or mechanical site preparation to expose mineral soils, aerial seeding), both with high impacts and complex effects.

The role of timber plantations in effective reforestation is often low. As Lamb (1998) commented for tropical forests, they 'usually restore the productive capacity of the landscape, but do little to recover biodiversity'. However, a number of measures proposed by Lamb might indeed have more positive effects: (1) the simple use of indigenous rather than alien species; (2) creating species mosaics by linking matching species and sites; (3) embedding plantation monocultures into a matrix of more natural vegetation; and (4) increasing variety by using species mixtures and/or encompassing diverse plant understoreys, are all likely to enhance wellbeing of native insects with relatively little loss of plantation productivity. The relative advantages and disadvantages of these, and the impediments to adopting them widely, are noted in Lamb's table (Table 10.3), as now widely advocated principles of wider sound landscape ecological management.

Table 10.3 Improving biodiversity in timber plantations. Advantages (A) and disadvantages (B) of various management options, and some impediments to their adoption (C) (from Lamb 1998)

Approach	Comments
Native species	(A) commercial value may be higher; social value may be higher; better adapted to environment. (B) slower growth; incomplete silvicultural knowledge; less suitable to degraded sites; less easily marketed. (C) ecological and silvicultural data often lacking.
Buffer strips	(A) greater watershed protection; improved fire control. (B) reduced plantation area; source of pests. (C) none.
Species mosaics	(A) improved overall productivity. (B) management more complicated; more than one species less easily marketed. (C) inadequate site descriptions; inadequate knowledge of species-site relationships.
Mixtures	(A) improved plantation productivity; reduced herbivory and disease; improved nutrition; earlier financial returns; social benefits from non-timber species. (B) management more complicated; reduced plantation productivity; marketing more complex. (C) finding complementary species; confirming benefits; developing silvicultural systems.
Understorey development	(A) reduced topsoil erosion; improved nutrient cycling; may include socially useful plant species. (B) reduction in overstorey tree growth; management more difficult;. (C) developing methods for fostering understorey at isolated sites; management of understories dominated by weeds.

Table 10.4 The categories of considerations for plantation management for biodiversity discussed by Hartley (2002)

Landscape context - larger spatial scale and context of the plantation determines how other aspects are decided and undertaken.

Harvest considerations – amount of native vegetation within a plantation is a key correlate of biodiversity conservation: design harvest approaches to retain some natural vegetation, as clumps, strips, and isolated trees.

Native versus alien species - from conservation viewpoint, use of alien tree species is the most contentious aspect of plantation forestry: native species should be favoured over alien trees if conservation of native biodiversity is to be optimised, and retention of native trees in other plantations can be highly beneficial.

Mixed species versus monoculture – mixed species plantations can result from retention of natural vegetation or deliberate plantings: debate continues over relative merits for conservation (mixed species equates to greater variety and heterogeneity) or harvest and economic returns.

Site preparation - decisions on site preparation can have considerable implications for biodiversity: they include removal of pre-existing or competing vegetation, and of woody debris, use of fire, and intensity of soil preparation and of machinery use.

Tending - management during first half of rotation especially important for biodiversity: extent of thinning, use of pesticides and fertilisers are relevant issues.

See also Table 3.1, p. 35 for further detail

Most discussions over conservation and management in plantations have emphasised need to enhance sites for forest species, so are largely restricted to plantations (such as many in Australia) that replace primary forests. Several measures to improve biodiversity values for those plantations (Hartley 2002) were noted earlier, with Hartley suggesting four main categories for management (Table 10.4, see also p. 35).

Focussing primarily on plant species richness, but likely to transfer easily to insect richness, Bremer and Farley (2010) addresssed the continuing debate over the possible values of plantations in enhancing biodiversity. Those values for plants varied considerably, with factors such as the form of the original land used (with plantations more likely to contribute to biodiversity when established on disturbed land rather than replacing natural ecosystems, whether these are forest, shrubland or grassland). Loss of plant diversity with afforestation of natural grasslands and shrublands flows from variety of causes – Bremer and Farley noted site preparation, progressive loss of shade-intolerant species as canopy developed, allelopathy, and physical barriers such as pine litter impeding germination. Their major conclusion was that primary forest and older non-forested vegetation should not be converted to plantations – as also recommended earlier by Brockerhoff et al. (2008). Biodiversity values of plantations are only one of the important benefits they furnish, together with carbon sequestration and more local benefits such as preventing soil erosion.

Transformation of monoculture plantations into mixed forests may be beneficial in increasing biological variety, and has been advocated for increasing the conservation values (and diminishing ecological impacts) of Norway spruce plantations replacing native beech forest in Bavaria (Huber and Baumgarten 2005), where small clear-cuts in the spruce forests are the templates for regeneration or for plantings of native trees. Responses of carabid beetles in Bavaria followed the general trend explored earlier by Niemela et al. (1993), with initial increase of open-habitat species in clear-cut areas and their later decline and replacement by forest-dwelling taxa. However, and again as in earlier studies, forest species simply disappeared or declined substantially in the clear-cuts, so that extensive clear-cutting may threaten those forest specialists. As another example, numbers of forest specialist Carabidae increased with plantation age (of several species of conifers) in Britain, with corresponding decrease in non-woodland species (Jukes et al. 2001). Whilst only 25 species (of the total 51) were found in both spruce and pine plantations, this pattern occurred in both over a chronosequence survey embracing pre-thicket, mid-rotation, mature and post-mature sample plots across the broad bioclimatic zones of upland, foothills and lowlands over a wide range in Scotland and England. In any such exercise, site replication is important in boosting confidence in the conclusions and, indeed, in discussing a further example of this trend amongst Carabidae of northern Sweden, Atlegrim et al. (1997) suggested that increasing the number of sites in such surveys may prove more informative than increasing the number of samples from a given area.

Because of advantages of sustaining the greatest levels of heterogeneity possible, a mosaic of different-aged forest stands, with regeneration planned to include a variety of ages and species may be advantageous over a more sudden and uniform widespread change.

Conservation-orientated strategies for forest biodiversity management are, fittingly, diverse (Gotmark 2013) and (for the intensively appraised northern hemisphere conifer forests) comprise two widely documented and extensively debated options. Thorn et al. (2015) distinguished these strategies as (1) 'benign

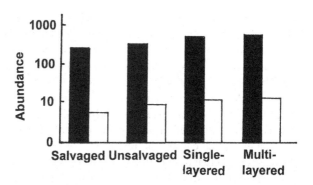

Fig. 10.4 Mean abundance of moth assemblages (non red-listed species, black; red-listed species, open) on forest stands with several treatments, based on 224 individual nights of light-trapping moth assemblages in Germany. (Thorn et al. 2015)

neglect' of stands disrupted by disturbances such as fire, storms or pest outbreaks, and (2) deliberate promotion of multi-layered forest stands assumed to be more natural and resilient against future disturbances. Comparisons of insect groups in multi-layered and 'single-layered' stands, however, are relatively scarce, but one such survey, in spruce-beech forest in Germany, compared Lepidoptera across these conditions, with additional light trap catches also comparing moths of salvaged and unsalvaged stands. Moth assemblages, especially if the ecologically diverse 'micro-lepidoptera' are included, can be good indicators of change in forest stands (Thorn et al. 2015). Catches included 291 'macromoth' species and 137 'micromoths', with an overall 97 species represented by singletons. Assemblages were compared both taxonomically and by larval feeding guild representation (Fig. 10.4), in which richness of species with larvae feeding on trees or shrubs (188 species) was the greatest. Higher overall moth abundance occurred in naturally disturbed unsalvaged forest than in salvaged forests, but no clear or significant effect of the two silvicultural contexts was evident.

In practice, 'benign neglect' usually includes some form of salvage logging to mitigate economic losses, and Thorn et al.'s study demonstrated values of benign neglect for saproxylic or detritus-feeding Lepidoptera, so that management components could embrace both kinds of stands, with suggestion that some naturally disturbed forest stands should not be salvaged, especially in protected areas. One other approach suggested tentatively by Thorn et al. (2015) is to allow natural disturbances in single-layered stands to generate sections of multi-layered stands.

Benign neglect was suggested also for use after bark beetle infestations and resultant large-scale tree mortality in a German national park (Muller et al. 2010), for which lack of active forest management improved habitat conditions for saproxylic beetles. To fully maintain the beetle assemblages in commercial montane forests, the amount of dead wood would need to be tripled (from a measured level of ca 15 m^3 ha^{-1}), with focus on spruce wood in open stands but on broadleaved tree wood in dense stands.

Intensive forest management poses many significant threats to native biodiversity, with 'natural disturbances' having largely given way in managed forests to regular linked disturbances such as clear-cutting, thinning, selective logging, and replanting. However, from a literature overview of natural disturbances to European forests (over the period 1850–2000), Schelhaas et al. (2003) clearly showed the relative frequency and impacts of some key factors. Storms were responsible for 53% of total damage, fire for 16%, snow for only 3%, and other abiotic causes accounted for 5%. Biotic agencies caused 16% of total damage, with half of this from bark beetles, and only 7% of the total damage could not be attributed clearly to any single cause, or its causes were combined. Collectively, an annual average of 35 million m² of wood was damaged by disturbances, with considerable variation between years. In a definition of wide relevance elsewhere, a 'natural disturbance' was 'an event that causes unforeseen loss of living forest biomass or an event that decreases the actual or potential value of the wood or forest stand'. Predicting when and where future disturbances will occur is, perhaps, impossible in detail – but much management in principle incorporates perceived risks, such as through fuel reduction burns used to reduce intensity of future wildfires in Australia. More generally, again from northern European forests, Halme et al. (2013) identified important challenges that must be heeded for effective restoration and benefits to biodiversity, as (1) coping with unpredictability; (2) maintaining connectivity in space and time; (3) assessment of functionality; (4) management of conflicting interests and restrictions; (5) social relations, such as attitudes of land owners; and (6) ensuring adequate funding for action and monitoring.

Legacy of land use can have have persistent influences in forest ecology. In Sweden, for example, large areas that were historically managed as meadows have been transformed into conifer forests, whilst other open areas are the outcome of clearing historical conifer forests. Comparisons of butterflies in the two categories of clearcut areas (Ibbe et al. 2011) showed the former meadows (after abandonment sufficient to allow a 70–90 year period for conifers to mature) supported more species and individual butterflies and were strongholds for some taxa threatened elsewhere in Europe.

Both extent and 'direction' of forest transformation are relevant in planning conservation management. Much of the forest given high conservation priority in the northern hemisphere has been largely transformed from being sustained by natural processes to becoming largely the outcomes of human influences. As Bernes et al. (2015) emphasised, this can affect well-intentioned conservation efforts and goals – such that the proportion of forest with high natural values, and are priority for set-aside and protection, may become far less than the stated target levels. Bernes et al. cited the 'western taiga' forest category of Sweden, for which only about 10% of the original forest extent satisfies criteria for the high quality category of the European Species and Habitat Directive. Old-growth forest, in particular, falls short of those ideals in many forest types, and many set-aside forests need active management to sustain or restore their previous high values for biodiversity. Especially since World War II, forest structures in much of northern Europe have been moved beyond their natural historical variability, in the interests of ensuring the supply of timber. Most ecosystem services were given low priority, or not considered specifically, in relation to maximising timber production by measures such as shortening rotation cycles. Both old-growth forests and natural early succession forests have become vulnerable and have largely disappeared over large areas. Indeed, Halme et al.

(2013) claimed that 'there are almost no truly natural areas remaining in Fennoscandian forests'. Forest conditions beyond the range of historical variability can still support considerable diversity of ant species and functional groups (Stephens and Wagner 2006). Their study in Ponderosa pine forest in Arizona (p. 105) showed differences in ant functional groups in relation to stand health and conditions, with different functional groups predominant under different disturbance intensities – and excluded or suppressed by other functional groups less suited to tolerating those disturbances.

Part of the outcome of increased scarcity of natural disturbances such as fires is that volumes of dead wood have decreased, affecting the wellbeing of numerous saproxylic invertebrates. Measures to restore such resources, in the contexts of wider disturbance regimes may have wide benefits for such taxa (Toivanen and Kotiaho 2007).

Restoring forest disturbance regimes to resemble those that occur naturally, together with establishing or retaining structural diversity, is a recognised management ideal for conserving the natural forest biodiversity. Those natural disturbances include fires and storms, both very variable in impacts and outcomes, reflecting their severity and area of influence. However, practical management for economic benefits may restrict the scope of those possible measures in the interests of expediency – and also ensure that some form of 'set aside' or protection from accidental imposed disturbances is intrinsic. Recognition that forest exploitation has led to severe changes in ecosystem processes and vegetation structure, with consequent changes (including declines and losses) of biodiversity has increased awareness of the needs to progressively but urgently modify many former practices. A formerly widespread desirable management target of creating forest stands of even ages and much reduced tree species diversity (in many cases, monocultures) and short rotation periods to create uniformity, is questioned increasingly. The complexity of forest conservation and development to serve either the local or the wider human community is a dynamic process. The biological condition and functions of the forest may largely determine its fundamental conservation value and significance, whilst Kusel (2001) suggested that other factors strongly influence the relationship between local people and forest. Discussed for the Mexican Monarch butterfly reserves (p. 195) (Manzo-Delgado et al. 2014), those features apply widely also elsewhere – in Australia, for example, a number of isolated 'timber towns' have grown up with, and continue to depend on, commercial forest use as the major or sole source of local livelihoods. Loss of the timber industry dictates the loss of those settlements. Kusel listed the following as important influences: (1) the extent to which commercial activities are allowed and promoted; (2) the extent to which forest-related jobs (including tourism) are promoted and maintained; (3) the terms stated for job availability; and (4) the extent of the market for the forest products, or for related products or tourism activities.

Natural disturbance regimes tend to preserve mosaics, with more local impacts rather than larger ('stand-replacing') ones, so that the forest remains diverse and 'patchy' with a variety of structure, species and age classes in and among stands. Debate continues over whether forest areas protected primarily for biodiversity conservation should be managed after disturbances (such as windthrows, below), and options have been discussed extensively in relation to bark beetle attacks. Interpretations for central European conifer-dominated National Parks with disturbances from a combination of fire, windthrows and bark beetle attack (*Ips typographus*, p. 249) (Lehnert et al. 2013) suggested that core zones of around 10,000 hectares can sustain all biological legacies of dense and early successional forests without losses of species adapted to any one stage. Such 'damaged forests' had high conservation value. Destructive effects caused by insects led to a variety of biological legacies, with effects rendered more complex by canopy opening and microclimate changes such as soil warming. The 'legacies' listed by Lehnart et al. were (1) the remaining living trees, with roles depending on the earlier stand characteristics; (2) the rapidly created supply of dead wood with successions from many snags to many logs; and (3) undisturbed understorey vegetation.

The gradation between extremes of intensive agriculture and reserved natural forests reflects many aspects of desirable restoration, in some contexts through the varying categories of plantation forests - each reflecting particular purpose and also that plantations may 'flow' from either resuming agricultural land or directly replacing felled/cleared natural forests. The stages, discussed by Brockerhoff et al. (2008), are summarised in Fig. 10.5, in which gradients of management intensity and conservation value are also shown, with numerous variations to consider successional rates and trajectories.

Gotmark (2013), soberingly, noted that 'the professional life of ecologists and managers (30-40 years) is much shorter than the development of a forest and the life span of most trees'. That reality emphasises the difficulty of undertaking continuous long-term studies on effects of disturbances, and it is perhaps inevitable that most 'conclusions' on impacts on insects (and on most other aspects of forest biology) are drawn from management documentation over relatively short periods – at most, of a few decades, and commonly far shorter. Reviewing past decisions and management trajectories and policies can be informative, but important components may easily be overlooked or have changed in scope or priority. Nevertheless, the purpose of forest conservation may have a very long-term foundation – as for the New Forest in England – and in many places this reflects historical and cultural traditions. Five questions for future research nominated by Gotmark (Table 10.5) may help understand how habitat management may be realised, each of them needing well-conducted long-term experiments focused on the values of that management in and to forests. Should they eventuate at all, those experiments are likely to occur in forests that have been changed by management of varying intensities, and based in the foundation belief that 'natural forests' are the reference condition for sustainable management (Paillet et al. 2009). From their meta-analysis of studies on European forests, for which Paillet et al. estimated only <1% to be truly 'natural', arguments were advanced for (1) conservation of unmanaged forests and (2) creation of forest reserves on a broad scale. The recovery of biodiversity in unmanaged forests might be long for some taxa, so that conservation policies need a long-term perspective.

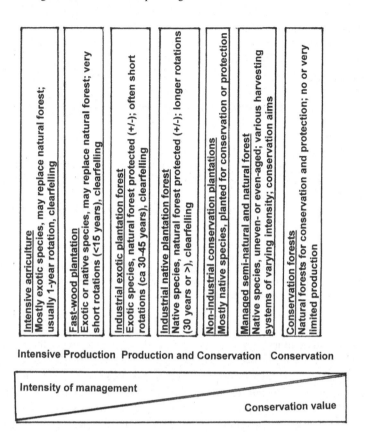

Fig. 10.5 A conceptual model of the relative conservation value of planted forest relative to conservation forest and agricultural land use forestry plantation stages (note that many plantation forests cannot be assigned clearly to one of the main categories outlined in the scheme). (Brockerhoff et al. 2008)

Table 10.5 Habitat management for biodiversity in temperate region forests: five important questions for future research, as nominated by Gotmark 2013

1. How do species-rich taxa respond to different management alternatives, such as traditional management versus minimal intervention in conservation forests?
2. To what extent are taxa of special interest (such as red-listed species) reduced or eliminated, or favoured or added, by one alternative compared to others?
3. How does active management versus minimal intervention change species composition of shrubs and trees, and regeneration of desirable woody plants?
4. How does active management change forest structure, including dead wood, light conditions and other factors, compared to minimal intervention?
5. How does active management for a single forest species (such as a red-listed or keystone taxon) influence this and other species, and could the species in the long run be favoured by disturbances and succession under minimal intervention?

Table 10.6 The treatments used to sample beetles in the Warra Silvicultural systems trial, Tasmania, to indicate management and structural differences (From Baker et al. 2009)

System name	Description of structure
Clearfell, burn and sow	
	Large openings with no structural retention, high intensity burns, applied seed.
As above, with understorey islands	
	As above, but with 40 m x 20 m machinery exclusion zones in up to 5 % of the coupe area.
Dispersed retention	
	10–15% of basal area retention of overstorey eucalypts, low intensity burn, natural seedfall.
Aggregated retention	
	30% of coupe area retained in aggregates of 0.5-1 ha within the majority of the harvested area within one tree height of retained forest, low intensity burn, natural seedfall.

Different silvicultural systems, as noted earlier, provide for different management alternatives with probable different consequences. Three families of pitfall-trapped beetles, all known to respond to forest management, were compared for four silvicultural systems in wet eucalypt forest at Warra, Tasmania (p. 164) (Baker et al. 2009). They comprised Carabidae (12,009 individuals, 43 species), Curculionidae (3004, 29) and Leiodidae (4226, 70), for which assemblage composition varied across systems and also with time since harvesting. Two of the systems sampled (Table 10.6), together with a long-term undisturbed native forest control site, all retained resources for beetles affiliated with mature forest, and the patterns reflected that that many species were strongly and specifically associated with either young regeneration or mature forest. Those 'preferences' were important determinants of species' responses to the alternative harvesting systems. Baker et al. suggested that, because most species were rarely entirely absent from one or other system, the alternatives did not create unique habitats for the beetles that do not usually occur in forested or clear-felled areas but, rather, changed the opportunity for the areas to be used. It seemed that mature forest refugia within harvested areas are important landscape features that facilitate colonisation of regenerating areas as these become suitable. A relevant management implication was that expanding the practice of 'aggregated retention' into production landscapes in place of some more widely used 'clear, burn and sow' treatments may provide general biodiversity benefits for mature forest inhabitants.

10.4 Gaps

Wind storms and similar events can create gaps in forest by 'windthrow', whereby toppled trees lead to locally changed conditions and microclimate, with exposed dead wood and that in some cases lead to perforation of closed forest canopy and a more heterogeneous mosaic stand. They can be a driver of forest succession, and also sources of regional biodiversity in forest ecosystems. Impacts on forest insect communities vary (Bouget and Duelli 2004). In essence, storms create gaps with increased sun exposure and substantial volumes of dead wood of all size categories from small twigs to large logs and snags (Bouget 2005a, b). The effects of increased exposure and insolation may be difficult to distinguish from those resulting simply from increased wood supply and diversity – and a high proportion of beetles living on some tree taxa in Sweden, for example, prefer open forests. Sunlit areas in those forests support characteristic saproxylic assemblages (aspen: Sverdrup-Thygeson and Ims 2002). Treefall gaps, defined fundamentally by one or more fallen trees and open sky above, have been likened to 'vertical holes' in the forest that may allow canopy-dwelling insects to descend to lower levels. The moth *Zunacetha annulata* (Dioptidae) feeding on the understorey shrub *Hybanthus prunifolius* (Violaceae) in Panama apparently focuses strongly on the food plants growing in recently-formed gaps (Wolda and Foster 1978, Harrison 1987). Predation on larvae was higher in gaps than in normal understorey, but nutritional quality of the open foliage probably also influenced occurrence (Harrison 1987). In short, the gaps provided a higher quality environment for the moth, in conditions of fortuitously created heterogeneity likely to affect many other insect species in similar ways.

As (1993) noted that strong wind, creating gaps in forest canopy, is 'the most important natural disturbance in modern forests', with potential to create a fine-scale mosaic of habitats. Although referring to northen boreal forest, in which forest fires have declined considerably, the principle of gaps creating mosaics is widespread even in forests where more predictable fires can augment that mosaic to even greater diversity. One feature on this heterogeneity, the focus of As' study in Sweden, is the incidence of deciduous forest remnants in large areas of conifer forest, in that those remnants might constitute 'islands' for native inhabitants. This study focused on beetles living in dead stems. Absence of clear 'island effects' (such as different species richness of beetle assemblages in large and small patches) may reflect simply the relatively recent use of this form of forestry, so that the deciduous matrix is still high quality and able to support many species – rather than being isolated, As (1993) suggested that the remnant patches could still be seen as 'incipient islands' able to exchange biota with surrounding areas.

Gap formation clearly creates heterogeneity, with closed canopy and treefall gaps of different sizes and ages contributing to a mosaic of resources on the forest floor and in the understorey layers. The changed conditions in recently formed gaps, notably higher temperatures and lower relative humidities, might necessitate some insects seeking refuge in areas where conditions are less stressful (Richards and Windsor 2007). Seasonal patterns of incidence and activity also occur, and Richards and Windsor reported (from Malaise trap catches in Panama) higher insect activity in gaps during the rainy season and higher activity in understorey during the dry season. Nevertheless, 'gap specialist' and 'shade tolerant' species may have different individualistic activity patterns that affect such inferences – different species of Odonata, for example, frequent one or other regime (Shelly 1982). Different species of fruit-feeding nymphalid butterflies utilise treefall gaps and forest understorey in Peru (Pardonnet et al. 2013), and the different assemblages facilitated by the heterogeneity caused by treefall gaps each contribute to regional richness. Comparisons of these butterflies in 15 gaps (5–10 years old) and paired understorey plots over 13 five-day trapping periods gave very similar numbers of individuals (gaps 756, understorey 775), but 71 species in gaps compared with 50 from understorey. The gaps apparently facilitated higher richness. Perhaps reflecting larval host plant needs, some nymphalids were clearly associated with gaps with specific vegetational features, such as dense vines. Pardonnet et al. suggested that a mosaic gradient including undisturbed understorey and gaps ranging from newly formed to older densely vegetated spaces may lead to coexistence of increased numbers of species. In some such cases, increased richness in 'gaps' represents colonisation by alien species, but mosaics of gaps largely encompassed within less disturbed forest may (at least partially) be immune from this, so that cases such as the above involve native species alone. Nevertheless, and as Ghazoul (2002) discussed, some forest butterflies that are considered specialists to understorey can indeed benefit from increased disturbance from gap establishment. Proliferation of larval food plants such as vines (above) may lead to concentrations of adults, and the varied forest butterfly faunas will normally include some species that 'prefer' the most open areas available and that can exploit gaps as either vagrants or transient populations. The gap assemblages of butterflies can thus be 'supplemented' by characteristically canopy species (Hill et al. 2001, for fruit-feeding Nymphalidae in Sabah), but may nevertheless be distinctive.

Forest gaps caused by selective logging from illegal tree poaching are a feature of the tropical Tam Dao Mountains, Vietnam, and impacts on forest butterflies in intact forest and gaps of 100-200 m^2, as relatively small-scale mosaic patterns, were studied by Spitzer et al. (1997). Species richness and abundance were higher in gaps, but the 'gap species' were mostly widespread opportunist taxa and many of the closed canopy species were absent from gaps. The high proportion of these localised specialists in closed forests reflects other such studies in endorsing conservation value of those forests, but Spitzer et al.'s data suggested that even small–scale localised damage through creation of anthropogenic gaps may change local butterfly assemblages, and led them to suggest that 'such small-scale but increasing destruction by local wood collectors seems to be one of the most serious conservation problems'.

Treefall gaps, commonly only small in area, are the most frequent natural disturbances in many forest systems and, although the resulting changes in plant communities – from high light penetration in the cleared gaps – have been well-studied, in general investigations of the responses of animals (including insects) have lagged somewhat (Patrick et al. 2012). 'Gap dynamics', however, has been a common theme for ecological study. Comparison of ant assemblages in Costa Rica by using Winkler bag sampling from leaf litter at a series of 12 intact forest sites and 12 large (>80 m^2) adjacent treefall gaps indicated that the gaps might impose little change. Thirty-seven species or morphospecies of ants were obtained, with richness similar in the two treatments - 31 in gaps, and 28 in intact forests – and no species was more common in gaps than in forest. The most abundant species occurred in most sites and across the two treatments: thus, *Pheidole monteverdensis* was found at all 24 sample sites, and several other taxa occurred at more than 20 sites. The gaps sampled in this study also represented a chronosequence, from <1–12 years old, but no successional changes of ant assemblages were detected, and cross-treatment similarity was high for all gap ages. Patrick et al.'s (2012) study contradicted some earlier suggestions that disturbances from gap creation may be an important influence on litter ant assemblages through changes in insolation, soil moisture and temperature and resulting influences on foraging activity. In this case, the possible short-lived impacts were thought likely to have little lasting effect on the assemblages present in intact forest. Nevertheless, interpreting impacts of treefall gaps on forest ants can be difficult. On Barro Colorado Island, Panama, Feener and Schupp (1998) commented that ant assemblages 'respond to the formation of treefall gaps in complex ways that do not give rise to simple, obvious differences between gap and forest habitats in numerical abundance, species richness, rates of resource discovery, or species composition'. This outcome was a marked contrast to some reported from other insects, and it was clear that the assemblages lacked obvious differences between gaps and nearby understorey.

The features of windthrow gaps in forests are very varied, but their formation 'results in a shifting mosaic of open early-successional patches rich in dead wood' inside the forest (Bouget 2005b). In oak-hornbeam forest in France, the initial response amongst saproxylic beetle assemblages was increased relative abundance of open habitat species and decrease of closed-canopy forest species in relation to undisturbed stands (Bouget 2005a), with some differences across beetle guilds in relation to gap size (Table 10.7). As an example, flower-feeding species were more abundant in large and moderate-sized gaps than in small gaps, presumed to be related to larger supplies of blossoms, but some other results were more anomalous. The abundance of secondary xylophagous species did not decrease with increasing gap area, despite the microclimate expected to be more destructive to forest specialists in the more exposed openings – so that sun-exposed harvest residues left in clear-felled areas might enhance diversity (Bouget 2005a).

Gaps allow concentrations of saproxylic species and of taxa adapted to clearings, both within mosaics because the gaps are created by chance events rather than any intent. Gaps can also help rare species, as indicated for Swedish red-listed beetles, of which about 20% are favoured by windthrow (Berg et al. 1994, Wermelinger et al. 2002). As for cleared corridors in forests, these small open spaces can enhance

Table 10.7 The abundance of major ecological groups of saproxylic beetles (catches from baited window pane traps given as 'mean/trap (+/−SD')) in gaps of different sizes (small, <0.3 ha; medium, 0.3 - <1 ha; large, >1 ha) and forest controls in managed oak-hornbeam forest in France Abbreviated from Bouget (2005b)

	Large	Medium	Small	Control
Pioneer xylophagous				
	129.8 (119.9)	113.9 (78.5)	98.6 (65.9)	108.4 (137.2)
Secondary xylophagous				
	11.2 (12.1)	9.1 (11.0)	7.0 (7.2)	11.5 (10.0)
Xylomycetophagous				
	3.7 (4.1)	3.7 (3.7)	2.2 (2.2)	1.7 (2.3)
Zoophagous				
	25.2 (26.7)	29.3 (34.6)	22.6 (27.9)	29.3 (25.8)
Floricolous				
	13.4 (12.9)	17.1 (20.6)	7.9 (7.5)	3.5 (3.9)
Total saproxylic species				
	166.4 (142.9)	155.3 (108.2)	129.9 (79.7)	148.2 (152.1)

supply of nectar or pollen by increased numbers of flowers, but also provide conditions attractive to early successional habitat specialist insects – even if over only relatively short periods whilst succession is rapid.

Management of windthrow gaps can contribute to both flower richness and supply of coarse woody debris, such as by clearing timber from part of the gap area (allowing it to flower prolifically) and leaving part of the gap unchanged (Bouget and Duelli 2004), so that fresh or sunlit dead wood is also provided, as the foundation of saproxylic insect successions that may last for many decades. The best method to enhance biodiversity after severe windthrow is likely to be this 'half and half' approach, as discussed by Duelli et al. (2002) and with this recommendation based on up to a decade of survey in three sites in alpine spruce forest in the eastern Swiss Alps following windthrow from the severe storm Vivian. Pairwise comparisons between cleared and uncleared windthrow areas and an intact managed forest control plot indicated that a combination of cleared and uncleared plots enhanced biodiversity considerably over that in either single treatment. Over the duration of the study, species richness in the new mosaic of forest and gaps doubled in the decade after the storm. However, the fauna in the two treatments did not converge but, instead, became less similar – so that 'conventional' resilience, taken commonly as faunal convergence toward similarity in species composition, did not occur.

The saproxylic beetles in the three windthrow areas were appraised in more detail from the 10-year survey of the three predominant families, Cerambycidae, Buprestidae and Scolytidae (Wermelinger et al. 2002) based on catches from a combination of window traps and yellow 'pan' traps. The total catches (about 68,000 individuals) included 86 species of these families (11 Buprestidae, 37 Cerambycidae, 38 Scolytidae). Abundance and richness from 1991–2000 is summarised in Fig. 10.6. Scolytidae peaked in the third year after the storm (1992) and declined

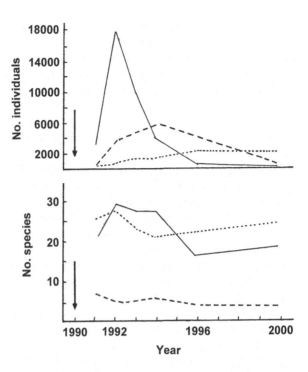

Fig. 10.6 Numbers of species and individual beetles of three groups (Scolytidae, solid lines; Cerambycidae, dotted lines; Buprestidae, dashed lines) for 10 years after storm in three windthrow areas from 1990 (year of windthrow, arrowed), in alpine spruce forests in Switzerland. (Wermelinger et al. 2002)

subsequently to low levels. Build-up of the other families was considerably slower, and abundance was not closely associated with richness. Only Cerambycidae richness increased gradually over the later years of the survey, and numbers of Buprestidae were generally low. Combined species richness of Cerambycidae and Buprestidae in forest (32 species) was far lower than in the uncleared windthrow areas (59 species.

No relationship between gap area and the proportion of trees colonised by the Spruce bark beetle, *Ips typographus*, was found in Sweden (Schroeder 2010), so suggesting that landscapes with many large gaps from storms should be those salvaged first, because the logging processes could be used most efficiently. This priority may help to optimise the three important management decisions that arise: as listed by Schroeder these are (1) optimising use of limited logging resources for greatest timber rewards over time; (2) the increased risk of tree mortality in later years in areas that are not salvaged; and (3) risk of degradation of fallen trees by beetles that are vectors of blue stain fungi. Harvesting is most efficient in areas where transport exists, and the higher value fallen trees can be salvaged before further degradation occurs.

Natural treefall gaps could influence insect assemblages in several ways, noted for ants in Panama by Feener and Schupp (1998). These included (1) altering microclimate variables such as insolation, and soil moisture and temperature – in turn affecting ant nest site quality and foraging behaviour; (2) inducing higher plant/herbivore insect productivity in gaps, possibly increasing food resources; and (3) that myrmecophytic plants are often more abundant and diverse in gaps than in closed forest. Comparison of ant assemblages in treefall gaps and closed forest (using baited pitfall traps) produced a minimum of 87 morphospecies (34 genera), but responses to gap formation were complex, with direct forest/gap differences clouded by other factors such as forest structure and season of sampling. It seemed that natural treefall gaps *per se* had rather little influence on the ant assemblages, despite the increased habitat heterogeneity they help to generate.

Accounts of insects in gaps attributed to specific documented natural events are exemplified by a study of impacts of Hurricane Hugo (September 1989, with maximum wind speeds around 225 Km/hour^{-1}) in Puerto Rico, in which canopy invertebrates were subsequently compared by examining branches from standing trees and nearby canopy gaps (Schowalter 1994). Some taxa were significantly more abundant on standing trees, but other taxa perhaps responded to changed microclimate or post-disturbance host plant condition, and were more abundant in gaps. Responses to the hurricane thereby differed across taxa, and also across the five major tree species that were individually sampled - broadly, the canopy invertebrate communities differed between disturbance levels and between tree taxa. Those trends, discussed further after a second evaluation some years later (Schowalter and Ganio 1999) persisted for at least six years after the hurricane, when the differences between functional groups were still evident.

Hurricane Hugo also facilitated the increased abundance, most pronounced at lower elevations where effects were most severe, of some Lepidoptera with larvae that feed on early successional vegetation developed after the disturbance (Torres 1992). The most conspicuous outbreak was of the Fall armyworm, *Spodoptera eridania* (Noctuidae), a highly polyphagous species whose larvae were recorded feeding on 56 plant species on hurricane-affected sites and whose abundance led to plant declines and later attack by natural enemies. In all, Torres (1992) noted 15 species benefitting from the early successional vegetation, some of these being previously unrecorded host plant associations.

Impacts of the hurricane on some individual species were also severe. Two snails and one phasmatid (*Agamemnon iphimedia*), for example, became so scarce that they were not detected ten months later on carefully searched 10 m diameter circular quadrats (Willig and Camilo 1991). Another stick insect, the endemic *Lamponius portoricensis*, also flightless, declined to very low densities, but the precise causes of these trends could not be determined, with both direct and indirect contributions likely.

Many other hurricanes, whilst noted as causing severe effects and damage have not been subsequently monitored in such detail. Lewis (2001) noted that Hurricane Hattie in 1961 'created a 50-Km wide band' in the Chiquibul Forest, Belize, and up to 90% of canopy trees were toppled, as one of the most severe reported impacts of hurricanes in forests in that region. His inferences that the butterfly species present there in his 1997–1998 surveys may be those able to survive such major disturbances – and which, in turn, may be relatively resilient to subsequent disturbances (such as selective logging) – may have wide application. Extending this, Lewis (2001) noted that selective logging (or other imposed disturbance) might have lesser impacts in areas in which natural disturbances are frequent.

The selective nature of logging operations in relation to gaps is also commonly overlooked. Thus, as noted by Willott et al. (1999), selective logging in Malaysian forests largely targets Dipterocarpaceae, a family lacking any records of butterfly larval feeding. Any effects of logging on butterflies are not due to loss of food plants but to changes in the physical environment, such as lack of canopy cover and changed understorey pioneer vegetation. Perhaps most notably, promotion of grasses provides host plants for numerous Hesperiidae and Nymphalidae: Satyrinae. In Willott et al.'s study, no local extinctions of butterflies in logged forest were evident. Both this study and that by Hamer et al. (2003) supported that selectively logged forests in Sabah contribute to conservation of 'biodiversity', notwithstanding the structural changes incurred. The principle of providing heterogeneity within forest environments may, indeed, support some creation of open areas if these are naturally sparse.

10.5 Modifying Forest Management

Forest management for insect conservation is a long-term need and endeavour and ideally must encompass all phases of the forestry cycle of production or silvicultural areas, undertaken in conjunction with increased needs to reserve undisturbed forests covering the full variety of global forest types. Thus, for *Nothofagus pumilio* forests in southern South America – an ecosystem that supports notable endemic insects, including southern 'relict' taxa of considerable conservation interest (and with Gondwanan parallels in Australia, where relatively small patches of *Nothofagus* persist only in pockets of southern temperate forests) – Spagarino et al. (2001) compared the insects present at different stages of forest development after harvest. They formulated a general suite of needs for ecologically sound and sustainable silvicultural practices under which insects may be better conserved. Those needs are clearly paralleled elsewhere, and the southern forests of Tasmania provide a generally similar evolutionary and conservation context. In Patagonia, variations in forest structure considerably modified the forest ecology, affecting the insect fauna and the interactions between species, with progressive loss of taxa. The relevant conservation plan would be needed over the 200 years of a forest management cycle from an immediately pre-harvest virgin stand (year 0) to regaining a mature even-aged stand (year 200) (Spagarino et al. 2001). That maturation time includes consideration of the deep natural litter layer accumulating over that period as having massive importance as a habitat for endemic invertebrates, both as residents and in providing refuges such as pupation or hibernation sites as a feature of the life cycles of numerous arboreal taxa. The needs formulated by Spagarino et al. are listed in Table 10.8. They are idealistic, logistically complex and improbable, and currently largely impracticable. They are also invaluable in suggesting an agenda that firmly incorporates insect conservation into forestry management and emphasises the needs for basic entomological documentation as a foundation for sustainable silviculture/forestry in conjunction with sustainable biodiversity. That documentation must also

Table 10.8 Needs in developing ecologically sound silvicultural practices in harmony with insect conservation
After Spagarino et al. 2001, based on studies of *Nothofagus pumilio* in Argentina

1. Quantify the insect diversity present
2. Identify and enumerate the insects present
3. Determine the vulnerable species
4. Study their ecology and environment relationships
5. Validate the effectiveness and practicability of the new silvicultural methodologies
6. Apply these to production areas
7. Where possible, enlarge studies to other taxa and relate these to changes observed among insects

consider sustainability of component habitats such as litter and coarse woody debris, and how these are supplied. Several species of *Nothofagus* dominate New Zealand's remaining broadleaved forests, and contribute substantially to the supply of coarse woody debris available to invertebrates. Analysed by Stewart and Burrows (1994), these large pools of nutrients (with impacts on community dynamics) are changed markedly by forest management – in this case removal of the woody debris is undertaken to eliminate breeding sites of a group of notable pest beetles (pinhole borers, *Platypus* spp.) that can increase to reach numbers sufficient to attack healthy trees. Removal of the debris thus contributes to minimising insect attack, and maintaining timber quality and supply through reducing stand mortality.

Table 10.8 provides a basis for comparing priority needs for Australia, and it is difficult to query the logic that led Spagarino et al. to formulate this embracing synopsis. Each parameter, however, is likely to remain highly incomplete without determined and well-supported agenda to address them. Even more restricted focus – such as concentrating on selected taxonomic groups for items 1 and 2 - is currently difficult to promote and coordinate in other than piecemeal fashion. Vulnerability of individual species is currently best reflected in the vulnerability of the forest systems they inhabit, because knowledge of the ecology and basic biology of many (most!) of Australia's rarer endemic insects is fragmentary or completely obscure. In some cases, characteristics may be inferred, with some caution, from knowledge of related taxa, but many of the forest insects of greatest concern are known from single localities or small areas, as putative 'narrow range endemics'. Election of insect species as 'vulnerable' currently draws largely on this paucity of records and knowledge to allocate this condition as precautionary, and any more formal equivalent to 'red-listing' is usually highly tentative. The first four criteria in Table 10.8, the very foundation of justifying insect conservation need, are all incomplete. Whilst many entomologists specialising in different insect taxa in Australia may have wise and probably accurate perceptions of vulnerability and scarcity, communicating and extending those perceptions effectively has far to go.

Similar limitations are evident for the principles espoused by Lindenmayer et al. (2006, p. 40), which are embedded firmly in ecological themes, each with series of 'checklist points' for off-reserve conservation in conjunction with large ecological reserves.

The long-term perspective also recognises that preventable disturbances to forests have long-term effects rather than solely the initial disturbances on which greater attention has focused. Many studies of impacts of logging on forest Lepidoptera, for example, have considered only relatively short-term trajectories of up to about 15 years after logging. In Uganda, Savilaakso et al. (2009) showed that assemblages of lepidopteran larvae compared across forest over 40 years of post-logging regeneration continued to differ significantly between logged and unlogged forest. That study was based on larvae found on foliage of *Neoboutonia macrocalyx* (Euphorbiaceae), a pioneer tree and host plant for numerous Lepidoptera. Larval density differed across treatments, as did levels of herbivory, but around 50 species were found in each treatment. However, Savilaakso et al. concluded that the assemblages had not fully recovered from the disturbance over this interval and, because of seasonal variations in the species present, only long-term surveys are suitable to encompass these and validate the findings.

The simultaneous sampling of 'chronosequence stages' provides one, strong, suite of comparative evaluations of the impact of disturbances and the progressive recovery inferred from the differences in assemblages found in different treatments. However, for 'sustainability' to be confirmed, operations such as selective logging must preserve biodiversity over more than one cycle of operations (Bawa and Siedler 1998), and also entails the species assemblages of logged forest approaching those of primary forest (Willott 1999) as regeneration progresses after disturbance. However, following and monitoring individual treatment plots over the time needed to confirm this is largely impracticable, and using 'space' as a surrogate for 'time' will inevitably continue to dominate such assessments, and interpretations improve as understanding of both the biota and the processes of change improve.

The major objective of the widely advocated 'Natural Forest Management' (NFM) principles is to harvest trees in a manner that allows the forest to regenerate naturally before the next round of extraction. In practice this term embraces a range of non-intensive management actions that differentiate them from the intensive nature of plantation management, and have varying impacts (Bawa and Siedler 1998). Achieving full NFM objectives may not be realistic, and Bawa and Siedler noted that the expectation of managing primary or little–disturbed tropical forest compatibly with full protection of biodiversity whilst also being economically profitable 'remains unsupported'. Improved management procedures are still needed, but Bawa and Siedler also suggested that NFM might be a means of 'buying time', filling some of the needs for wood and wood products whilst imposing only minimal effects on biodiversity.

Saura et al. (2014) summarised part of their review by writing 'having extensive forests that are well connected is fundamental to conserving forest biodiversity', so that existing stands should be maintained, and landscape management incorporate minimising fragmentation, patch formation and isolation. The 'quality' of the forests is of major concern as management for commercial production and alternative uses continues, with wide recognition that pristine old-growth forests are increasingly rare, often increasingly vulnerable to direct exploitation or changes to surrounding landscapes, and support numerous insect species that do not occur

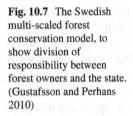

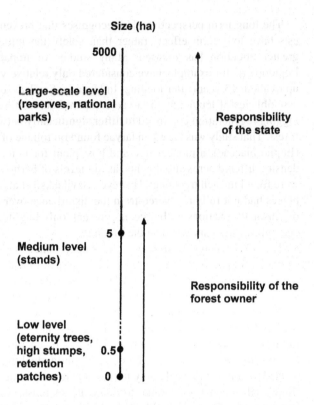

Fig. 10.7 The Swedish multi-scaled forest conservation model, to show division of responsibility between forest owners and the state. (Gustafsson and Perhans 2010)

elsewhere. Insects often occur there in considerable diversity, reflecting the structural diversity and complexity of those forests, with their high living and dead biomass. However, again from Saura et al., the aim of maintaining 'landscapes of well-connected, extensive and high quality forests … is a simple goal to articulate, but a challenging one to implement'.

Long-term efforts for forest conservation and sustainability are exemplified by the farsighted model for forest biodiversity conservation introduced in Sweden from 1978 and developed further by policy executed in 1994. A stated aim of this was to give equal emphasis to environmental and production issues. Reviewed by Gustafsson and Perhans (2010), that multi-scaled model involved set aside spanning the levels from individual trees to large (thousands of hectares) areas, with different management responsibilities (Fig. 10.7) along that continuous gradient of sizes, which is simplified to three stages in the figure. At the smallest scales, dead and live trees are left in harvested stands. The 'medium scale' incorporates the normally used range of stand sizes; and 'large scale' includes areas such as protected nature reserves, with some bias also toward low productivity forests. Despite some shortcomings, and allowing for flexibility to accomodate future needs, the principles embodied form a tested model for emulation elsewhere.

Despite long-term and widespread calls for forest sustainability involving reduced impacts and intensity of forest exploitation, the practical adoption of those

Table 10.9 The reasons listed to help explain why poor logging practices persist in the tropics, and why reduced-impact logging (RIL) has not progressed further
From section headings in Putz et al. (2000)

1. RIL is too expensive.
2. In part due to high profits, many loggers unlikely to admit that there is anything wrong with current logging practices.
3. Lack of government incentives to change logging practices.
4. 'The forest will be converted anyway'.
5. Available logging equipment is unsuitable for RIL.
6. Lack of training and guidance by RIL experts.
7. Lack of focused pressure for better logging from environmental groups

measures over much of the tropics, in particular, has lagged considerably. Putz et al. (2000) commented that 'good logging practices are still the exception rather than the rule in most of the tropics'. Trends in commercial forestry in Europe include using thinning and rapid harvesting over short rotations, reducing the period of canopy closure. Open forests are reduced or eliminated by rapid planting after clearfelling or other harvesting operations and, for protected areas, changes in the gradient from closed to open canopy are a major influence on conservation values and insect diversity. Reduced impact logging (RIL) guidelines span a range of measures to reduce damage to the forests, and adoption of those techniques and approaches would be substantial progress towards more sustainable forest management. RIL is predicated in both pre-logging and post-logging measures, such as protecting regenerating vegetation from injury preventing or minimising damage to non-target species, and protecting ecosystem processes. Putz et al. discussed seven major reasons, the first six of them based on discussions with commercial loggers on the operating constraints they face (Table 10.9). The most serious concerns devolved around the first of these, that RIL is simply too expensive to be adopted widely in commercial operations.

Current changes, or suggested changes in forest management, draw on long historical traditions. Historical differences between, for example, the northern European forests and Atlantic forests of Brazil cannot be overlooked in exploring general patterns or 'recipes'. As Schiegg (2000) pointed out, many notable forests have been managed for centuries, so that many of the insects present have survived the pressures of intensive forestry, or immigrated from elsewhere. Such pressures have arisen only more recently in Australia, so that the legacies of impacts are themselves more recent and, probably, have not yet 'stabilised' in any comparable way. Nevertheless, despite the seemingly enormous amount of information on northern forest saproxylic insects, Langor et al. (2008) still regarded them as 'quite poorly understood'. Information on the causes of their spatial variations and their specific dead wood resource needs is still required to assess responses to forest management. Knowledge of host tree species, for example, is most complete in Canada for Buprestidae and some Scolytidae (Fig. 10.8) in which about half the species (182 of 384) are known from one tree genus, many from single tree species.

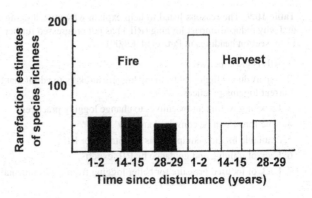

Fig. 10.8 Rarefaction estimates of species richness of saproxylic beetles in boreal forest *Populus* stands in Alberta, Canada, of three stages of fire age and time since harvest. (Langor et al. 2008)

Table 10.10 Some silvicultural options to increase the regeneration of rainforest tree species when harvesting mixed forests
As listed for Tasmania by Tabor et al. (2007)

Reserve and protect mixed forest adjacent to coupe edges.
Add seed of dominant rainforest tree species to aerial sowing mix.
Understorey islands: machinery exclusion areas within clear-felled coupes.
Reduce coupe size.
Reduce burn intensity.
Longer rotation length.
Aggregated retention: retain remnant patches of forest throughout an otherwise completely felled coupe.
Strip felling: fell alternate strips around 80 m wide.
Single tree/ small group selection; retain >75 % forest cover, harvest 40 m³/ha every 20 years, mechanical disturbance, no burning.

Commonwealth and State/Territory Governments in Australia have made commitments to maintain the floristic diversity of native forests used for conservation, and three main components contribute to this. Those measures (Tabor et al. 2007) are (1) creation of new reserves; (2) applying silvicultural measures that maintain dominant species with local provenances in the forests used for production; and (3) reserving appropriate corridors and patches of vegetation with careful planning of harvesting options. Nevertheless, concerns persist over regeneration potential for all native flora affected after coupe-level harvesting, and have led to queries over the suitability of some current practices such as the 'clearfell, burn and sow' processes widespread in Tasmania, across a variety of forest types. Factors such as size of coupes and proximity to forest edge (as sources of seeds) may be important, and a variety of silvicultural options may each have advantages and disadvantages (Table 10.10) - but collectively demonstrate the wide range of contexts and considerations that influence the regeneration of rainforest trees in mixed forests and may benefit associated insects.

The fate of plantation forests usurping non-forested ecosystems is coming under increasing scrutiny, with calls elsewhere to re-create the systems they have replaced, together with their associated biodiversity, and in some contexts as part of a wider 'benefits package'. The example of pine plantations on former heathland discussed by Vangansbeke et al. (2017) included finding the balance between wood production, biodiversity conservation and human recreational activities such as trail walking, with their models exploring how this might be optimised. In that example, viable populations of some specialist heathland or grassland species can persist within the pine plantation matrix because of a network of clearcut and natural open spaces. Recreational activities – including human traffic on trails – could have adverse impacts on key species, but trails can be designated to largely confine or direct people to areas of lesser importance for conservation, and reduce the damage of tradeoffs to accommodate the various needs for the plantation areas. One outcome is safeguarding most wood harvesting whilst also facilitating other benefits. Multiple purpose land use and integration of those purposes without undue sacrifice is likely to be an important future driver of enlightened forest management.

Conversion of forest plantations to restore networks and landscapes more suited for conservation than large monoculture stands is becoming more common. In parts of Belgium, some former coniferous plantations have been converted in order to comply with European Union regulations for restoration of native habitats, and also to provide benefits for recreation and amenity. Because such processes involve clearcuts and subsequent landscape alterations, local controversies can arise through differing opinions on aesthetics and significance/worth of the changes (De Valck et al. 2014). Public preferences vary enormously with context, often with appreciation for landscape diversity favouring conservation of small areas, such as to form open glades within the plantation forests.

10.6 Needs and Prospects

Australia's forests, together with their denizens and enveloping environments, have undergone much change over only about two centuries since European colonisation. Over that period, (1) much forest has been cleared for agriculture, urban settlement snd related needs, and (2) forests and woodlands have been exploited widely for timber and latterly increasingly for timber products such as wood pulp for both local use and export. The forestry industry continues to be an important component of Australia's economy, and the enduring needs for high quality saw logs from native hardwoods (predominantly of *Eucalyptus* and its allies) continue to provoke debates and conflict over how these uses can be sustained in the face of increasing pressures on native forests in harmony with assuring the wellbeing of the vast numbers of insects and other species that are restricted to, and depend on, forest environments. Many of these are very restricted in distribution, and have specialised ecological needs.

As emphasised in this book, insects are major components of forest fauna. They collectively occupy a wide range of feeding guilds and·ecological roles, and are taxonomically diverse. Many are actually or potentially under threat as forest extent and condition changes, and their losses diminish future understanding of the Australian fauna. Some phytophagous insect groups have co-evolved and radiated with their related plant hosts over long periods and, whilst the diversity of many such radiations is accepted broadly as high, precise data on the numbers of species present are often approximations, pending further study. Recent taxonomic studies of some groups have almost invariably revealed numerous undescribed taxa or distinctive local 'forms' or genotypes. Some such radiations (such as the oecophorid moths whose larvae feed on dead *Eucalyptus* foliage on the forest floor – p. 59) are globally unusual. Numerous species are endemic, many very local in incidence, and other evolutionarily intriguing manifestations of diversity include (1) parallel radiations of related taxa on different host plant groups (such as psyllids on *Eucalyptus* and *Acacia*), and (2) specialisations on only one of these predominant plant genera (such as gall-forming thrips on *Acacia*: Crespi et al. 2004). Such native insect complexes can contribute answers to many fundamental questions on the origins and processes that have generated Australia's insect diversity.

However, not unexpectedly, individual species from amongst many such diverse lineages are amongst the major pests of their hosts in production forestry or amenity plantings. Outbreaks or persistently high populations of taxa from several orders can defoliate, weaken and contribute to deaths of numerous native and plantation trees. Others degrade wood quality through their feeding and tunneling activities. Collectively, those species may command strenuous suppression or 'control', often by methods with undocumented or unanticipated effects on co-occurring or related non-target species.

Besides destruction of forests (either selectively or more generally by clear-cutting to release land for agriculture or urban developments), the other major trend of change has been to increase the plantation estate, both to augment supply of harvestable timber and to supply the demand for softwood products. Native *Eucalyptus* and alien conifers (predominantly *Pinus radiata*) are the major components of the highly managed forests, in which silvicultural methods are modified continuously to increase production – by means such as shortening rotation times, promoting genetic and physical uniformity, and minimising potential pest damage. Whilst the 'plantation estate' comprises largely alien conifers as potential hosts to which rather few Australian native insects have become adapted, a rather different ecological situation prevails for eucalypt plantations. Use of *E. globulus* (p. 37) in regions of Australia far beyond its true native range presents local insects with a superabundant novel host with which they have had no earlier history, but may live normally on local hosts closely related to this alien eucalypt, to which they might switch readily. Collectively, the varied forests – from tropical to cool temperate and the many distinct floristic categories spanning these – provide an equally broad array of resources that insects can exploit. Many have done so, but interpreting and inventorying the taxa present and the interactions in which they participate remain daunting and highly incomplete.

The sheer richness of insect species in any forest – the beetles of cool wet forest of *Eucalyptus obliqua* at Warra, Tasmania (p. 164) are one example – manifests through almost every survey undertaken. Less species-rich groups than beetles are also notable: unusual species of ants, cockroaches, lacewings and many others are known only from very restricted forest sites, in many cases despite extensive searches elsewhere.

There is no doubt that (1) Australia's native forests harbour enormous numbers of insect species, many of them restricted to those forests and with specialised resource needs; (2) many of those species are endemic to Australia, and many to small parts of the continent or particular forest types; so that (3) the unique collective associations are sufficiently unusual and globally significant to merit conservation, and (4) conservation is increasingly urgent in the face of continuing losses and changes to forests, largely from tangible and direct human interference – including forest removal and exploitation, and the novel interactions wrought through introductions of alien species. Less tangible influences such as climate change are also evident, and their impacts difficult to predict.

That conservation need embraces protection and management of the forests themselves, and their characteristic resident insects (and, of course, other biota!) that occur nowhere else on Earth, to sustain the complex forest environments, the array of ecological processes that contribute to that sustainability, and the species that participate in and drive those processes.

Practical conservation, also, must necessarily proceed with highly incomplete knowledge and documentation. Not only is a considerable proportion of forest insects currently undescribed, and many not diagnosed or recognised clearly in institutional collections (Taylor 1983), but functional interpretation is hampered by lack of basic ecological knowledge for many species, and the paucity of inventories ('ecological collections', advocated for Australia by Yen 1993) against which changes may be evaluated. The impediments to practical insect conservation discussed by Cardoso et al. (2011) are perhaps nowhere more evident than amongst the massive and mosaic variety of Australia' forests and their inhabitants.

These complex systems illustrate needs for wide conservation approaches that can extend well beyond the most conventional approach of considering needs of individual threatened insect species to embrace diverse assemblages, whose size and conservation status cannot be evaluated fully. For these, the best possible pragmatic arguments needed to garner political and public support may appear unconvincing as necessarily based on 'ideals' rather than full documentation. That conservation advocacy may also conflict with commercial interests in which regional community livelihoods are at stake.

It is self-evident that (1) reserving key forest areas from exploitation and ensuring the management needed to sustain their communities is a key need; (2) defining those priority areas in part by presence of threatened flagship vertebrates can be persuasive; (3) reliance on this without attention to invertebrates provides only incomplete pictures of biodiversity values; (4) simply 'locking up' forest areas without management to protect them effectively, reduce hazards to forest settlements from wildfire and other disturbances, and regenerate the sucessional stages on which numerous insects depend, is not effective; and (5) harmonising the entire process within wider societal interests may not be wholly practicable. Beyond the laudable principle of reserving forests in order to protect native biodiversity, honing priorities from amongst the numerous potential candidates (at scales from country or region to site or patch) is a complex exercise, restricted in many cases by practicality and competing demands. As discussed earlier, relevant parameters include uniqueness, size and condition, land tenure, diversity, numbers of known threatened species, the incidence of notable or unique species or lineages, political advocacy for particular taxa, ecosystem values, and many others.

Flagship insect species and known 'endangered' or 'critically endangered' species can play important roles in conservation advocacy – as elsewhere, formal listing of such species in Australia conveys some obligation to act to protect them. However, relatively few forest insects amongst those intrinsically worthy taxa have been recognised formally. Most forest insect species, even though they are demonstrably rare, and many known from single localities, can only be inferred to truly need conservation. For them, sustaining the forests from which they have been recorded, incorporating any available biological or ecological knowledge of those species (perhaps drawing, with caution, also on any information available on their close relatives), is likely to be the most useful – even, only – possible option.

Each reasonably undisturbed forest area is likely to contain one or more insect species that are sufficiently unusual and/or poorly known to imply conservation significance, and to give the forest 'conservation value'. In Australia, ability to designate the 'most important' insect species and faunas lags far behind capability in the better-documented landscapes of Europe or North America: saproxylic beetles in Europe, for example, are an important focal umbrella group for forest conservation. A suite of 168 such species in central Europe and described as 'primeval forest relict species' were selected for priority interest because they depend on the continuing presence of primeval forest areas with features such as over-mature trees, large amounts of dead wood, and the diversity of dead wood (Eckelt et al. 2017). These species are also absent from managed forests in the region, reflecting ecological disruption of those forests (Siebold et al. 2015), so their presence clearly signals forest stands that merit prompt conservation for that biodiversity. However, Eckelt et al. also commented that identification of those sites 'is of particular importance in landscapes with a long cultural and agricultural history', and that the connotation of 'primeval' has valuable public appeal for advocacy. Even in Europe, though, true 'primeval forest remnants' are now scarce. Different individual country assessments contributed to the total for Central Europe, and most of the eight countries harbour at least 100 such species. Only Switzerland, with 95 species, supported fewer, and the common standard to rank forest conservation value through the entire region and expand public awareness and interest in the problems faced may be a useful model for emulation elsewhere. The 16 saproxylic beetle species acknowledged by listing in the European Habitats Directive are important individual flagship taxa.

The far shorter modern history of forest change in Australia suggests that parallel 'primeval remnants' may still occur in the more remote and least accessible forest areas, such as the Tasmanian wilderness and parts of south east mainland highlands, and the northern wet tropics, but lack of fundamental documentation renders comprehensive selection of such suites of species very tentative. However, a somewhat broader election of the better-known potential umbrella insect taxa in forests is an important theme to develop – perhaps selecting Gondwanan relicts in the south, and evolutionary outliers and southern relicts further north for some preliminary emphasis.

The importance of Australia's forests for numerous restricted and endemic insects is not acknowledged or recognised widely in much conservation discussion. A well-reasoned representative national selection of Australian forest insects to be incorporated into wider forest conservation policy, and to form the basis for monitoring programmes, would ensure that insects gain a higher practical profile, to the benefit of far more invertebrate taxa. Likewise, the difficulties of making a representative selection of the larger forest areas with secure core areas for conservation priority and reservation, as a sound principle, might be aided by such justifiable tools. 'Optimal' reservation of Australia's forests demands some agreement on priorities and the means to attain these. Two rather different but strongly held perspectives are widespread in nominating land priorities for reserves, and were discussed for Australian woodlands by Lindenmayer et al. (2010) in relation to the wisest allocation of limited support and resources available. On the one hand (1) priority should be given for the 'best' patches in which the native vegetation is most intact and general disturbance is least, so that they are presumed to be of high quality and represent the condition closest to 'natural'. On the other hand (2) other advocates hold that attention should instead be given to sites where greatest gains ('returns for effort') may be made, so action is most likely on sites in only poor or moderate condition. This approach thus relies to a far greater extent on restoration and rehabilitation, rather than putatively simple reservation alone. Intact, or near-pristine, sites are the benchmarks against which other conditions can be compared, and the trajectories of management toward those conditions tracked. The diversity of insects and their responses to forest change gives many opportunities to involve them in that tracking, as specific indicators or from more general trends of richness and abundance.

The levels of forest protection claimed by ABARES (2017) and the potential review of Australia's Regional Forest Agreements (p. 229) to increase effective protection of biodiversity (as suggested by Lindenmayer et al. 2015) provide an indicative template of needs. They also illustrate priorities for augmentation to create a truly representative system of forest reserves that can be managed to sustain their biological legacies.

Expansion of the plantation estate and of the wider forestry industry, provides numerous opportunities to enhance practical conservation. Although not commonly viewed in that context, the planning, establishment and management of plantations, and the silvicultural and extraction methods employed in both plantations and natural forests progressively incorporate measures that reduce harmful impacts and benefit native biota. The importance of retaining dead wood, of mosaic treatments and cutting schedules, management of fire regimes, sustaining a variety of understorey vegetation, and avoiding excessive fragmentation and edge effects are all practices illuminated by studies on insect wellbeing. There is clear potential for increased appreciation and understanding of the roles of Australia's forest insects in informing wise use and sustainability of the country's unique forest biomes. In achieving this, the future of high numbers of endemic forest insects may also be secured.

References

ABARES (Australian Bureau of Agricultural and Resource Economics and Sciences) (2017) Australia's forests at a glance, with data to 2015–2016. Department of Agriculture and Water Resources, Canberra

Andam KS, Ferraro PJ, Pfaff A, Sanchez-Azofeita GA, Robalino JA (2008) Measuring the effectiveness of protected area networks in reducing deforestation. Proc Nat Acad Sci 105:16089–16094

As S (1993) Are habitat islands islands? Woodliving beetles (Coleoptera) in deciduous forest fragments in boreal forest. Ecography 16:219–228

Atlegrim O, Sjoberg K, Ball JP (1997) Forestry effects on a boreal beetle community in spring: selective logging and clear-cutting compared. Entomol Fennica 8:19–26

Baker SC, Grove SJ, Forster L, Bonham KJ, Bashford D (2009) Short-term responses of ground-active beetles to alternative silvicultural systems in the Warra Silvicultural systems trial, Tasmania, Australia. For Ecol Manag 258:444–459

Barlow J, Gardner TA, Araujo IS, Avila-Pires TC, Bonaldo AB et al (2007) Quantifying the biodiversity value of tropical primary, secondary and plantation forests. Proc Nat Acad Sci 104:18555–18560

Bawa KS, Siedler R (1998) Natural forest management and conservation of biodiversity in tropical forests. Conserv Biol 12:46–55

Berg A, Ehnstrom B, Gustafsson L, Hallingback T, Jonsell M, Weslien J (1994) Threatened plant, animal and fungus species in Swedish forests; distribution and habitat associations. Conserv Biol 8:718–731

Bernes C, Jonsson BG, Junninen K, Lohmus A, Macdonald E, Muller J, Sandstrom J (2015) What is the impact of active management on biodiversity in forest set aside for conservation or restoration? A systematic review protocol. Environ Evid 4:25. https://doi.org/10.1186/s13750-015-0050-7

Bouget C (2005a) Short-term effect of windstorm disturbance on saproxylic beetles in broadleaved temperate forests. Part I: Do environmental changes induce a gap effect? For Ecol Manage 216:1–14

Bouget C (2005b) Short-term effect of windstorm disturbance on saproxylic beetles in broadleaved temperate forests. Part II: Effects of gap size and gap isolation. For Ecol Manage 216:15–27

Bouget C, Duelli P (2004) The effects of windthrow on forest insect communities: a literature review. Biol Conserv 118:281–299

Bradshaw CJA (2012) Little left to lose: deforestation and forest degradation in Australia since European colonization. J Plant Ecol 5:109–120

Bremer LL, Farley KA (2010) Does plantation forestry restore biodiversity or create green deserts? A synthesis of the effects of land-use transitions on plant species richness. Biodivers Conserv 19:3893–3915

Brockerhoff EG, Jactel H, Parotta JA, Quine CP, Sayer J (2008) Plantation forests and biodiversity: oxymoron or opportunity? Biodivers Conserv 17:925–951

Cale JA, Klutsch JG, Erbilgin N, Negron JF, Castello JD (2016) Using structural sustainability for forest health monitoring and triage: case study of a mountain pine beetle (Dendroctonus ponderosae) - impacted landscape. Ecol. Indic 70:451–459

Cardoso P, Erwin TL, Borges PAV, New TR (2011) The seven impediments to invertebrate conservation and how to overcome them. Biol Conserv 144:2647–2655

Carron LT (1985) A history of forestry in Australia. Australian National University Press, Canberra

Commonwealth of Australia (1992) National Forest Policy Statement: a new focus for Australia's forests. Australian Government Publishing Service, Canberra

Commonwealth of Australia (1995) Regional forest agreements: the commonwealth position. Australian Government Publishing Service, Canberra

Coulson RN, Stephen FM (2008) Impacts of insects in forest landscapes: implications for forest health management. In Paine TD, Lieutier F (eds) Insects and diseases of Mediterranean forest ecosystems. Springer, Cham, pp. 101–125

Crespi BJ, Morris DC, Mound LA (2004) Evolution of ecological and behavioural diversity: Australian *Acacia* thrips as model organisms. Australian Biological Resources Study and Australian National Insect Collection, Canberra

De Valck J, Vlaeminck P, Broekx S, Liekens I, Aertsens J, Chen W, Vranken L (2014) Benefits of clearing forest plantations to restore nature? Evidence from a discrete choice experiment in Flanders, Belgium. Landsc Urb Plann 125:65–75

Duelli P, Obrist MK, Wermelinger B (2002) Windthrow-induced changes in faunistic biodiversity in alpine spruce forests. For Snow Landsc Res 77:117–131

Eckelt A, Muller J, Bense U, Brustel H, Bussler H et al. (2017) "Primeval forest relict beetles" of Central Europe: a set of 168 umbrella species for the protection of primeval forest remnants. J Insect Conserv https://doi.org/10.1007/s10841-017-0028-6

Evans MC (2016) Deforestation in Australia: drivers, trends and policy responses. Pac Conserv Biol 22:130–150

Feener DH, Schupp EW (1998) Effect of treefall gaps on the patchiness and species richness of Neotropical ant assemblages. Oecologia 116:191–201

Ghazoul J (2002) Impact of logging on the richness and diversity of forest butterflies in a tropical dry forest in Thailand. Biodivers Conserv 11:521–541

Gibb H, Hjalten J, Ball JP, Atlegrim O, Pettersson RB et al (2006) Effects of landscape composition and substrate availability on saproxylic beetles in boreal forests: a study using experimental logs for monitoring assemblages. Ecography 29:191–204

Gotmark F (2013) Habitat management alternatives for conservation forests in the temperate zone: review, synthesis, and implications. For Ecol Manag 306:292–307

Gough LA, Birkemoe T, Sverdrup-Thygeson A (2014) Reactive forest management can also be proactive for wood-living beetles in hollow oak trees. Biol Conserv 180:75–83

Grimbacher PS, Catterall CP (2007) How much do site age, habitat structure and spatial isolation influence the restoration of rainforest beetle species assemblages? Biol Conserv 135:107–118

Gustafsson L, Perhans K (2010) Biodiversity conservation in Swedish forests: ways forward for a 30-year-old multi-scaled approach. Ambio 39:546–554

Halme P, Allen KA, Aunins A, Bradshaw RHW, Beunelkis G et al (2013) Challenges of ecological restoration: lessons from forest in northern Europe. Biol Conserv 167:248–256

Hamer KC, Hill JK, Benedick S, Mustaffa N, Sgerratt TN, Maryatis M, Chey VK (2003) Ecology of butterflies in natural and selectively logged forests of northern Borneo: the importance of habitat heterogeneity. J Appl Ecol 40:150–162

Hanski I (2008) Insect conservation in boreal forests. J Insect Conserv 12:451–454

Harrison S (1987) Treefall gaps versus forest understorey as environments for a defoliating moth on a tropical forest shrub. Oecologia 72:65–68

Hartley MJ (2002) Rationale and methods for conserving biodiversity in plantation forests. For Ecol Manag 155:81–95

Hill JK, Hamer KC, Tangah J, Dawood M (2001) Ecology of tropical butterflies in rainforest gaps. Oecologia 128:294–302

Huber C, Baumgartner M (2005) Early effects of forest regeneration with selective and small scale clear-cutting on ground beetles (Coleoptera, Carabidae) in a Norway spruce stand in southern Bavaria (Hoglwald). Biodivers Conserv 14:1989–2007

Ibbe M, Milberg P, Tuner A, Bergman K-O (2011) History matters: impact of historical land use on butterfly diversity in clear-cuts in a boreal landscape. For Ecol Manag 261:1885–1891

Jacobs DF, Oliet JA, Aronson J, Bolte A, Bullock JM et al (2015) Restoring forests: what constitutes success in the twenty-first century? New Forests 46:601–614

Jactel H, Nicoll BC, Branco M, Gonzalez-Olabarria JR, Grodzki W et al (2009) The influences of forest stand management on biotic and abiotic risks of damage. Ann For Sci 66:701 18 pp

Jukes MR, Peace AJ, Ferris R (2001) Carabid beetle communities associated with coniferous plantations in Britain: the influence of site, ground vegetation and stand structure. For Ecol Manag 148:271–286

Kirkpatrick JB (1998) Nature conservation and the Regional Forest Agreement process. Austr J Environ Manage 5:31–37

Kusel J (2001) Assessing well-being in forest dependent communities. J Sust For 13:359–384

Lamb D (1998) Large-scale ecological restoration of degraded tropical forest lands: the potential role of timber plantations. Restor Ecol 6:271–279

Langor DW, Hammond HEJ, Spence JR, Jacobs J, Cobb TP (2008) Saproxylic insect assemblages in Canadian forests: diversity, ecology, and conservation. Canad Entomol 140:453–474

Larsson S, Ekbom B, Schroeder LM, McGeoch MA (2006) Saproxylic beetles in a Swedish boreal forest landscape managed according to 'new forestry. In Grove SJ, Hanula JL (eds) Insect biodiversity and dead wood. Proceedings of a symposium for the 22nd international congress of entomology. USDA General and Technical Report SRS-93, Ashville, NC, pp 75–82

Lehnert LW, Bassler C, Brandl R, Burton PJ, Muller J (2013) Conservation value of forests attacked by bark beetles: highest number of indicator species is found in early successional stages. J Nat Conserv 21:97–104

Lewis OT (2001) Effect of experimental selective logging on tropical butterflies. Conserv Biol 15:389–400

Lindenmayer DB, Franklin JF, Fischer J (2006) General management principles and a checklist of strategies to guide forest biodiversity conservation. Biol Conserv 131:433–445

Lindenmayer D, Bennett AF, Hobbs R (2010) How far have we come? Perspectives on ecology, management and conservation of Australia's temperate woodlands (Chapter 42). In Lindenmayer D, Bennett A, Hobbs R (eds) Temperate woodland conservation and management CSIRO Publishing, Melbourne, 14 pp

Lindenmayer DB, Blair D, McBurney L, Banks SC (2015) The need for a comprehensive reassessment of the Regional Forest Agreements in Australia. Pac Conserv Biol 21:266–270

Lovei GL, Cartellieri M (2000) Ground beetles (Coleoptera, Carabidae) in forest fragments of the Manawatu, New Zealand: collapsed assemblages? J Insect Conserv 4:239–244

Manzo-Delgado L, Lopez-Garcia J, Alcantara-Ayala I (2014) Role of forest conservation in lessening land degradation in a temperate region: the Monarch Butterfly Biosphere Reserve, Mexico. J Environ Manag 138:55–66

Muller J, Noss RF, Bussler H, Brandl R (2010) Learning from a "benign neglect strategy" in a national park: response of saproxylic beetles to dead wood accumulation. Biol Conserv 143:2559–2569

Nakamura A, Proctor H, Catterall CP (2003) Using soil and litter arthropods to assess the state of rainforest restoration. Ecol Manage Restor 4:S20–S28

Neumann FG (1992) Responses of foraging ant populations to high-intensity wildfire, salvage logging and natural regeneration processes in *Eucalyptus regnans* regrowth forest of the Victorian Central Highlands. Aust For 55:29–38

Niemela J, Langor D, Spence JR (1993) Effects of clear-cut harvest on boreal ground-beetle assemblages (Coleoptera: Carabidae) in western Canada. Conserv Biol 7:551–561

Norton TW (1996) Conserving biological diversity in Australia's temperate eucalypt forests. For Ecol Manag 85:21–33

Norton TW (1997) Conservation and management of eucalypt ecosystems. In Williams JA, Woinarski JCZ (eds) Eucalypt ecology: individuals to ecosystems. Cambridge University Press, Cambridge, pp 373–401

Nunez-Mir GC, Iannone BV III, Curtis K, Fei S (2015) Evaluating the evolution of forest restoration research in a changing world: a "big literature" review. New For 46:669–682

Paillet Y, Berges L, Hjalten J, Odor P, Avon C et al (2009) Biodiversity differences between managed and unmanaged forests: meta-analysis of species richness in Europe. Conserv Biol 24:101–112

Pardonnet S, Beck H, Milberg P, Bergman K-O (2013) Effect of tree-fall gaps on fruit-feeding nymphalid butterfly assemblages in a Peruvian rain forest. Biotropica 45:612–619

Patrick M, Fowler D, Dunn RR, Sanders NJ (2012) Effects of treefall gap disturbances on ant assemblages in a tropical montane cloud forest. Biotropica 44:472–478

Putz FE, Dykstra DP, Heinrich R (2000) Why poor logging practices persist in the tropics. Conserv Biol 14:951–956

Rhodes JR, Cattarino L, Seabrook L, Maron M (2017) Assessing the effectiveness of regulation to protect threatened forests. Biol Conserv 216:33–42

Richards LA, Windsor DM (2007) Seasonal variation of arthropod abundance in gaps and the understorey of a lowland moist forest in Panama. J Trop Ecol 23:169–176

Routley R, Routley V (1975) The fight for the forests: the takeover of Australian forests for pines, wood chips and intensive forestry. Australian National University Press, Canberra

Saura S, Martin-Quellen E, Hunter ML (2014) Forest landscape change and biodiversity conservation. In Azevedo JC, Perera AH, Pinto MA (eds) Forest landscapes and global change: challenges for research and management. Springer, New York, pp. 167–198

Savilaakso S, Koivisto J, Veteli TO, Pusenius J, Roininen H (2009) Long lasting impact of forest harvesting on the diversity of herbivorous insects. Biodivers Conserv 18:3931–3948

Schelhaas M-J, Nabuurs G-J, Schuck A (2003) Natural disturbances in the European forests in the 19th and 20th centuries. Glob Change Biol 9:1620–1633

Schiegg K (2000) Effects of dead wood volume and connectivity on saproxylic insect species diversity. Ecoscience 7:290–298

Schowalter TD (1994) Invertebrate community structure and herbivory in a tropical rain forest canopy in Puerto Rico following Hurricane Hugo. Biotropica 26:312–319

Schowalter TD, Ganio LM (1999) Invertebrate communities in a tropical rain forest canopy in Puerto Rico following Hurricane Hugo. Ecol Entomol 24:191–201

Schroeder LM (2010) Colonization of storm gaps by the spruce bark beetle: influence of gap and landscape characteristics. Agric For Entomol 12:29–39

Shelly TE (1982) Comparative foraging behavior of light-versus shade-seeking adult damselflies in a lowland Neotropical forest (Odonata: Zygoptera). Physiol Zool 55:335–343

Siebold S, Brandl R, Buse J, Hothoen T, Schmidl J, Thorn S, Muller J (2015) Association of extinction risk of saproxylic beetles with ecological degradation of forest in Europe. Conserv Biol 29:382–390

Spagarino C, Pastur GM, Peri PL (2001) Changes in *Nothofagus pumilio* forest biodiversity during the forest management cycle. 1. Insects. Biodivers Conserv 10:2077–2092

Spies TA (2004) Ecological concepts and diversity of old-growth forests. J For 102:14–20

Spitzer K, Jaros J, Havelka J, Leps J (1997) Effect of small-scale disturbance on butterfly communities of an Indochinese montane rainforest. Biol Conserv 80:9–15

Stanturf JA (2015) Future landscapes: opportunities and challenges. New For 46:615–644

Stephens SS, Wagner MR (2006) Using ground foraging ant (Hymenoptera: Formicidae) functional groups as bioindicators of forest health in northern Arizona ponderosa pine forests. Environ Entomol 35:937–949

Stewart GH, Burrows LE (1994) Coarse woody debris in old-growth temperate beech (*Nothofagus*) forests of New Zealand. Can J For Res 24:1988–1996

Stork NE, Srivastava DS, Watt AD, Larsen TB (2003) Butterfly diversity and silvicultural practice in lowland rainforests of Cameroon. Biodivers Conserv 12:387–410

Sverdrup-Thygeson A, Ims RA (2002) The effect of forest clearcutting in Norway on the community of saproxylic beetles on aspen. Biol Conserv 106:347–357

Tabor J, McElhinny C, Hickey J, Wood J (2007) Colonisation of clearfelled coupes by rainforest tree species from mature mixed forest edges, Tasmania, Australia. For Ecol Manag 240:13–23

Taki H, Makihara H, Matsumura T, Hasegawa M, Matsuura T et al (2013) Evaluation of secondary forests as alternative habitats to primary forest for flower-visiting insects. J Insect Conserv 17:549–556

Taylor RW (1983) Descriptive taxonomy: past, present, and future. In Highley E, Taylor RW (eds) Australian systematic entomology; a bicentenary perspective. CSIRO Publishing, Melbourne, pp 91–134

Taylor RJ, Doran N (2001) Use of terrestrial invertebrates as indicators of the ecological sustainability of forest management under the Montreal process. J Insect Conserv 5:221–231

Teale SA, Castello JD (2011) Regulators and terminators: the importance of biotic factors to a healthy forest. In: Castello JD, Teale SA (eds) Forest health: an integrated perspective. Cambridge University Press, Cambridge, pp 81–114

Thorn S, Hacker HH, Seibold S, Jehl H, Bassler C, Muller J (2015) Guild-specific responses of forest Lepidoptera highlight conservation-oriented forest management – implications from conifer-dominated forests. For Ecol Manag 337:41–47

Toivanen T, Kotiaho JS (2007) Burning of logged sites to protect beetles in managed boreal forests. Conserv Biol 21:1562–1572

Torres JA (1992) Lepidoptera outbreaks in response to successional changes after the passage of Hurricane Hugo in Puerto Rico. J Trop Ecol 8:285–298

Vangansbeke P, Blondeel H, Landuyt D, De Frenne P, Gorissen L, Verheyen K (2017) Spatially combining wood production and recreation with biodiversity conservation. Biodivers Conserv 26:3213–3239

Watt AD, Stork NE, Bolton B (2002) The diversity and abundance of ants in relation to forest disturbance and plantation establishment in southern Cameroon. J Appl Ecol 39:18–30

Wermelinger B, Duelli P, Obrist MK (2002) Dynamics of saproxylic beetles (Coleoptera) in windthrow areas in alpine spruce forests. For Snow Landsc Res 77:133–148

Willig MR, Camilo GR (1991) The effect of Hurricane Hugo on six invertebrate species in the Luquillo Experimental Forest of Puerto Rico. Biotropica 23:455–461

Willott SJ (1999) The effects of selective logging on the distribution of moths in a Bornean rainforest. Phil Trans R Soc Lond B 354:1783–1790

Willott SJ, Lim DC, Compton SG, Sutton SL (1999) Effects of selective logging on the butterflies of a Bornean rainforest. Conserv Biol 14:1055–1065

Wolda H, Foster R (1978) *Zunacetha annulata* (Lepidoptera: Dioptidae), an outbreak insect in a neotropical forest. Geo-Eco-Trop 2:443–454

Yaffee SL (1999) Three faces of ecosystem management. Conserv Biol 13:713–725

Yen AL (1993) Some practical issues in the assessment of invertebrate biodiversity. In: Beattie AJ (ed) Rapid biodiversity assessment. Macquarie University, Sydney, pp 21–25

Zermeno-Hernandez I, Pingarroni A, Martinez-Ramos M (2016) Agricultural land-use diversity and forest regeneration potential in human-modified tropical landscapes. Agric Ecosyst Environ 230:210–220

Zou Y, Sang W, Warren-Thomas E, Liu Y, Yu Z, Wang C, Axmacher JC (2015) Diversity patterns of ground beetles and understorey vegetation in mature, secondary, and plantation forest regions of temperate northern China. Ecol Evol 5:531–542

Zou Y, Sang W, Warren-Thomas E, Axmacher JC (2016) Geometrid moth assemblages reflect high conservation value of naturally regenerated secondary forest in temperate China. For Ecol Manag 374:111–118

Appendix

Australian Forest Insects: Candidate Taxa for Conservation Priority and Use in Conservation Management

The following brief comments and listing of Australian terrestrial insect groups is to suggest those groups that are of greatest interest and value in conservation activities related to forests, and augment the perspective given in earlier examples. Many insects, despite their diversity and abundance in forests, play no part in practical conservation and, other than as 'impressive numbers', are essentially passive passengers in management planning. In the absence of any defined roles as pests or tangible ecological benefits (other than those advanced optimistically by specialists seeking to promote 'their' group: in many cases valid, but seen more widely as of only limited relevance), their conservation value in forests is largely in contributing to 'numbers' in surveys documenting insect richness and representation. For example, Rentz (2014) commented that the most diverse family of cockroaches (Blattodea: Ectobiidae) are 'to be found in every handful of leaf litter whether it be in rainforest or dry sclerophyll woodland', where they are important contributors to recycling. Many species are undescribed and many known from few specimens or localities. Abundance of termites, some Hemiptera, and others in forests implies importance as food resources for vertebrates, and the loss of insect biomass may have ramifying impacts in local food webs. Likewise, the values of Diptera as pollinators are likely to be substantial, but remain largely unexplored.

Specialists in several of those orders continue to demonstrate ocurrence of rare, localised and evolutionarily significant taxa within their groups: those taxa clearly merit conservation just as much as do the more popular insects, but levels of knowledge and interest simply preclude that involvement, and their wellbeing can best be assured by continuing to shelter them under the umbrella of better-documented taxa for the forseeable future. The public perception of conserving cockroaches, for example, is commonly one of ridicule rather than sympathy, and contrasts markedly with attitudes toward another group of orthopteroids, Phasmida. The tropical forests

© Springer International Publishing AG, part of Springer Nature 2018
T. R. New, *Forests and Insect Conservation in Australia*,
https://doi.org/10.1007/978-3-319-92222-5

of Queensland support some of the largest of all stick insects, and many species appear to be rare – a status perhaps confused by their crypsis and infrequent discovery (Brock and Hasenpusch 2009) – but with popularity enhanced by widespread hobbyist interests.

The insect groups noted below include those that fill one or more of the following criteria: (1) are restricted to and depend on native forest environments; (2) comprise ecological and taxonomic radiations that constitute evolutionary lineages of wider interest in documenting the evolution of Australia's insect fauna; (3) are ancient or otherwise notable and endemic elements of Australia's insect fauna; (4) fulfil criteria that can give them status as umbrella, flagship or indicator taxa; (5) are taxonomically 'tractable', in that many of the species can be recognised and identified by non-specialists, or from recent synopses or field guides; and (6) with values that can be communicated effectively to people, and demonstrate them worthy of conservation. About 15 orders contain possible candidates, excluding groups of ectoparasites on vertebrates, whose wellbeing will clearly be affected by loss of their forest bird or mammal hosts.

Those orders are Blattodea (including Isoptera), Mantodea, Dermaptera, Orthoptera, Phasmida, Embioptera, Psocoptera, Hemiptera, Thysanoptera, Neuroptera, Coleoptera, Mecoptera (only marginally in forests) Diptera, Lepidoptera, and Hymenoptera. Direct involvements in forest conservation are almost wholly confined to members of only four of these, Hemiptera, Coleoptera, Lepidoptera and Hymenoptera, in each of which conservation concerns devolve on vulnerability of individual species and values of insects in wider environmental assessments through undergoing changes in assemblage composition, relative abundance and richness related to forest conservation, as predictably responsive to imposed changes. In each of these orders, serious pest species of forest trees are related closely to other native species that are far scarcer. The longhorn beetle genus *Phoracantha* (Cerambycidae), for example, contains some of the most serious wood-boring pests of eucalypts, the most frequently addressed species being *P. semipunctata*. Some other species of *Phoracantha* are relatively scarce (Wang 1995). As for many other forest insects, concern over a few pest species of Cerambycidae has generated much information on the family worldwide (Wang 2016), but little direct interest in their conservation, or that of most other saproxylic beetles, has yet developed in Australia, despite substantially augmented taxonomic knowledge of the family (Slipinski and Escalona 2013, 2016).

Isolated endemic lineages occur in all four orders, some being highly characteristic of forest environments. The archaic moss-bugs (Peloridiidae) are a Gondwanan group of bugs, with Australian species found amongst the epiphytic coverings of *Nothofagus* trees in wet forests in the south, and flat-bugs (Aradidae, p. 152) are more widely distributed under bark. Many such endemic taxa are very scarce, and some endemic families comprise single or very few species: two such notable families are (1) Lamingtoniidae (Coleoptera), with both larvae and adults of the only species (*Lamingtonia binnaburense*) associated with polyporous fungi in rainforests of southern Queensland and northern New South Wales, and (2) Anomosetidae (Lepidoptera), containing only *Anomoses hylecoetes*, with similar distribution and probably associated with rotting logs. In common with many other scarce and inconspicuous insects, these taxa are unlikely to become specific foci for conservation action, but their detection at a site may help provide support for site protection.

The following families and other higher categories, some of which need much further investigation, may have practical values in developing forest conservation practices for insects and related biodiversity, and in monitoring impacts of anthropogenic changes.

Hemiptera: Psylloidea, Eurymelidae (radiations on eucalypts), Coccoidea (especially the gall-forming *Apiomorpha* [Eriococcidae]), Peloridiidae, Aradidae.

Coleoptera: Carabidae, Scarabaeidae (Rutelinae, Melolonthinae), Lucanidae, Geotrupidae, Buprestidae, Elateridae, Dermestidae, Melyridae, Tenebrionidae, Cerambycidae, Chrysomelidae, Curculionidae. More general attention to saproxylic beetles, with many families represented, and Staphylinidae, which are very diverse but mostly difficult to identify, is also important.

Lepidoptera: Oecophoridae (as important in breakdown of leaf litter: most other groups of 'microlepidoptera' are not sufficiently tractable for regular evaluation), Geometridae (often amongst the most diverse large moths in forests), Anthelidae (a characteristic forest group in Australia, occurring also in New Guinea), Limacodidae (cup moths, with some pest species), Cossidae (wood borers, including a large relict genus, *Xyleutes*, with some of the largest moths in Australia); Hepialidae (with highly characteristic and spectacular genera, notably *Zelotypia* and *Aenetus*), Papilionoidea (butterflies, as the most popular group of insects, are frequently noticed in forests, and background biology is relatively well-known, so that novelties can be recognised easily). Individual species of several families of Lepidoptera have become significant flagships for conservation: examples, some mentioned earlier, are *Ornithoptera richmondia* (Papilionidae), *Phyllodes imperialis smithersi* (Erebidae) and *Attacus wardi* (Saturniidae).

Hymenoptera: Symphyta (some sawflies are important defoliators, but larvae of some ['spitfires'] garner general interest), Formicidae (ants, diverse and widely tested as indicators, reflecting their ecological diversity and division amongst distinct functional groups), Apoidea (bees, as a central group of pollinators in forest ecosystems).

References

Brock PD, Hasenpusch JW (2009) The complete field guide to stick and leaf insects of Australia. CSIRO Publishing, Collingwood

Rentz D (2014) A guide to the cockroaches of Australia. CSIRO Publishing, Collingwood

Slipinski A, Escalona HE (2013) Australian longhorn beetles (Coleoptera: Cerambycidae). Volume I, Introduction and subfamily Lamiinae. CSIRO Publishing, Collingwood

Slipinski A, Escalona HE (2016) Australian longhorn beetles (Coleoptera: Cerambycidae). Volume 2, Subfamily Cerambycinae. CSIRO Publishing, Melbourne

Wang Q (1995) A taxonomic revision of the Australian genus *Phoracantha* Newman (Coleoptera: Cerambycidae). Invertebr Taxon 9:865–958

Wang Q (ed) (2016) Cerambycidae of the world. Biology and pest management. CRC Press, Boca Raton

Index

Printed in the United States
By Bookmasters

Printed in the United States
By Bookmasters